Lecture Notes in Computer Science 7877

Commenced Publication in 1973
Founding and Former Series Editors:
Gerhard Goos, Juris Hartmanis, and Jan van Leeuwen

Walter G. Kropatsch   Nicole M. Artner
Yll Haxhimusa   Xiaoyi Jiang (Eds.)

Graph-Based Representations in Pattern Recognition

9th IAPR-TC-15 International Workshop, GbRPR 2013
Vienna, Austria, May 15-17, 2013
Proceedings

 Springer

Volume Editors

Walter G. Kropatsch
Nicole M. Artner
Yll Haxhimusa
Vienna University of Technology
Department of Pattern Recognition and Image Processing
Favoritenstr. 9-11, 186-3, 1040 Vienna, Austria
E-mail: {krw, artner, yll}@prip.tuwien.ac.at

Xiaoyi Jiang
University of Münster
Department of Computer Science
Einsteinstr. 62, 48149 Münster, Germany
E-mail: xjiang@uni-muenster.de

ISSN 0302-9743   e-ISSN 1611-3349
ISBN 978-3-642-38220-8   e-ISBN 978-3-642-38221-5
DOI 10.1007/978-3-642-38221-5
Springer Heidelberg Dordrecht London New York

Library of Congress Control Number: 2013937223

CR Subject Classification (1998): I.5, I.4, I.3, I.2.10, G.2.2, I.2

LNCS Sublibrary: SL 6 – Image Processing, Computer Vision, Pattern Recognition,
and Graphics

Typesetting: Camera-ready by author, data conversion by Scientific Publishing Services, Chennai, India

Printed on acid-free paper

Springer is part of Springer Science+Business Media (www.springer.com)

Preface

These proceedings present the papers accepted for the 9th IAPR-TC-15 Workshop on Graph-based Representations in Pattern Recognition (GbR) 2013. For more than 15 years, GbR has been providing a forum for researchers from the fields of pattern recognition, image processing, and computer vision who build their works on the basis of graph theory. This year it was a great pleasure for us to organize the GbR 2013 in the heart of Europe – Vienna, Austria.

The Technical Committee 15 (TC15) of the International Association for Pattern Recognition (IAPR) was created in 1996. It encourages elaboration of graph-based research works, is an integral partner in organizing biennial GbR workshops, sponsors related special sessions at conferences, and promotes special issues in journals.

Traditionally the work presented at GbR covers a wide range of topics. The scope of the papers varies from theoretical contributions to applications, from discovering the new properties of a single graph (graph edit distance, maximum cut, graph characteristics derived from Schrödinger equation) to developing algorithms for sets of graphs, maximum subgraph problems and graph matching. A great interest was shown in the problems of graph kernels and topology.

Besides the regular research papers, this workshop featured two highlights: the IAPR distinguished speakers Mario Vento and Herbert Edelsbrunner. Mario Vento was among the founders of TC15 some 16 years ago. He summarized the development of our field starting with the original motivation. It allowed the younger generation of our community to compare the goals and expectations of the early years with the current state. Herbert Edelsbrunner created a new bridge from TC15 to the area of topology and persistence.

Overall, GbR 2013 attracted 27 submissions from 10 countries. Each paper went through a critical reviewing process by at least two members of the international Program Committee. Finally, 24 papers including the contributions of the invited speakers were accepted for oral presentation and publication in these proceedings.

On behalf of the organizers, we would like to thank the members of the Program Committee for their timely and competent reviews; the authors of the submitted papers for their work and their abidance to all deadlines. Finally, we would like to thank the IAPR for sponsoring our workshop and the IAPR distinguished speakers for their contributions.

May 2013

Walter G. Kropatsch
Nicole M. Artner
Yll Haxhimusa
Xiaoyi Jiang

Organization

Program Co-chairs

Walter G. Kropatsch   Vienna University of Technology, Austria
Nicole M. Artner   Vienna University of Technology, Austria
Yll Haxhimusa   Vienna University of Technology, Austria
Xiaoyi Jiang   University of Münster, Germany

Program Committee

Nicole Artner (Austria)
Antonio Bandera (Spain)
Csaba Beleznai (Austria)
Isabelle Bloch (France)
Luc Brun (France)
Wilhelm Burger (Austria)
Donatello Conte (Italy)
Francisco Escolano (Spain)
Rocio Gonzalez-Diaz (Spain)
Edwin Hancock (UK)
Yll Haxhimusa (Austria)
Xiaoyi Jiang (Germany)
Dimosthenis Karatzas (Spain)
Yukiko Kenmochi (Japan)
Walter Kropatsch (Austria)
Tetsuji Kuboyama (Japan)
Christoph Lampert (Austria)
Cheng-Lin Liu (China)

Josep Lladós (Spain)
Bin Luo (China)
Rebeca Marfil (Spain)
Jean-Marc Ogier (France)
Marcello Pelillo (Italy)
Pedro Real (Spain)
Radim Sara (Czech Republic)
Christian Schellewald (Norway)
Francesc Serratosa (Spain)
Ali Shokoufandeh (USA)
Robin Strand (Sweden)
Peter Sturm (France)
Salvatore Tabbone (France)
Andrea Torsello (Italy)
Antoine Vacavant (France)
Ernest Valveny (Spain)
Mario Vento (Italy)

Local Organizing Committee

Walter Kropatsch
Nicole Artner
Yll Haxhimusa
Elfriede Oberleitner
Aysylu Gabdulkhakova

Sponsoring Institutions

Vienna University of Technology
PRIP Club
International Association for Pattern Recognition (IAPR)

Table of Contents

A One Hour Trip in the World of Graphs, Looking at the Papers of the Last Ten Years

Mario Vento

Dept. of Computer Eng. and Electrical Eng. and
Applied Mathematics University of Salerno
Via Ponte Don Melillo, Fisciano (SA), Italy
mvento@unisa.it

1 Motivations of the Trip

The use of a graph-based pattern representation induces the need to formulate the main operations required in Pattern Recognition in terms of operations on graphs: classification, usually intended as the comparison between an object and a set of prototypes, and learning, which is the process for obtaining a model of a class starting from a set of known samples, are among the key issues that must be addressed using graph-based techniques.

Forty years have passed since the first papers on this topic appear in Pattern Recognition literature: a lot of research effort has been devoted to explore this challenging field and some approaches have been meanwhile consolidated. These notes aren't a scientific paper but some considerations inspiring my future talk at gbr 2013 conference, a little trip in the word of graphs aimed at better knowing treasures and outstanding locations.

2 Trip Diary

The use of graphs in Pattern Recognition dates back to the early seventies, and the paper Thirty years of graph matching in Pattern Recognition" [14] reports a survey of the literature on graph-based techniques since the first years and up to the early 2000's. In the last decade we have assisted to a growing interest in graphs, as confirmed by the number of papers using graphs for different aspects of Pattern Recognition.

We have surely assisted to a maturation of the classical techniques for graph comparison, either exact or inexact; contemporarily we are assisting to a rapid growth of many alternative approaches, such as graph embedding and graph kernels, aimed at making possible the application to graphs of vector-based techniques for classification and learning (such as the ones derived from the statistical classification and learning theory).

The trip is devoted to analyze the main advances registered in graph-based methodologies in the last ten years, looking at the main recent literature on this topic; the aim is to reconstruct an unifying view of these approaches when used in the context of Pattern Recognition tasks.

W.G. Kropatsch et al. (Eds.): GbRPR 2013, LNCS 7877, pp. 1–10, 2013.

The analysis starts from the above mentioned survey [14] and enriches the discussion by considering a selection of the most recent main contributions; consequently, the talk, for the sake of conciseness, will mainly focus on the papers published during the last ten years. At the beginning, the interests of Pattern Recognition researchers on graphs were mainly concentrated on graph matching, either exact or inexact. While for the exact methods the attention was concentrated on the definition of novel algorithms attempting to progressively reduce the computational burden, the approaches used in the inexact methods were inspired to some different rationales:

- Optimal inexact matching algorithms, able to find a solution minimizing the matching cost; it is guaranteed that, if an exact solution exists, it will be found. The algorithms ascribed to this class essentially concentrate on dealing with the input graph variability; the optimality of the solution requires an exploration of the solution space, usually making the algorithms fairly more expensive than the exact ones. Most of the algorithms are based on some forms of Tree search with backtracking, but also other techniques based on group theory or other mathematical properties of the graphs used in the matching process have been proposed.
- Suboptimal or approximate matching algorithms, able to ensure only a local minimum of the matching cost: it is expected that the obtained minimum is close to the global minimum, even if an upper bound of this distance is often unknown, so limiting their applicability to cases in which the maximum error value assumes a secondary importance. The big advantage of this class of algorithms is the polynomial matching time. While tree searching methods of this kind have been developed, the most common approach is based on continuous optimization, by replacing the matching problem, inherently discrete, with a continuous one, usually not linear; the advantage is that it is possible the use of a well established theoretical framework. Another important class of algorithms, although not as common as continuous optimization, is based on the exploitation of the eigenvalues of the adjacency matrix, which are invariant to node permutations, which can help to reduce the computational complexity in the matching process. Also a wide array of other techniques have been used less frequently.
- Error correcting graph matching algorithms, based on the definition of an explicit model of the errors (missing nodes and/or edges, changes on the attributes, etc.); the cheapest sequence of operations needed to transform one graph into the other is used to evaluate a similarity between the two graphs. So, with respect to exact and inexact graph matching, the discussion will be finalized to the presentation of the main advances on these kinds of graph matching algorithms.

Of course what is happening in the recent past, see the journal papers reported in the Section References, cannot be left out of the discussion: in fact, in the last decade we have assisted to the birth and growth of methods facing learning and classification in a rather innovative scientific vision: the computational burden of matching algorithms together with their intrinsic complexity, in opposition to the well established world of statistical Pattern recognition methodologies, suggested new paradigms for the graph-based methods: why don't we try to reduce graph matching and learning to vector-based operations, so as to make it possible the use of statistical approaches?

Two opposite ways of facing the problem, each with its pros and cons: graphs from the beginning to the end", with a few heavy algorithms, but the exploitation of all the information contained into the graphs; on the other side, the risk of loosing discriminating power during the conversion of graphs into vectors (by selecting suitable properties), counterbalanced by the immediate access to all the theoretically assessed achievements of the statistical framework.

These two opposite factions are now simultaneously active, each hoping to overcome the other; ten years ago these innovative methods were in the background, but now they are gaining more and more attention in the scientific literature on graphs.

Graph embedding, intended as the technique that map whole graphs onto points in a vector space, in such a way that similar graphs are mapped onto close points is perhaps the most significant novelty in graph-based in Pattern Recognition in the recent years. Although seminal works on these fields were already present in earlier literature, it is in the last decade that these techniques have gained popularity in the Pattern Recognition community. Bunke et al. [10] present a survey on the topic of graph kernels and graph embeddings, and in [11] extend this review and present these techniques as a way to unify the statistical and structural approaches in Pattern Recognition.

Graph kernels represent a sort of generalization of graph embedding; if we denote with G the space of all the graphs, a graph kernel is a function that maps a couple of graphs onto a real number, and holds similar properties to the dot product defined on vectors. More formally they can be seen as a measure of the similarity between two graphs; however its formal properties allow a kernel to replace the vector dot product in several vector-based algorithms that use this operator (and other functions related to dot product, such as the Euclidean norm). Among the many Pattern Recognition techniques that can be adapted to graphs using kernels we mention Support Vector Machine classifiers and Principal Component Analysis.

Kernels have been used for a long time to extend to the nonlinear case linear algorithms working on vector spaces, thanks to the Mercer's theorem: given a kernel function defined on a compact space X, there is a vector space V and a mapping between X and V such that the value of the kernel computed on two points in X is equal to the dot product of the corresponding points in V. Thus, for compact spaces, a kernel can be seen as an implicit way of performing an embedding into a vector space. Although Mercer's theorem do not apply to graph kernels, in practice these latter can be used as a theoretically sound way to extend a vector algorithm to graphs. Of course, the actual performance of these algorithms strongly depend on the appropriateness (with respect to the task at hand) of the notion of similarity embodied in the graph kernel.

3   Trips Souvenirs

What have we experienced from the trip? The analysis of the recent literature of graph-based techniques shows there is still a warm interest toward the use of this important data structure for facing Pattern Recognition problems. However, a definite interpretation of the best promising future directions seems to be still a bit uncertain: on one hand, we have surely assisted to a maturation of the classical techniques for graph comparison, either exact or inexact; on the other hand, we are assisting to a rapid growth of many alternative approaches, such as graph embedding and graph kernels, whose rationale is to reduce graphs to vectors so as to make it possible the use of the well established statistical theory of classification and learning.

The main questions posed by researchers advocating the graphs from beginning to end" approach could be: Is it really effective to solve a problem starting with graph representations, and going back to vectors, risking to lose important chunks of discriminative power? If so, why don't you renounce to use graphs, and directly use vector-based descriptions from the start?"

The opposite faction could reply: Why do you insist on describing the world by graphs if there is still a lack of completely assessed and computationally acceptable algorithms for classifying and for learning graph prototypes?"

The conclusion? We will discuss!

References

1. Auwatanamongkol, S.: Inexact graph matching using a genetic algorithm for image recognition. Pattern Recognition Letters (PRL) 28(12), 1428–1437 (2007)
2. Bagdanov, A.D., Worring, M.: First order gaussian graphs for efficient structure classification. PR 36(6), 1311–1324 (2003)
3. Bai, X., Latecki, L.: Path similarity skeleton graph matching. IEEE Trans. on PAMI 30(7), 1282–1292 (2008)
4. Bengoetxea, E., Larrañaga, P., Bloch, I., Perchant, A., Boeres, C.: Inexact graph matching by means of estimation of distribution algorithms. PR 35(12), 2867–2880 (2002)
5. Bergamasco, F., Albarelli, A.: A graph-based technique for semi-supervised segmentation of 3D surfaces. PRL (2012) (in press)
6. Borzeshi, E.Z., Piccardi, M., Riesen, K., Bunke, H.: Discriminative prototype selection methods for graph embedding. PR (2012)
7. Bourbakis, N., Yuan, P., Makrogiannis, S.: Object recognition using wavelets, L-G graphs and synthesis of regions. PR 40(7), 2077–2096 (2007)
8. Bunke, H., Dickinson, P., Irniger, C., Kraetzl, M.: Recovery of missing information in graph sequences by means of reference pattern matching and decision tree learning. PR 39(4), 573–586 (2006)
9. Bunke, H., Riesen, K.: Improving vector space embedding of graphs through feature selection algorithms. PR 44(9), 1928–1940 (2011)
10. Bunke, H., Riesen, K.: Recent advances in graph-based pattern recognition with applications in document analysis. PR 44(5), 1057–1067 (2011)
11. Bunke, H., Riesen, K.: Towards the unification of structural and statistical pattern recognition. PRL 33(7), 811–825 (2012)

12. Caelli, T., Kosinov, S.: Inexact graph matching using eigen-subspace projection clustering. IJPRAI 18(3), 329–354 (2004)
13. Caetano, T., McAuley, J., Cheng, L., Le, Q., Smola, A.: Learning graph matching. IEEE Trans. on PAMI 31(6), 1048–1058 (2009)
14. Conte, D., Foggia, P., Sansone, C., Vento, M.: Thirty years of graph matching in Pattern Recognition. IJPRAI 18(3), 265–298 (2004)
15. Conte, D., Foggia, P., Jolion, J.M., Vento, M.: A graph-based, multi-resolution algorithm for tracking objects in presence of occlusions. PR 39(4), 562–572 (2006)
16. Culp, M., Michailidis, G.: Graph-based semisupervised learning. IEEE Trans. on PAMI 30(1), 174–179 (2008)
17. Czech, W.: Invariants of distance k-graphs for graph embedding. PRL 33(15), 1968–1979 (2012)
18. Dhillon, I., Guan, Y., Kulis, B.: Weighted graph cuts without eigenvectors: A multilevel approach. IEEE Trans. on PAMI 29(11), 1944–1957 (2007)
19. Dickinson, P.J., Kraetzl, M., Bunke, H., Neuhaus, M., Dadej, A.: Similarity measures for hierarchical representations of graphs with unique node labels. IJPRAI 18-3(3), 425–442 (2004)
20. Dickinson, P.J., Bunke, H., Dadej, A., Kraetzl, M.: Matching graphs with unique node labels. Pattern Analysis & Applications 7, 243–254 (2004)
21. Duchenne, O., Bach, F., Kweon, I.S., Ponce, J.: A tensor-based algorithm for high-order graph matching. IEEE Trans. on PAMI 33(12), 2383–2395 (2011)
22. Ducournau, A., Bretto, A., Rital, S., Laget, B.: A reductive approach to hypergraph clustering: An application to image segmentation. PR 45(7), 2788–2803 (2012)
23. Emms, D., Wilson, R.C., Hancock, E.R.: Graph matching using the interference of continuous-time quantum walks. PR 42(5), 985–1002 (2009)
24. Felzenszwalb, P., Zabih, R.: Dynamic programming and graph algorithms in computer vision. IEEE Trans. on PAMI 33(4), 721–740 (2011)
25. Fernandez-Madrigal, J.A., Gonzalez, J.: Multihierarchical graph search. IEEE Trans. on PAMI 24(1), 103–113 (2002)
26. Ferrer, M., Karatzas, D., Valveny, E., Bardaji, I., Bunke, H.: A generic framework for median graph computation based on a recursive embedding approach. CVIU 115(7), 919–928 (2011)
27. Ferrer, M., Valveny, E., Serratosa, F.: Median graph: A new exact algorithm using a distance based on the maximum common subgraph. PRL 30(5), 579–588 (2009)
28. Ferrer, M., Valveny, E., Serratosa, F.: Median graphs: A genetic approach based on new theoretical properties. PR 42(9), 2003–2012 (2009)
29. Ferrer, M., Valveny, E., Serratosa, F., Riesen, K., Bunke, H.: Generalized median graph computation by means of graph embedding in vector spaces. PR 43(4), 1642–1655 (2010)
30. Foggia, P., Percannella, G., Sansone, C., Vento, M.: A graph-based algorithm for cluster detection. IJPRAI 22(5), 843–860 (2008)
31. Fränti, P., Virmajoki, O., Hautamaki, V.: Fast agglomerative clustering using a k-nearest neighbor graph. IEEE Trans. on PAMI 28(11), 1875–1881 (2006)
32. Gao, X., Xiao, B., Tao, D., Li, X.: A survey of graph edit distance. Pattern Analysis & Applications 13, 113–129 (2010), doi:10.1007/s10044-008-0141-y
33. Gauzere, B., Brun, L., Villemin, D.: Two new graphs kernels in chemoinformatics. PRL (2012) (in press)
34. Gerstmayer, M., Haxhimusa, Y., Kropatsch, W.: Hierarchical interactive image segmentation using irregular pyramids. In: Jiang, X., Ferrer, M., Torsello, A. (eds.) GbRPR 2011. LNCS, vol. 6658, pp. 245–254. Springer, Heidelberg (2011)

35. Gibert, J., Valveny, E., Bunke, H.: Feature selection on node statistics based embedding of graphs. PRL 33(15), 1980–1990 (2012)
36. Gibert, J., Valveny, E., Bunke, H.: Graph embedding in vector spaces by node attribute statistics. PR 45(9), 3072–3083 (2012)
37. Gonzalez-Diaz, R., Ion, A., Iglesias-Ham, M., Kropatsch, W.G.: Invariant representative cocycles of cohomology generators using irregular graph pyramids. CVIU 115(7), 1011–1022 (2011)
38. Gori, M., Maggini, M., Sarti, L.: Exact and approximate graph matching using random walks. IEEE Trans. on PAMI 27(7), 1100–1111 (2005)
39. Guigues, L., Le Men, H., Cocquerez, J.P.: The hierarchy of the cocoons of a graph and its application to image segmentation. PRL 24(8), 1059–1066 (2003)
40. Günter, S., Bunke, H.: Self-organizing map for clustering in the graph domain. PRL 23(4), 405–417 (2002)
41. Günter, S., Bunke, H.: Validation indices for graph clustering. PRL 24(8), 1107–1113 (2003)
42. Hagenbuchner, M., Gori, M., Bunke, H., Tsoi, A.C., Irniger, C.: Using attributed plex grammars for the generation of image and graph databases. PRL 24(8), 1081–1087 (2003)
43. Hancock, E.R., Wilson, R.C.: Pattern analysis with graphs: Parallel work at bern and york. PRL 33(7), 833–841 (2012)
44. He, L., Han, C.Y., Everding, B., Wee, W.G.: Graph matching for object recognition and recovery. PR 37(7), 1557–1560 (2004)
45. Hidović, D., Pelillo, M.: Metrics for attributed graphs based on the maximal similarity common subgraph. IJPRAI 18(3), 299–313 (2004)
46. Hu, W., Hu, W., Xie, N., Maybank, S.: Unsupervised active learning based on hierarchical graph-theoretic clustering. IEEE Trans. on SMC-B 39(5), 1147–1161 (2009)
47. Jain, B.J., Obermayer, K.: Graph quantization. CVIU 115(7), 946–961 (2011)
48. Cesar Jr., R.M., Bengoetxea, E., Bloch, I., Larrañaga, P.: Inexact graph matching for model-based recognition: Evaluation and comparison of optimization algorithms. PR 38(11), 2099–2113 (2005)
49. Justice, D., Hero, A.: A binary linear programming formulation of the graph edit distance. IEEE Trans. on PAMI 28(8), 1200–1214 (2006)
50. Kammerer, P., Glantz, R.: Segmentation of brush strokes by saliency preserving dual graph contraction. PRL 24(8), 1043–1050 (2003)
51. Kang, H.W.: G-wire: A livewire segmentation algorithm based on a generalized graph formulation. PRL 26(13), 2042–2051 (2005)
52. Kim, D.H., Yun, I.D., Lee, S.U.: Attributed relational graph matching based on the nested assignment structure. PR 43(3), 914–928 (2010)
53. Kim, J.S., Hong, K.S.: Colortexture segmentation using unsupervised graph cuts. PR 42(5), 735–750 (2009)
54. Kokiopoulou, E., Frossard, P.: Graph-based classification of multiple observation sets. PR 43(12), 3988–3997 (2010)
55. Kokiopoulou, E., Saad, Y.: Enhanced graph-based dimensionality reduction with repulsion laplaceans. PR 42(11), 2392–2402 (2009)
56. Kostin, A., Kittler, J., Christmas, W.: Object recognition by symmetrised graph matching using relaxation labelling with an inhibitory mechanism. PRL 26(3), 381–393 (2005)
57. Lezoray, O., Elmoataz, A., Bougleux, S.: Graph regularization for color image processing. CVIU 107(12), 38–55 (2007)

58. Lin, L., Liu, X., Zhu, S.C.: Layered graph matching with composite cluster sampling. IEEE Trans. on PAMI 32(8), 1426–1442 (2010)
59. Liu, J., Wang, B., Lu, H., Ma, S.: A graph-based image annotation framework. PRL 29(4), 407–415 (2008)
60. Lladós, J., Sánchez, G.: Graph matching versus graph parsing in graphics recognition: A combined approach. IJPRAI 18(3), 455–473 (2004)
61. Luo, B., Wilson, R.C., Hancock, E.R.: Spectral embedding of graphs. PR 36(10), 2213–2230 (2003)
62. Luo, B., Wilson, R.C., Hancock, E.R.: A spectral approach to learning structural variations in graphs. PR 39(6), 1188–1198 (2006)
63. Ma, F., Bajger, M., Slavotinek, J.P., Bottema, M.J.: Two graph theory based methods for identifying the pectoral muscle in mammograms. PR 40(9), 2592–2602 (2007)
64. Macrini, D., Dickinson, S., Fleet, D., Siddiqi, K.: Bone graphs: Medial shape parsing and abstraction. CVIU 115(7), 1044–1061 (2011)
65. Macrini, D., Dickinson, S., Fleet, D., Siddiqi, K.: Object categorization using bone graphs. CVIU 115(8), 1187–1206 (2011)
66. Mantrach, A., van Zeebroeck, N., Francq, P., Shimbo, M., Bersini, H., Saerens, M.: Semi-supervised classification and betweenness computation on large, sparse, directed graphs. PR 44(6), 1212–1224 (2011)
67. Martínez, A.M., Mittrapiyanuruk, P., Kak, A.C.: On combining graph-partitioning with non-parametric clustering for image segmentation. CVIU 95(1), 72–85 (2004)
68. Massaro, A., Pelillo, M.: Matching graphs by pivoting. PRL 24(8), 1099–1106 (2003)
69. Maulik, U.: Hierarchical pattern discovery in graphs. IEEE Trans. on SMC-C 38(6), 867–872 (2008)
70. de Mauro, C., Diligenti, M., Gori, M., Maggini, M.: Similarity learning for graph-based image representations. PRL 24(8), 1115–1122 (2003)
71. Neuhaus, M., Bunke, H.: Self-organizing maps for learning the edit costs in graph matching. IEEE Trans. on SMC-B 35(3), 503–514 (2005)
72. Neuhaus, M., Bunke, H.: Edit distance-based kernel functions for structural pattern classification. PR 39(10), 1852–1863 (2006)
73. Neuhaus, M., Bunke, H.: Automatic learning of cost functions for graph edit distance. Information Sciences 177(1), 239–247 (2007)
74. Qiu, H., Hancock, E.R.: Graph matching and clustering using spectral partitions. PR 39(1), 22–34 (2006)
75. Qiu, H., Hancock, E.R.: Graph simplification and matching using commute times. PR 40(10), 2874–2889 (2007)
76. Raveaux, R., Adam, S., Héroux, P., Trupin, É.: Learning graph prototypes for shape recognition. CVIU 115(7), 905–918 (2011)
77. Raveaux, R., Burie, J.C., Ogier, J.M.: A graph matching method and a graph matching distance based on subgraph assignments. PRL 31(5), 394–406 (2010)
78. Riesen, K., Bunke, H.: Graph classification by means of lipschitz embedding. IEEE Trans. on SMC-B 39(6), 1472–1483 (2009)
79. Riesen, K., Bunke, H.: Approximate graph edit distance computation by means of bipartite graph matching. Image and Vision Computing 27(7), 950–959 (2009)
80. Riesen, K., Bunke, H.: Graph classification based on vector space embedding. IJPRAI 23, 1053–1081 (2009)

81. Riesen, K., Bunke, H.: Reducing the dimensionality of dissimilarity space embedding graph kernels. Engineering Applications of Artificial Intelligence 22, 48–56 (2009)
82. Robles-Kelly, A., Hancock, E.: Graph edit distance from spectral seriation. IEEE Trans. on PAMI 27(3), 365–378 (2005)
83. Robles-Kelly, A., Hancock, E.R.: String edit distance, random walks and graph matching. IJPRAI 18(3), 315–327 (2004)
84. Robles-Kelly, A., Hancock, E.R.: A graph-spectral method for surface height recovery. PR 38(8), 1167–1186 (2005)
85. Robles-Kelly, A., Hancock, E.R.: A riemannian approach to graph embedding. PR 40(3), 1042–1056 (2007)
86. Rohban, M.H., Rabiee, H.R.: Supervised neighborhood graph construction for semi-supervised classification. PR 45(4), 1363–1372 (2012)
87. Rota Bulò, S., Pelillo, M., Bomze, I.M.: Graph-based quadratic optimization: A fast evolutionary approach. CVIU 115(7), 984–995 (2011)
88. Ruberto, C.D.: Recognition of shapes by attributed skeletal graphs. PR 37(1), 21–31 (2004)
89. da, S., Torres, R., Falcão, A., da, F., Costa, L.: A graph-based approach for multiscale shape analysis. PR 37(6), 1163–1174 (2004)
90. Sanfeliu, A., Alquézar, R., Andrade, J., Climent, J., Serratosa, F., Vergés, J.: Graph-based representations and techniques for image processing and image analysis. PR 35(3), 639–650 (2002)
91. Sanfeliu, A., Serratosa, F., Alquezar, R.: Second-order random graphs for modeling sets of attributed graphs and their application to object learning and recognition. IJPRAI 18(3), 375–396 (2004)
92. Sanromà, G., Alquézar, R., Serratosa, F.: A new graph matching method for point-set correspondence using the em algorithm and softassign. CVIU 116(2), 292–304 (2012)
93. Santo, M.D., Foggia, P., Sansone, C., Vento, M.: A large database of graphs and its use for benchmarking graph isomorphism algorithms. PRL 24(8), 1067–1079 (2003)
94. Scheinerman, E.R., Tucker, K.: Modeling graphs using dot product representations. Computational Statistics 25, 1–16 (2010)
95. Sebastian, T., Klein, P., Kimia, B.: Recognition of shapes by editing their shock graphs. IEEE Trans. on PAMI 26(5), 550–571 (2004)
96. Serratosa, F., Alquezar, R., Sanfeliu, A.: Synthesis of function-described graphs and clustering of attributed graphs. IJPRAI 16(6), 621–655 (2002)
97. Serratosa, F., Alquézar, R., Sanfeliu, A.: Function-described graphs for modelling objects represented by sets of attributed graphs. PR 36(3), 781–798 (2003)
98. Shang, F., Jiao, L., Wang, F.: Graph dual regularization non-negative matrix factorization for co-clustering. PR 45(6), 2237–2250 (2012)
99. Shiga, M., Mamitsuka, H.: Efficient semi-supervised learning on locally informative multiple graphs. PR 45(3), 1035–1049 (2012)
100. Skomorowski, M.: Syntactic recognition of distorted patterns by means of random graph parsing. PRL 28(5), 572–581 (2007)
101. Solé-Ribalta, A., Serratosa, F.: Models and algorithms for computing the common labelling of a set of attributed graphs. CVIU 115(7), 929–945 (2011)
102. Solnon, C.: AllDifferent-based filtering for subgraph isomorphism. Artificial Intelligence 174, 850–864 (2010)
103. Sumengen, B., Manjunath, B.: Graph partitioning active contours (gpac) for image segmentation. IEEE Trans. on PAMI 28(4), 509–521 (2006)

104. Tang, H., Fang, T., Shi, P.F.: Nonlinear discriminant mapping using the laplacian of a graph. PR 39(1), 156–159 (2006)
105. Tang, J., Jiang, B., Zheng, A., Luo, B.: Graph matching based on spectral embedding with missing value. PR 45(10), 3768–3779 (2012)
106. Tao, W., Chang, F., Liu, L., Jin, H., Wang, T.: Interactively multiphase image segmentation based on variational formulation and graph cuts. PR 43(10), 3208–3218 (2010)
107. Torsello, A., Hancock, E.R.: Graph embedding using tree edit-union. PR 40(5), 1393–1405 (2007)
108. Ullmann, J.R.: Bit-vector algorithms for binary constraint satisfaction and subgraph isomorphism. J. Exp. Algorithmics 15, 1.6:1.1–1.6:1.64 (2011)
109. Wan, M., Lai, Z., Shao, J., Jin, Z.: Two-dimensional local graph embedding discriminant analysis (2dlgeda) with its application to face and palm biometrics. Neurocomputing 73(13), 197–203 (2009)
110. Wang, B., Pan, F., Hu, K.M., Paul, J.C.: Manifold-ranking based retrieval using k-regular nearest neighbor graph. PR 45(4), 1569–1577 (2012)
111. Wang, J.T., Zhang, K., Chang, G., Shasha, D.: Finding approximate patterns in undirected acyclic graphs. PR 35(2), 473–483 (2002)
112. Wilson, R., Hancock, E., Luo, B.: Pattern vectors from algebraic graph theory. IEEE Trans. on PAMI 27(7), 1112–1124 (2005)
113. Wilson, R.C., Zhu, P.: A study of graph spectra for comparing graphs and trees. PR 41(9), 2833–2841 (2008)
114. van Wyk, B., van Wyk, M.: Kronecker product graph matching. PR 36(9), 2019–2030 (2003)
115. van Wyk, B., van Wyk, M.: A pocs-based graph matching algorithm. IEEE Trans. on PAMI 26(11), 1526–1530 (2004)
116. van Wyk, M.A., van Wyk, B.J.: A learning-based framework for graph matching. IJPRAI 18(3), 355–374 (2004)
117. Xiao, B., Hancock, E.R., Wilson, R.C.: Graph characteristics from the heat kernel trace. PR 42(11), 2589–2606 (2009)
118. Xiao, Y., Dong, H., Wu, W., Xiong, M., Wang, W., Shi, B.: Structure-based graph distance measures of high degree of precision. PR 41(12), 3547–3561 (2008)
119. Xu, N., Ahuja, N., Bansal, R.: Object segmentation using graph cuts based active contours. CVIU 107(3), 210–224 (2007)
120. Yan, F., Christmas, W., Kittler, J.: Layered data association using graph-theoretic formulation with application to tennis ball tracking in monocular sequences. IEEE Trans. on PAMI 30(10), 1814–1830 (2008)
121. Yan, S., Xu, D., Zhang, B., Zhang, H.J., Yang, Q., Lin, S.: Graph embedding and extensions: A general framework for dimensionality reduction. IEEE Trans. on PAMI 29(1), 40–51 (2007)
122. Yang, F., Kruggel, F.: A graph matching approach for labeling brain sulci using location, orientation, and shape. Neurocomputing 73(13), 179–190 (2009)
123. Yang, L.: Building k-edge-connected neighborhood graph for distance-based data projection. PRL 26(13), 2015–2021 (2005)
124. You, Q., Zheng, N., Gao, L., Du, S., Wu, Y.: Analysis of solution for supervised graph embedding. IJPRAI 22(7), 1283–1299 (2008)
125. Yu, G., Peng, H., Wei, J., Ma, Q.: Mixture graph based semi-supervised dimensionality reduction. Pattern Recognition and Image Analysis 20, 536–541 (2010)
126. Zampelli, S., Deville, Y., Solnon, C.: Solving subgraph isomorphism problems with constraint programming. Constraints 15, 327–353 (2010)

127. Zanghi, H., Ambroise, C., Miele, V.: Fast online graph clustering via Erdos-Rényi mixture. PR 41(12), 3592–3599 (2008)
128. Zanghi, H., Volant, S., Ambroise, C.: Clustering based on random graph model embedding vertex features. PRL 31(9), 830–836 (2010)
129. Zaslavskiy, M., Bach, F., Vert, J.P.: A path following algorithm for the graph matching problem. IEEE Trans. on PAMI 31(12), 2227–2242 (2009)
130. Zhang, C., Wang, F.: A multilevel approach for learning from labeled and unlabeled data on graphs. PR 43(6), 2301–2314 (2010)
131. Zhang, F., Hancock, E.R.: Graph spectral image smoothing using the heat kernel. PR 41(11), 3328–3342 (2008)
132. Zhao, H., Robles-Kelly, A., Zhou, J., Lu, J., Yang, J.Y.: Graph attribute embedding via riemannian submersion learning. CVIU 115(7), 962–975 (2011)
133. Zhi, R., Flierl, M., Ruan, Q., Kleijn, W.: Graph-preserving sparse nonnegative matrix factorization with application to facial expression recognition. IEEE Trans. on SMC-B 41(1), 38–52 (2011)

A Unified Framework for Strengthening Topological Node Features and Its Application to Subgraph Isomorphism Detection

Nicholas Dahm[1,3], Horst Bunke[4], Terry Caelli[2,5], and Yongsheng Gao[3]

[1] Queensland Research Laboratory, National ICT Australia
[2] Victoria Research Laboratory, National ICT Australia
terry.caelli@nicta.com.au
[3] School of Engineering, Griffith University, Brisbane, Australia
{n.dahm,yongsheng.gao}@griffith.edu.au
[4] Institute of Computer Science and Applied Mathematics,
University of Bern, Switzerland
bunke@iam.unibe.ch
[5] Electrical and Electronic Engineering, University of Melbourne, Australia
tcaelli@unimelb.edu.au

Abstract. This paper presents techniques to address the complexity problem of subgraph isomorphism detection on large graphs. To overcome the inherently high computational complexity, the problem is simplified through the calculation and strengthening of topological node features. These features can be utilised, in principle, by any subgraph isomorphism algorithm. The design and capabilities of the proposed unified strengthening framework are discussed in detail. Additionally, the concept of an n-neighbourhood is introduced, which facilitates the development of novel features and provides an additional platform for feature strengthening. Through experiments performed with state-of-the-art subgraph isomorphism algorithms, the theoretical and practical advantages of using these techniques become evident.

Keywords: Graph Matching, Subgraph Isomorphism, Topological Node Features.

1 Introduction

Identifying subgraph isomorphisms between a pair of graphs is a key problem in structural pattern recognition with an exponential worst case complexity. Subgraph isomorphism algorithms can be either exact, or inexact, with the latter dealing with incomplete or noisy graphs. In this paper we focus solely on exact subgraph isomorphism, where the graphs are complete and error-free. The concepts presented in this paper however, may apply equally to inexact algorithms. Applications for exact subgraph isomorphism include chemical substructure and protein-protein interaction network matching, social network analysis and VLSI design [1]. The most widely used algorithms for subgraph isomorphism are Ullmann's algorithm [11] and the VF2 algorithm [2]. These tree-search algorithms

W.G. Kropatsch et al. (Eds.): GbRPR 2013, LNCS 7877, pp. 11–20, 2013.

are able to obtain practical runtime speeds by pruning branches from the search tree that contain incompatible node matchings. However due to the exponentiality of subgraph isomorphism, as the number of nodes in the graphs increases, the matching times of both algorithms can quickly become infeasible [4].

To identify more node incompatibilities, and hence further reduce matching times, local topological information about a node can be encoded into a *topological node feature* (TNF). VF2 for example, uses the simplest TNF, namely the vertex degree (number of adjacent nodes) to identify incompatible node matchings. Topological node features like this are also known as *subgraph isomorphism consistents* or, in the case of graph isomorphism, *invariants*. On the simpler problem of graph isomorphism, a number of algorithms exist that utilise complex structural information. An early example of this is the Nauty algorithm by McKay [6], which uses TNFs and a strengthening procedure similar to the tree-index method discussed in this paper. Using this structural information, Nauty is able to identify nodes which have identical topological structure, so becoming acceptable isomorphic mappings. Recently, this idea was extended by Sorlin & Solnon to create the IDL algorithm [10]. In another recent paper Riesen et al. [5] ignores the search tree entirely and uses only TNFs to determine isomorphisms. Their method has a polynomial runtime, but in some cases cannot resolve the matching due to outstanding permutations.

When extending these concepts to subgraph isomorphism, any pair of matched nodes is less likely to have identical topological structure. Despite this, it is still possible to exploit the fact that a node in the full graph will contain the same topological structure as a node from the subgraph, however with some extra structure possibly added. From this observation, a rule can be defined for each TNF as to when a node mapping is considered invalid. For example, if a subgraph node has a degree of 5, a mapping to any full graph node with a degree less than 5 is invalid. A recent algorithm utilising TNFs is the ILF algorithm by Zampelli et al. [12]. This algorithm uses TNF values strengthened through a similar procedure to Nauty to eliminate incompatibilities. In the paper, the authors show that ILF can outperform VF2 in many cases even while only using a simple TNF like degree. The recent LAD algorithm by Solnon [9] uses a local *all different* constraint which ensures that for each mapping, the nodes adjacent to the subgraph node can be uniquely mapped to nodes adjacent to the full graph node. When combined with the generalised arc consistency (GAC) *all different* constraint that is commonly used in constraint programming, LAD has been shown to be even faster than ILF, on most cases.

In this paper we present a number of techniques that can be used to simplify the subgraph isomorphism process. Similar to [4], the techniques described here are not designed to challenge existing methods. On the contrary, they are designed so that they can be utilised as an enhancement to any subgraph isomorphism algorithm. In Section 2, we describe the concept of a node's *n*-neighbourhood and propose some novel topological node features which utilise it. Section 3 presents a unified framework consisting of three strengthening techniques. These strengthening techniques, introduced in Sections 3.1 to 3.3, can be applied to both topological node features as well as application-specific node labels. Section 3.4 then shows how these concepts can be combined to create strengthened features that are resistant to noise. All of these techniques can be calculated independently on each graph, allowing them to be computed ahead of time and stored. This makes matching against a large database of graphs particularly effective. Finally in Section 4, we empirically show the performance of these techniques to determine when they are best applied.

2   Topological n-Neighbourhood Features

The graphs dealt with in this paper are simple (no self loops, no duplicate edges) unlabelled graphs, both directed and undirected. A graph is defined as an ordered pair $G = (V, E)$, where $V = \{v_1, \ldots, v_n\}$ is a set of vertices and $E = \{\{v_x, v_y\}, \ldots\}$ is a set of edges.

A topological node feature is defined as any feature that is calculated solely on topological information, as viewed from a particular node. Some traditional TNFs used in graph matching are:

- degree. The number of adjacent nodes.
- clusterc (clustering coefficient). The number of edges between adjacent nodes (not including edges to the node being evaluated).
- ncliques$_k$. The number of cliques of size k that include a particular node.
- nwalksp$_k$. The number of walks of length k that pass through a particular node.

Both ncliques$_k$ and nwalksp$_k$ are vectors, holding values for each different k.

An n-neighbourhood (nN) of a node v is an induced subgraph formed from all the nodes that can be reached within n steps from v. This induced subgraph is centered around node v and contains all nodes up to n steps away, and all edges between those nodes. It is denoted as nN(v, n). For any single node v, a unique nN may be created for each value $n = 1, 2, \ldots, m$, where nN$(v, m) = G$ (the entire graph can be reached in m steps).

There are a number of TNFs that can be calculated from each nN of a node. Firstly we have the node count, or number of nodes in the nN, denoted by nN-ncount. Likewise we have the edge count, denoted by nN-ecount. Next we have the number of walks of length k in the nN, denoted nN-nwalks$_k$. Lastly we have the number of walks of length k in the nN, that pass through the main node, denoted nN-nwalksp$_k$. Each of these TNFs will give a different result for each nN of a node, giving n values, or $n * p$ values for nN-nwalks$_k$ and nN-nwalksp$_k$ where $k = 1, 2, \ldots, p$. The primary benefit of calculating TNFs on nNs is the reduced likelihood of *noise* (topological structure not present in the subgraph) from distant nodes being encoded in the feature. For small values of n, the features contain less information but also less noise. On larger values of n, the amount of information encoded is higher, but so is the likelihood that noise will prevent the feature from detecting mismatches.

3 Node Label Strengthening Framework

Our *strengthening framework* consists of the summation, listing, and tree indices, SI, LI and TI, respectively. In this order, there is a natural progression from one index to the next as they provide more resolution but take longer to compare. Each of these indices can be applied iteratively and works by incorporating the indices of neighbouring nodes. One or all of these indices may be applied for each different value created by a TNF or even an application-specific label (listing and tree indices only). A detailed description of these indices is given in the following subsections. Table 1 in Section 3.3 compares the strengths and weaknesses of each index.

3.1 Summation Index

The *Morgan index* [7] is an effective TNF, originally used to characterise chemical structures, and more recently to assist in graph isomorphism detection [5]. Despite its success in graph isomorphism, it has limited effectiveness on subgraph isomorphism.

Derived from the calculation procedure of the Morgan index, we propose the *summation index* (SI). The summation index propagates TNFs through the graph, allowing nodes to encode neighbouring structural information into their own *strengthened* TNFs. An example is shown in Figure 1. In this example, we show how SI can strengthen the node degree. Initially (iteration 0), the SI values of nodes $A - E$ are their TNF (degree in this case) values: 1, 2, 3, 2, and 2. For all subsequent iterations, the SI values are the sum of the neighbouring SI values from the last iteration. After iteration 1, these are 2, 4, 6, 5, and 5. After iteration 2, these are 4, 8, 14, 11, and 11. This process continues for a user-defined number of iterations, or as required by a particular matching algorithm. Note that on iteration 0, nodes B, D, and E all had an equal degree, and hence SI, but after iteration 1, B could be separated from the others.

Definition 1 (Summation Index (SI)).

$$SI_i(v) = \begin{cases} feature(v) & if\ i = 0, \\ \sum_u SI_{i-1}(u) & otherwise. \end{cases}$$

where u is a vertex adjacent to v.

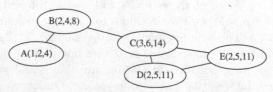

Fig. 1. Calculating the summation indices. Here we see a simple graph with three iterations of (degree) SI values shown for each node (iterations 0, 1, and 2).

As summation requires features to be added (+) and ranked (≤), it cannot be used for many application-specific labels. To strengthen such labels, the listing or tree indices below can be used.

3.2 Listing Index

The second feature strengthening technique in our framework is the *listing index* (LI). The listing index is a natural progression from summation, containing more information but also requiring more complex comparisons. Fankhauser et al. [5] presented this technique for graph isomorphism under the name *neighbourhood information*. A node's neighbourhood information is essentially a list (formally a multiset) of all feature values of the neighbouring nodes.

The key difference with LI is that features are evaluated separately. This provides more resolution, at the cost of increased comparison time. As with summation, this process can be repeated to include more information. The listing index of a node, at iteration i, is equal to the union of the listing indices of all neighbouring nodes at $i - 1$. For Figure 1, iteration 0 would be $\{1\}$, $\{2\}$, $\{3\}$, $\{2\}$, and $\{2\}$. This would then become $\{2\}$, $\{1, 3\}$, $\{2, 2, 2\}$, $\{2, 3\}$, and $\{2, 3\}$ for iteration 1, and $\{1, 3\}$, $\{2, 2, 2, 2\}$, $\{1, 2, 2, 3, 3, 3\}$, $\{2, 2, 2, 2, 3\}$, and $\{2, 2, 2, 2, 3\}$ for iteration 2.

The listing index also follows the same convention as summation, in that on each iteration, only the previous values of the neighbours are considered, with no regard to the node's own previous value.

Definition 2 (Listing Index (LI)).

$$LI_i(v) = \begin{cases} \{feature(v)\} & if\ i = 0, \\ \cup_u LI_{i-1}(u) & otherwise. \end{cases}$$

One advantage of listing over summation is that there is no requirement for the feature values to be numerical. The only requirement is that they can be compared for equality (=), unless they are TNFs being used on subgraph isomorphism, in which case they must be able to be ranked (≤).

3.3 Tree Index

The final feature strengthening technique we present is the *tree index* (TI). We show this technique in its pure form as a natural progression from the other indices, and discuss some alternative versions. This technique can be thought of as a second interpretation of the listing technique. The initial step is identical to listing, however the iterations are performed differently. In our listing technique, each iteration takes the union of neighbouring lists from the last iteration. Instead of taking the union of neighbouring lists from the last iteration, we simply create a list with those lists as elements. This creates an iteratively deeper list which can be thought of as a tree, beginning from the node and branching i layers, where i is the iteration number. Using the tree index, the second iteration indices for nodes $A - E$ would be $\{\{1, 3\}\}$, $\{\{2\}, \{2, 2, 2\}\}$, $\{\{1, 3\}, \{2, 3\}, \{2, 3\}\}$, $\{\{2, 2, 2\}, \{2, 3\}\}$, and $\{\{2, 2, 2\}, \{2, 3\}\}$ respectively.

Definition 3 (Tree Index (TI)).

$$TI_i(v) = \begin{cases} feature(v) & \text{if } i = 0, \\ \cup_u \{TI_{i-1}(u)\} & \text{otherwise.} \end{cases}$$

This provides us with a rich description of the node's local structure, resulting in more complex comparison challenges. In the worst case, where feature values are not ranked, this leaves us with a tree of linear assignments for each node comparison. Since the tree index effectively creates a tree of the graph starting from a node, there is no need to store the resulting values. Instead, we can simply traverse the graph during the matching.

Alternative versions of this have been presented for graph isomorphism [6,10] and subgraph isomorphism [12]. These alternative versions precompute the values and use a renaming step in an attempt to limit the size that must be stored. The effectiveness of this renaming step depends on the type of graph and can vary greatly.

Table 1. A naive comparison of the strengths and weaknesses of each index

	SI	LI	TI
Preprocessing Time	Very Low	Moderate	Zero
Preprocessing Space	Very Low	High	Zero
Matching Time	Very Low	Low	Very High
Pruning Effectiveness	Moderate	High	Very High

3.4 Strengthening in n-Neighbourhood

As mentioned in Section 2, TNFs calculated on an nN can be thought of as less noisy than their counterparts obtained on the main graph. This same concept applies equally (if not more) to the indices introduced in Sections 3.1 to 3.3. Instead of propagating the indices through the original graph, we can propagate them through the nN of each node. Although this means that each node's strengthened TNF values are calculated on a unique nN graph, these values are still valid to compare in subgraph isomorphism. The benefit of propagating through nNs is that it allows us to construct a very distinct picture of the local structure without being distorted by structural information many steps from the node. The downside to this is that structure many steps away is ignored completely, regardless of how useful such information could have been. Since nN propagation requires propagating information through each nN separately, the computation required is far more than propagation on the main graph, however the storage cost is not significantly higher. This makes nN propagation ideal for databases where graphs can be preprocessed once and matched many times.

4 Experimental Results

To evaluate the effectiveness of the techniques discussed in this paper, we first perform some analytical tests, followed by practical tests using a state-of-the-art subgraph isomorphism algorithm. For our testing data, we use positive subgraph isomorphism instances created using the geometric random graph generator from the igraph library [3]. This generator was chosen as it generates edges using a geometric method, resulting in graphs likely to be found in computer vision applications. Each test instance contains a 100 node full graph and a 90 node subgraph.

4.1 Evaluating Pruning Techniques by ABF

Subgraph isomorphism detection is most commonly performed using a search tree and some pruning techniques. Given a full graph with N nodes and a subgraph with M nodes, we have $\frac{N!}{(N-M)!}$ permutations in the search tree. The depth of the search tree is determined by the number of nodes in the subgraph. This value is static as solutions are found only at the leaves. The number of branches at each search tree node is the number of valid mappings possible for a particular subgraph node. This value, called the node's branching factor, starts as N, but may be pruned to be significantly lower.

To compare the pruning effectiveness of the techniques discussed in this paper, we use the average branching factor (ABF). We define the ABF as the average of the branching factors of subgraph nodes. To ensure the order of nodes does not skew the results, we calculate each node's branching factor as if it were at the root of the search tree. More information regarding ABF can be found in Section 3.5 of [8].

4.2 Analytical Experiments

We evaluate each of the techniques shown in this paper using the average branching factor (ABF) defined in Section 4.1, averaged over 10 test instances. For comparison with our techniques in the following figures, we provide the ABFs of the traditional TNFs that were described in Section 2. These ABF values are: 39.5 (degree), 39.7 (clusterc), 57.0 (ncliques$_k$, $k = 3, 4, \ldots, 8$), 48.1 (nwalksp$_k$, $k = 1, 2, \ldots, 8$), and 29.5 (the combination of all four). These ABF values are based solely on the traditional TNFs, without the use of any nN or index-related strengthening. As such, they can be compared *as is* with every value reported in the following figures.

In Figure 2 we compare the results from the different n-neighbourhood (nN) features defined in Section 2. For each feature, we show the ABF and comparison time for nNs up to a depth of 10 steps. Note that for a maximum nN depth of x in the figure, nNs are calculated and compared for $n = 1, 2, \ldots, x$. The comparison time here is the average time required to determine all compatibilities between the subgraph and full graph nodes on each test instance.

Comparing these figures to the traditional TNFs listed above, we can clearly see the pruning effectiveness of these nN features. At a maximum nN depth of just two steps, each nN feature has achieved more pruning than all four of the traditional TNFs. While not performing as well as some others, the nN-nwalksp feature here is particularly noteworthy, as it is the only feature that has a non-nN

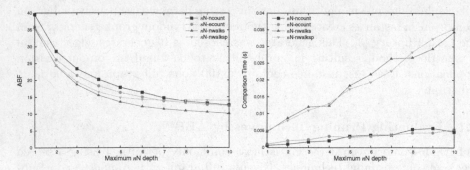

Fig. 2. ABF and comparison times for n-neighbourhood features

counterpart. A comparison between the non-nN nwalksp (ABF: 48.1) and nN-nwalksp (ABF: 35.3 $\rightarrow$ 14.0) shows the effectiveness of nNs. This effectiveness comes from the noise-reduction inherent in nNs, as discussed in Section 3.4.

Figure 3 compares the results for each of the strengthening indices detailed in Sections 3.1 to 3.3, and their nN-strengthened counterparts from Section 3.4. These figures show how the ABF and comparison times change as the maximum iteration number of the indices increases. The TNF strengthened here is degree, as this simple TNF will best show the capabilities of the strengthening indices. Note that due to the significant time required for tree-index comparisons, all times are shown on a logarithmic scale.

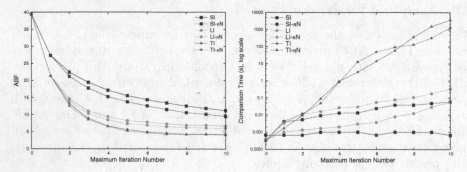

Fig. 3. ABF and comparison times for SI/LI/TI and their nN-strengthened counterparts, using only the degree TNF

Starting from the standard degree ABF of 39.5, it takes only a single iteration before all indices are below the 29.5 ABF of the combined traditional TNFs. After four iterations, both LI and TI have achieved greater pruning with just degree than any single TNF, including nN TNFs. Given a more complex feature, or combination of features, these strengthening indices can achieve even greater pruning. Of course as we include more features and strengthening techniques, the comparison time may also increase.

4.3   Practical Experiments

The ABF reductions in the previous section show that our techniques can simplify the matching problem, allowing matching algorithms to perform faster. In this section, we determine whether this reduction in matching time is worth the increase in feature comparison time. To achieve this, we perform subgraph isomorphism detection (searching for all solutions) using the VF2 algorithm, as this is the most commonly used benchmark.

Each full graph in our test set contains exactly 100 nodes. However in order to show how the number of edges affects the performance of certain techniques, we run our tests five times, each time with an increasing number of edges. For each algorithm and number of edges, we report the average matching time for 100 test instances. Figure 4 shows subgraph isomorphism detection times for VF2 when combined with the techniques from this paper.

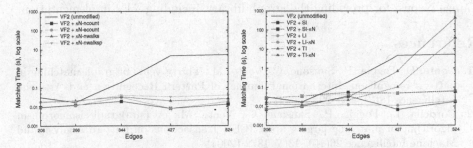

Fig. 4. Subgraph Isomorphism times for VF2 using the nN TNFs and strengthening techniques (again only strengthening degree) from this paper. Regular VF2 is included for comparison.

Due to the exponentiality of the graph matching problem, VF2 often takes significantly longer on certain instances, giving its results a high standard deviation. In addition to reducing the average matching times as shown in Figure 4, our techniques also help VF2 overcome these hard instances. This results in a more consistent matching time, which is advantageous in real-time applications.

5   Conclusions

In this paper we proposed a number of strengthening techniques that can greatly increase the pruning power of both TNFs and application-specific labels. By iteratively encoding the neighbouring topological information, these techniques were able to significantly reduce matching times while only using the simple TNF of degree. Both the summation and listing indices were able to perform well in all cases, whereas the tree index only proved useful on graphs with fewer edges. With index type, TNFs used, and iteration depth configurable, this framework can be tailored to suit particular problems or classes of graphs as required.

Additionally we presented some new TNFs based on the n-neighbourhoods of nodes. These TNFs use the reduced noise inherent in nNs to identify otherwise-hidden topological incompatibilities between mismatched nodes.

Through analytical and practical experiments, the effectiveness of these techniques has been shown to achieve significant gains over the standard VF2 algorithm. Subsequent testing has also shown that by adding these strengthening techniques to the *iterative node elimination* technique from [4], matching times can be over twice as fast as those reported in that paper, on the same data.

It should be noted that certain constraint programming algorithms, such as LAD, achieve significantly lower gains from these techniques. This is due to additional requirements to integrate TNFs and how such algorithms utilise this information in the matching process. The creation of alternative integration techniques to circumvent this issue remains an open research problem. Another interesting research problem is the creation and matching of topological edge features.

Acknowledgements. The authors would like to thank Gábor Csárdi and Tamás Nepusz for providing the igraph library, including VF2 implementation.

References

1. Conte, D., Foggia, P., Sansone, C., Vento, M.: Thirty years of graph matching in pattern recognition. International Journal of Pattern Recognition and Artificial Intelligence 18(3), 265–298 (2004)
2. Cordella, L., Foggia, P., Sansone, C., Vento, M.: A (sub)graph isomorphism algorithm for matching large graphs. IEEE Transactions on Pattern Analysis and Machine Intelligence 26(10), 1367–1372 (2004)
3. Csardi, G., Nepusz, T.: The igraph software package for complex network research. Inter. Journal Complex Systems 1695, 1–9 (2006),
   http://igraph.sourceforge.net
4. Dahm, N., Bunke, H., Caelli, T., Gao, Y.: Topological features and iterative node elimination for speeding up subgraph isomorphism detection. In: Proceedings of the 21st International Conference on Pattern Recognition (2012)
5. Fankhauser, S., Riesen, K., Bunke, H., Dickinson, P.: Suboptimal graph isomorphism using bipartite matching. International Journal of Pattern Recognition and Artificial Intelligence (accepted for publication)
6. McKay, B.B.: Practical graph isomorphism. Congressus Numerantium 30, 45–87 (1981)
7. Morgan, H.L.: The generation of a unique machine description for chemical structures - a technique developed at chemical abstracts service. Journal of Chemical Documentation 5(2), 107–113 (1965)
8. Russell, S., Norvig, P.: Artificial Intelligence: A Modern Approach, 1st edn. Prentice Hall Press, Upper Saddle River (1995)
9. Solnon, C.: AllDifferent-based filtering for subgraph isomorphism. Artificial Intelligence 174(12–13), 850–864 (2010)
10. Sorlin, S., Solnon, C.: A parametric filtering algorithm for the graph isomorphism problem. Constraints 13, 518–537 (2008)
11. Ullmann, J.R.: An algorithm for subgraph isomorphism. Journal of the ACM 23(1), 31–42 (1976)
12. Zampelli, S., Deville, Y., Solnon, C.: Solving subgraph isomorphism problems with constraint programming. Constraints 15(3), 327–353 (2010)

On the Complexity of Submap Isomorphism

Christine Solnon[1,2], Guillaume Damiand[1,2], Colin de la Higuera[3],
and Jean-Christophe Janodet[4]

[1] INSA de Lyon, LIRIS, UMR 5205 CNRS, 69621 Villeurbanne, France
[2] Université de Lyon, France
[3] LINA, UMR CNRS 6241, Université de Nantes, France
[4] IBISC, Université d'Evry, France

Abstract. Generalized maps describe the subdivision of objects in cells, and incidence and adjacency relations between cells, and they are widely used to model 2D and 3D images. Recently, we have defined submap isomorphism, which involves deciding if a copy of a pattern map may be found in a target map, and we have described a polynomial time algorithm for solving this problem when the pattern map is connected. In this paper, we show that submap isomorphism becomes NP-complete when the pattern map is not connected, by reducing the NP-complete problem Planar-4 3-SAT to it.

1 Motivations

Combinatorial maps and generalized maps [1] are very nice data structures to model the topology of nD objects subdivided in cells (*e.g.*, 0D vertices, 1D edges, 2D faces, 3D volumes, ...) by means of incidence and adjacency relationships between these cells. In 2D, maps may be used to model the topology of an embedding of a planar graph in a plane. In particular, these models are very well suited for scene modeling [2], and for 2D and 3D image segmentation [3].

In [4], we have defined a basic tool for comparing 2D maps, *i.e.*, submap isomorphism (which involves deciding if a copy of a pattern map may be found in a target map), and we have proposed an efficient polynomial-time algorithm for solving this problem when the pattern map is connected. This work has been generalized to nD maps in [5]. The subisomorphism defined in [5] is based on induced submap relations, such that submaps are obtained by removing some darts and all their seams, just like induced subgraphs are obtained by removing some vertices and all their incident edges. In [6], we have introduced a new kind of submap relation, called partial submap: partial submaps are obtained by removing not only some darts (and all their seams), but also some other seams, just like partial subgraphs are obtained by removing not only some vertices (and their incident edges), but also some other edges. The polynomial time algorithm described in [5] for solving the induced submap isomorphism problem may be extended to the partial case in a very straightforward way. However, this algorithm still assumes that the pattern map is connected. In this paper, we show that the submap isomorphism problem becomes NP-complete when the pattern map is not connected, both for partial and induced submaps.

W.G. Kropatsch et al. (Eds.): GbRPR 2013, LNCS 7877, pp. 21–30, 2013.
© Springer-Verlag Berlin Heidelberg 2013

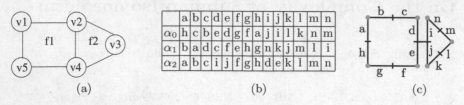

	a	b	c	d	e	f	g	h	i	j	k	l	m	n
α_0	h	c	b	e	d	g	f	a	j	i	l	k	n	m
α_1	b	a	d	c	f	e	h	g	n	k	j	m	l	i
α_2	a	b	c	i	j	f	g	h	d	e	k	l	m	n

(a)   (b)   (c)

Fig. 1. (a) A plane graph. (b) The corresponding 2G-map. (c) Its graphical representation: darts are represented by segments labeled with letters, consecutive darts separated with a little segment are 0-sewn (*e.g.*, $\alpha_0(b) = c$ and $\alpha_0(c) = b$), consecutive darts separated with a dot are 1-sewn (*e.g.*, $\alpha_1(a) = b$ and $\alpha_1(b) = a$), parallel adjacent darts are 2-sewn (*e.g.*, $\alpha_2(d) = i$ and $\alpha_2(i) = d$).

Outline of the Paper. In Section 2, we recall definitions related to generalized maps. In Section 3, we define the submap isomorphism problem and recall some complexity results about this problem. In Section 4, we describe the planar-4 3-SAT problem, which is NP-complete. In Section 5, we describe a polynomial-time reduction of planar-4 3-SAT to submap isomorphism, thus showing that submap isomorphism is NP-complete.

2   Recalls and Basic Definitions on Generalized Maps

In this work we consider generalized maps, which are more general than combinatorial maps, and we refer the reader to [1] for more details.

Definition 1. *(nG-map) Let $n \geq 0$. An n-dimensional generalized map (or nG-map) is defined by a tuple $G = (D, \alpha_0, \ldots, \alpha_n)$ such that (i) D is a finite set of darts; (ii) $\forall i \in \{0, \ldots, n\}$, α_i is an involution[1] on D; and (iii) $\forall i, j \in \{0, \ldots, n\}$ such that $i + 2 \leq j$, $\alpha_i \circ \alpha_j$ is an involution.*

2G-maps may be used to model the embedding of a planar graph into a plane. For example, Fig. 1 displays a plane graph and the corresponding 2G-map. We say that a dart d is *i-sewn* with a dart d' whenever $d = \alpha_i(d')$ and $d \neq d'$, whereas it is *i-free* whenever $d = \alpha_i(d)$. A *seam* is a tuple (d, i, d') such that d' is i-sewn to d. For example, $(a, 0, h)$ is a seam of the map displayed in Fig. 1 because $\alpha_0(a) = h$.

Definition 2. *(seams of a set of darts in an nG-map) Let $G = (D, \alpha_0, \ldots, \alpha_n)$ be an nG-map and $E \subseteq D$ be a set of darts. The set of seams associated with E in G is: $seams_G(E) = \{(d, i, \alpha_i(d)) | d \in E, i \in \{0, \ldots, n\}, \alpha_i(d) \in E, \alpha_i(d) \neq d\}$.*

A map is connected if any pair of darts is connected by a path of sewn darts.

Definition 3 (Connected map). *A generalized map $G = (D, \alpha_0, \ldots, \alpha_n)$ is connected if $\forall d \in D, \forall d' \in D$, there exists a path between d and d', i.e., a sequence of darts $(d_1, \ldots, d_k)$ such that $d_1 = d$, $d_k = d'$ and $\forall i \in \{1, \ldots, k-1\}$, $\exists j_i \in \{0, \ldots, n\}, d_{i+1} = \alpha_{j_i}(d_i)$.*

---

[1] An involution f on D is a bijective mapping from D to D such that $f = f^{-1}$.

Map isomorphism [1] allows us decide of the equivalence of two maps.

Definition 4. *(nG-map isomorphism [1]) Two nG-maps $G = (D, \alpha_0, \ldots, \alpha_n)$ and $G' = (D', \alpha'_0, \ldots, \alpha'_n)$ are isomorphic if there exists a bijection $f : D \to D'$, such that $\forall d \in D, \forall i \in [0, n], f(\alpha_i(d)) = \alpha'_i(f(d))$.*

In [4], induced submaps have been defined: G is an induced submap of G' if G preserves all seams of G', i.e, for every couple of darts (d_1, d_2) of G, d_1 is i-sewn to d_2 in G' if and only if d_1 is i-sewn to d_2 in G.

Definition 5. *(induced submap) A map $G' = (D', \alpha'_0, \ldots, \alpha'_n)$ is an induced submap of $G = (D, \alpha_0, \ldots, \alpha_n)$ if $D' \subseteq D$ and $seams_{G'}(D') = seams_G(D')$.*

In [6], we have introduced another submap relation, called partial submap by analogy with existing work on graphs. Indeed, induced subgraphs are obtained by removing some nodes (and all their incident edges) whereas partial subgraphs are obtained by removing not only some nodes (and all their incident edges) but also some edges. In our map context, partial submaps are obtained by removing not only some darts (and all their seams) but also some other seams.

Definition 6. *(partial submap) A map $G' = (D', \alpha'_0, \ldots, \alpha'_n)$ is a partial submap of $G = (D, \alpha_0, \ldots, \alpha_n)$ if $D' \subseteq D$ and $seams_{G'}(D') \subseteq seams_G(D')$.*

3   The Submap Isomorphism Problem

The submap isomorphism problem involves deciding if a pattern map is isomorphic to a submap of a target map, and it is formally defined as follows:

> **Problem:** Partial (resp. induced) submap isomorphism
> **Instance:** A triple (n, G, G') such that $n > 0$, and G and G' are nG-maps.
> **Question:** Does there exist a partial (resp. induced) submap of G' which is isomorphic to G?

We note $G \sqsubseteq^p G'$ (resp. $G \sqsubseteq^i G'$) when the answer is yes. Note that $G \sqsubseteq^i G' \Rightarrow G \sqsubseteq^p G'$. Fig. 2 displays examples of submap isomorphisms.

The complexity of the submap isomorphism problem depends on the connectedness of the pattern map. For example, the map G_1 of Fig. 2 is not connected, and is composed of two connected components, whereas the maps G_2, G_3 and G_4 are connected. In [5], we have described a polynomial-time algorithm which solves the submap isomorphism problem when the pattern map G is connected. When the pattern map G is not connected, we may use this algorithm to search for all occurrences of each connected component of G in the target map G'. Let us consider, for example, the map G_1 of Fig. 2. Its left hand side component occurs once in G_2 and twice in G_3 and G_4, and its right hand side component occurs twice in G_2, G_3 and G_4 (as it is automorphic). To solve the submap isomorphism problem from these occurrence lists, we have to solve the following combinatorial problem: Can we select one occurrence in G' of each connected component of G so that the selected occurrences do not overlap in G'?

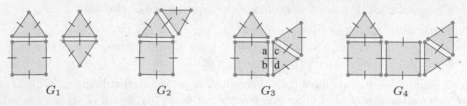

Fig. 2. Submap isomorphism examples. G_1 is not isomorphic to a submap of G_2 (i.e., $G_1 \not\sqsubseteq^p G_2$ and $G_1 \not\sqsubseteq^i G_2$), though each connected component of G_1 is a submap of G_2. G_1 is isomorphic to a partial submap of G_3, but not to an induced one (i.e., $G_1 \sqsubseteq^p G_3$ and $G_1 \not\sqsubseteq^i G_3$), because the seams $(a, 2, c)$ and $(b, 2, d)$ of G_3 are not preserved in G_1. G_1 is isomorphic to an induced submap of G_4 and, therefore, it is also isomorphic to a partial submap of G_4 (i.e., $G_1 \sqsubseteq^p G_4$ and $G_1 \sqsubseteq^i G_4$).

Theorem 1 claims that this combinatorial problem is NP-complete.

Theorem 1. *The partial (resp. induced) submap isomorphism problem is $\mathcal{NP}$-complete.*

The problem trivially belongs to NP since one can check that a given partial (resp. induced) submap of the target map G' is isomorphic to the pattern map G in polynomial time. We may use for example the polynomial algorithm of [5], which has been defined for non connected maps.

To prove that it is NP-complete, we show in Section 5 that Planar-4 3-SAT, which is known to be NP-complete, may be reduced to it in polynomial time.

4   Planar-4 3-SAT

Planar-4 3-SAT is a special case of the SAT problem, which involves deciding if there exists a truth assignment for a set X of variables such that a boolean formula F over X is satisfied [7]. We assume that F is in Conjunctive Normal Form (CNF), *i.e.*, it is a conjunction of clauses such that each clause is a disjunction of literals which are either variables of X or negations of variables of X.

The formula-graph associated with a CNF formula F over a set of variables X is the bipartite graph $G_{X,F} = (V, E)$ such that V associates a vertex with every variable $x_i \in X$ and every clause c_j of F, and E associates an edge (x_i, c_j) with every variable/clause couple such that variable x_i occurs in clause c_j.

The planar-4 3-SAT problem is formally defined as follows.

Problem: Planar-4 3-SAT

Instance: A couple (X, F) such that X is a set of boolean variables and F is a CNF formula over X such that (i) every clause of F is a disjunction of 3 literals, (ii) the formula-graph $G_{X,F}$ is planar, and (iii) the degree of every vertex of $G_{X,F}$ is bounded by 4 (*i.e.*, each variable occurs in at most 4 different clauses)

Question: Does there exist a truth assignment for X which satisfies F?

$X = \{x, y, z, u, w\}$
$F = (\bar{x} \vee y \vee u) \wedge$
$\quad (\bar{x} \vee y \vee \bar{z}) \wedge$
$\quad (\bar{y} \vee z \vee u) \wedge$
$\quad (\bar{z} \vee u \vee \bar{w}) \wedge$
$\quad (x \vee w \vee \bar{u})$

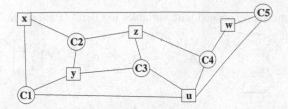

Fig. 3. An instance of Planar-4 3-SAT and its associated formula graph (clauses correspond to circles, and variables to squares)

Planar-4 3-SAT has been shown to be NP-complete in [8]. Fig. 3 displays an instance of Planar-4 3-SAT and its associated formula-graph.

To reduce an instance (X, F) of Planar-4 3-SAT to an instance (n, G, G') of submap isomorphism, we first perform a preprocessing: We iteratively eliminate from (X, F) every variable $x_i \in X$ which occurs in only one clause c_j of F (those whose degree is equal to 1 in the formula-graph), set x_i to the truth value which satisfies c_j, and eliminate c_j from F, until either X and F become empty (thus showing that the answer is trivially yes), or all variables in X occur in 2, 3, or 4 clauses of F.

5   Reduction of Planar-4 3-SAT to Submap Isomorphism

Let us first show that planar-4 3-SAT can be reduced to induced submap isomorphism in polynomial time: The partial case will be studied at the end of this section. We consider an instance (X, F) of planar-4 3-SAT and we show how to build an instance (n, G, G') such that $G \sqsubseteq^i G'$ iff the answer to (X, F) is yes. We consider 2G-maps, so that $n = 2$, and the 2G-maps G and G' are constructed by assembling building blocks which are 2G-maps. Fig. 4 displays building blocks associated with variables: For each variable $x_i \in X$ such that the degree of x_i in the formula-graph $G_{X,F}$ is equal to k with $2 \leq k \leq 4$ (as the preprocessing step has removed any variable whose degree is equal to 1), we build two variable patterns V'_k and V_k which will respectively occur in G' and G. Each variable pattern V'_k (resp. V_k) looks like a flower whose core is a $2k$-edge face and which have $2k$ petals (resp. k petals), where each petal is a 6-edge face. For each petal in each variable pattern V'_k, the edge opposite to the core of the flower is a connecting edge which may be 2-sewn with clause patterns to define G'.

For each clause, we build two clause patterns C' and C which will respectively occur in G' and G. The clause pattern C' is composed of a 3-edge central face which has 3 adjacent 4-edge faces, whereas the clause pattern C is composed of a 3-edge face which has 1 adjacent 4-edge face, as displayed below:

Clause pattern C':   Clause pattern C: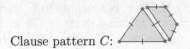

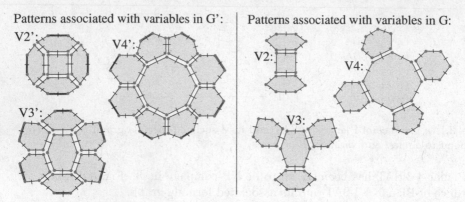

Fig. 4. Variable patterns used as building blocks to define G' and G. Connecting edges in G' are displayed in bold.

Edges of C' displayed in bold are connecting edges which are 2-sewn with variable patterns to define G'.

Definition of the 2G-map G'. For each variable $x_i \in X$ such that the degree of x_i in the formula-graph $G_{X,F}$ is equal to k, G' contains an occurrence of the variable pattern V'_k. Each petal of this occurrence of V'_k is alternatively labeled with x_i and $\bar{x}_i$. For each clause c_j of F, G' contains an occurrence of the clause pattern C'. Each 4-edge face of this occurrence of C' is labeled with a different literal of c_j. Variable and clause patterns are 2-sewn to define a connected 2G-map: every connecting edge of each clause pattern is 2-sewn with a different connecting edge of a variable pattern such that the two faces which become adjacent by this seam are labeled with the same literal. We can easily check that this 2G-map can always be built in polynomial time as the formula-graph $G_{X,F}$ is planar, and there exist polynomial-time algorithms for embedding a planar graph in a plane [9]: We can use the same embedding for constructing G'. Fig. 5 displays the 2G-map associated with the formula displayed in Fig. 3.

Definition of the 2G-map G. If the SAT instance has n variables and c clauses, then G is composed of $n + c$ different components: a component V_k is associated with every variable $x_i \in X$, where k is the degree of x_i in $G_{X,F}$; a component C is associated with every clause. For example, the 2G-map G associated with the formula displayed in Fig. 3 contains 10 components: 3 occurrences of V_3, 1 occurrence of V_4, 1 occurrence of V_2, and 5 occurrences of C.

Proof of $(G \sqsubseteq^i G') \Rightarrow (\exists \text{ truth assignment of } X \text{ which satisfies } F)$. Let us first assume that there exists an induced submap G'' of G' which is isomorphic to G, and let us show that there exists a truth assignment of X which satisfies F.

If G'' is isomorphic to G then, according to Def. 4, there exists a bijection f which matches darts of G'' with darts of G and which preserves all seams. By extension, we say that f matches faces of G'' with faces of G. As we consider

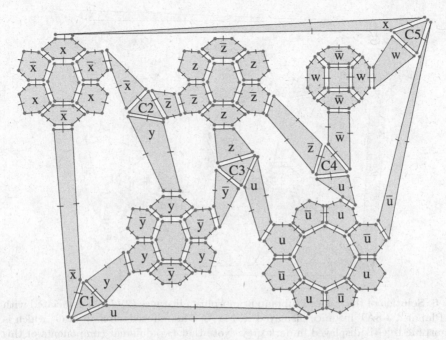

Fig. 5. 2G-map G' associated with the SAT instance displayed in Fig. 3. Note that this map contains holes (corresponding to white parts in the figure): each dart d adjacent to these holes is 2-free so that $\alpha_2(d) = d$.

induced submap isomorphism, two faces of G which belong to two different connected components cannot be matched by f with faces which are 2-sewn in G'' (according to Def. 5). Fig. 6 displays an example of such a solution for the instance $(2, G, G')$ of the induced submap isomorphism problem associated with the instance (X, F) of Planar-4 3-SAT displayed in Fig. 3.

G contains c occurrences of C, where c is the number of clauses of F. Each occurrence of C has a 3-edge face adjacent to a 4-edge face. These faces can only be matched with faces which belong to occurrences of C' in G' as 3-edge faces in G' only come from C' patterns. As there are c occurrences of C in G, each occurrence of C' in G' is matched with a different occurrence of C in G. For the same reasons, each occurrence of a variable pattern V_k in G is matched with a different occurrence of a variable pattern V_k' in G': Petal and core faces in G can only be matched with petal and core faces in G', and an occurrence of V_i cannot be matched with faces of an occurrence of V_j' if $i \neq j$. For each variable pattern V_k', the label of the petals of V_k' which are not matched with petals of variable patterns of G gives the truth assignment for the corresponding variable. For each clause pattern C', the label of the 4-edge face of C' which is matched with a 4-edge face of C corresponds to a literal which satisfies the clause associated with C'. As two faces of G which belong to two different connected components cannot be matched by f with faces which are 2-sewn in G'', we ensure that when a 4-edge face of a clause pattern is matched, then the adjacent petal is not

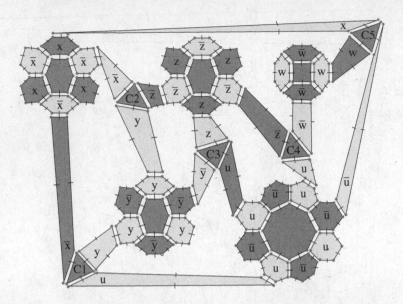

Fig. 6. Solution of the induced submap isomorphism instance $(2, G, G')$ associated with the Planar-4 3-SAT instance displayed in Fig. 3. The induced submap of G' which is isomorphic to G is displayed in dark grey. Note that two different components of this submap cannot be 2-sewn in G' as we consider induced submap isomorphism.

matched, *i.e.*, when a clause is satisfied by a literal l, then no other clause can be satisfied by the negation of this literal so that the truth assignment deduced from the flower matching actually satisfies all clauses of F. For example, the truth assignment corresponding to the solution displayed in Fig. 6 is $\{\bar{x}, y, \bar{z}, u, w\}$.

Proof of ($\exists$ truth assignment of X which satisfies F) $\Rightarrow$ ($G \sqsubseteq^i G'$). Let us assume that there exists a truth assignment of X which satisfies F and let us show that there exists an induced submap G'' of G' which is isomorphic to G, i.e., that there exists a dart matching which preserves all seams of G. For each variable pattern V_k in G associated with a variable x_i, we match the darts of the core face with the darts of the core face of the variable pattern associated with x_i in G' and we match the darts of the k 6-edge petals of V_k with the darts of the k 6-edge petals which are labeled with the negation of the truth value of x_i. For each clause pattern C in G associated with a clause c_j, we match the darts of the 3-edge face of C with the darts of the 3-edge face of the clause pattern associated with c_j in G' and we match the darts of the 4-edge face of C with the darts of one of the three 4-edge faces: We choose a 4-edge face which is labeled with a literal which is satisfied by the truth assignment (this 4-edge face cannot be 2-sewn with a matched 6-edge petal).

Proof for the Partial Case. Let us now consider the partial case: We consider an instance (X, F) of planar-4 3-SAT and we show how to build an instance (n, G, G') such that $G \sqsubseteq^p G'$ iff the answer to (X, F) is yes. The proof is similar

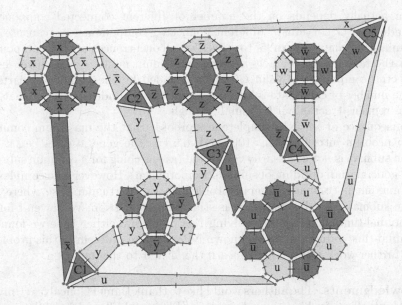

Fig. 7. Solution of the partial submap isomorphism instance $(2, G, G')$ associated with the Planar-4 3-SAT instance displayed in Fig. 3. The partial submap of G' which is isomorphic to G is displayed in dark grey.

to the induced case. The difference between the induced and the partial cases is that, when considering induced submap isomorphism, two faces which belong to two different components in G cannot be matched with faces of G' which are 2-sewn whereas, when considering partial submap isomorphism, two faces which belong to two different components in G may be matched with faces of G' which are 2-sewn. Therefore, we modify the clause pattern C so that the 4-edge face is adjacent to a 3-edge face, on one side, and to a 6-edge face on the opposite side, as displayed below:

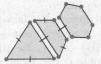

These 6-edge faces can only be matched with petals. The label of the petal which is matched with the 6-edge face of a clause pattern corresponds to the literal which satisfies the clause. Fig. 7 displays an example of solution for partial submap isomorphism.

6   Conclusion

We have shown that submap isomorphism is NP-complete when the pattern map G is not connected. This implies that there does not exist a polynomial-time algorithm for this problem, unless P=NP. The practical tractability of this problem actually depends on the number of different connected components of G. Indeed, if G contains k different connected components, we can use the polynomial-time algorithm of [5] to search for all occurrences of each component of G in the target map G'. Let m be the maximum number of occurrences of a connected component of G in G' (m is bounded by the number of darts of G'). The number of candidate solutions to explore is bounded by m^k so that the problem remains tractable if k is small enough.

A consequence of our NP-completeness proof is that the maximum common submap problem introduced in [10] is NP-hard in the general case, i.e., if the common submap is not necessarily connected (as searching for a common submap is more general than deciding of submap isomorphism). However, the complexity of the maximum common submap problem in the particular case where the common submap must be connected is still an open question: We haven't found a polynomial-time algorithm for solving this problem, neither have we found a polynomial-time reduction from a known NP-complete problem to this problem. Hence, further work will mainly concern the answer to this question.

Acknowledgments. The authors would like to thank Daniel Goncalves (University of Montpellier) for his pointer to problem Planar-3SAT, and fruitful remarks.

References

1. Lienhardt, P.: N-dimensional generalized combinatorial maps and cellular quasi-manifolds. Computational Geometry and Applications 4(3), 275–324 (1994)
2. Fradin, D., Meneveaux, D., Lienhardt, P.: A hierarchical topology-based model for handling complex indoor scenes. Computer Graphics Forum 25(2), 149–162 (2006)
3. Braquelaire, J.P., Brun, L.: Image segmentation with topological maps and inter-pixel representation. Visual Communication and Image Representation 9(1), 62–79 (1998)
4. Damiand, G., de la Higuera, C., Janodet, J.-C., Samuel, É., Solnon, C.: A polynomial algorithm for submap isomorphism: Application to searching patterns in images. In: Torsello, A., Escolano, F., Brun, L. (eds.) GbRPR 2009. LNCS, vol. 5534, pp. 102–112. Springer, Heidelberg (2009)
5. Damiand, G., Solnon, C., de la Higuera, C., Janodet, J.C., Samuel, E.: Polynomial algorithms for subisomorphism of nd open combinatorial maps. Computer Vision and Image Understanding (CVIU) 115(7), 996–1010 (2011)
6. Combier, C., Damiand, G., Solnon, C.: From maximum common submaps to edit distances of generalized maps. Pattern Recognition Letters 33(15), 2020–2028 (2012)
7. Cook, S.A.: The complexity of theorem-proving procedures. In: ACM Symposium on Theory of Computing, pp. 151–158 (1971)
8. Jansen, K., Müller, H.: The minimum broadcast time problem for several processor networks. Theoretical Computer Science 147(1-2), 69–85 (1995)
9. Mohar, B.: A linear time algorithm for embedding graphs in an arbitrary surface. SIAM Journal on Discrete Mathematics 12(1), 6–26 (1999)
10. Combier, C., Damiand, G., Solnon, C.: Measuring the distance of generalized maps. In: Jiang, X., Ferrer, M., Torsello, A. (eds.) GbRPR 2011. LNCS, vol. 6658, pp. 82–91. Springer, Heidelberg (2011)

Flooding Edge Weighted Graphs

Fernand Meyer

CMM-Centre de Morphologie Mathématique,
Mathématiques et Systèmes, MINES ParisTech, France
fernand.meyer@mines-paristech.fr

Abstract. This paper characterizes floodings on edge weighted graphs. Of particular interest are the highest floodings of a graph below a ceiling function defined on the nodes. Two classes of algorithms for their construction are presented. The first are applied on the dendrogram representing the hierarchy associated to the edge weighted graph. The second consist in shortest distance algorithms on the graph itself.

1 Introduction

Edge weighted graphs are ubiquitous in the field of classification and image processing. A hierarchy is easily derived from an edge weighted graph: cutting all edges with a weight above some threshold produces a number of connected subgraphs, representing each one scale of a taxonomy. For higher thresholds less edges are cut, resulting in larger subgraphs, obtained by the union of smaller ones. If the nodes represent the different tiles of a partition, and the edge weights represent a dissimilarity between adjacent tiles, the hierarchy is a series of nested partitions which are coarser and coarser, each tile at a given level, being obtained by the union of tiles at lower levels. In the region adjacency graph for instance, the nodes represent the catchment basins of a topographic surface, edges link neighboring basins and their weights represents the altitudes of the pass points between neighboring basins. If the topographic surface is flooded, the flood passes from basin to basin through these pass points. The progression of the flood is thus the same on the topographic surface or on the RAG. The present paper defines floodings on arbitrary edge weighted graphs. Criteria are given characterizing physically valid floodings. We then study the extension of a lake containing a given node when its flooding level increases. The highest flooding below a ceiling function defined on the nodes is unique. It has a great interest in image segmentation and filtering. Various algorithms are proposed for its construction.

2 The Laws of Hydrostatics and Floodings

2.1 Criteria Characterizing a Flooding

Consider a non oriented node and edge weighted graph $G = [E, N]$, E representing the edges and N the nodes.

W.G. Kropatsch et al. (Eds.): GbRPR 2013, LNCS 7877, pp. 31–40, 2013.

Fig. 1. Tank and pipe network:
- a and b form a regional minimum with $\tau_a = \tau_b = \lambda$; $e_{ab} \leq \lambda$; $e_{bc} > \lambda$
- b and c have unequal levels but are separated by a higher pipe.
- d and e form a full lake, reaching the level of its lowest exhaust pipe e_{cd}
- e and f have the same level ; however they do not form a lake, as they are linked by a pipe which is higher.

The distribution in the last four tanks is not compatible with the laws of hydrostatics.

In order to give a physical interpretation to our graph, we consider the nodes as vertical tanks of infinite height and depth. The weight τ_i represents the level of water in the tank i, equal to $-\infty$ if no water is present. Two neighboring tanks i and j are linked by a pipe at an altitude e_{ij} equal to the weight of the edge. We call such an edge weighted graph a tank network. Edge weights e and flooding levels τ take their values in $[-\infty, +\infty]$. We suppose that the laws of hydrostatics apply to our network of tanks and pipes:

* if the level τ_i in the tank i is higher than the pipe e_{ij}, then $\tau_i = \tau_j$.
* the level τ_i in the tank i cannot be higher than the level τ_j, unless $e_{ij} \geq \tau_i$.

In fact, this second condition implies the first one. We adopt it as a criterion defining valid floodings on a tank network.

Definition 1. *The distribution τ of water in the tanks of the graph $G = [E, N]$ is a flooding of this graph, i.e. is a stable distribution of fluid if it verifies the criterion { for any couple of neighboring nodes (p, q) : $(\tau_p > \tau_q \Rightarrow e_{pq} \geq \tau_p)$ (criterion 1) }*

Figure 1 presents a number of configurations compatible with the laws of hydrostatics and others which are not.

The following equivalences yield other useful criteria for recognizing flood distributions on tank networks:

$$(\tau_p > \tau_q \Rightarrow e_{pq} \geq \tau_p) \Leftrightarrow (\text{not } (\tau_p > \tau_q) \text{ or } e_{pq} \geq \tau_p) \Leftrightarrow$$
$$(\tau_p \leq \tau_q \text{ or } \tau_p \leq e_{pq}) \Leftrightarrow (\tau_p \leq \tau_q \vee e_{pq})   (criterion\ 2)$$

2.2   The Algebra of Floodings

Lemma 1. *If τ and ν are two floodings of a tank network G, then $\tau \vee \nu$ and $\tau \wedge \nu$ also are floodings of G.*

2.3   Creation of Lakes

We first define the ultrametric flooding distance $ud(p, q)$ between two nodes p and q on an edge weighted graph as the lowest value λ such that there exists a connected path between p and q with no edge higher than λ. The highest edge along the path has a weight λ. For a node p the closed ball of centre p and radius ρ is defined by $\overline{\text{Ball}}(p, \rho) = \{q \in N \mid \text{ud}(p, q) \le \rho\}$. Such balls have strange properties:

- two closed balls with the same radius are either disjoint or identical.
- each element of a closed ball is a centre of this ball.
- the radius of a ball is equal to its diameter, that is the longest distance between two nodes in the ball.

Open balls $\text{Ball}(p, \rho) = \{q \in N \mid \text{ud}(p, q) < \rho\}$ have similar properties.

The following lemma presents the basic mechanism generating lakes.

Lemma 2. *If (p, q) are neighboring nodes of the flooded graph G, linked by an edge with weight $e_{pq} < \tau_p$, then $\tau_p = \tau_q$.*

Proof. Indeed the criterion $(\tau_p > \tau_q \Rightarrow e_{pq} \ge \tau_p)$ is equivalent with $(e_{pq} < \tau_p \Rightarrow \tau_p \le \tau_q)$. Hence if $e_{pq} < \tau_p$, we have $\tau_p \le \tau_q$; so we also have $e_{pq} < \tau_q$ implying $\tau_q \le \tau_p$; finally $\tau_p = \tau_q$.

Consider now a node p with a flood level λ. In the open ball $\text{Ball}(p, \lambda)$ all neighboring nodes (s, t) are connected by an edge $e_{st} < \lambda$, hence $\tau_s = \tau_t = \lambda$ and the whole ball X is a lake with the same altitude λ as p.

Lemma 3. *If an open ball $\text{Ball}(p, \rho)$ has one node with a floding level $\lambda > \rho$, then its flooding level is uniform and equal to λ.*

By definition of an open ball, all edges in the cocycle of X have weights $\ge \lambda$. the smallest of them has a weight $\mu > \lambda$ or $\mu = \lambda$. Consider both cases separately.

Creation of a Lake Zone: $\mu = \lambda$. There exists an edge with weight λ in the cocycle of X, linking a node s of X with a node t outside X. The node t does not belong to the open ball $\text{Ball}(p, \lambda)$ but to the closed ball $Y = \overline{\text{Ball}}(p, \lambda) \supset \text{Ball}(p, \lambda)$. What is the level of the flooding within Y ? Each node s of Y is linked with p by a path whose edges are lower or equal than λ. The criterion 2 characterizing floodings may then be applied to all pairs (u, v) of edges along this path: $\tau_v \le \tau_u \vee e_{uv}$. If $\tau_u, e_{uv} \le \lambda$, then also $\tau_v \le \lambda$. This proves:

Lemma 4. *If there is at least one node with a weight $\le \lambda$ in a closed ball $\overline{\text{Ball}}(p, \lambda)$ of level λ, all other nodes in this ball have a flooding level $\le \lambda$.*

The diameter of Y is λ. Such a closed ball is called lake zone.

Creation of a Regional Minimum Lake: $\mu > \lambda$. If the smallest edge of the cocycle is higher than λ, so are all edges of the cocycle. Hence X forms a lake with a uniform flooding level λ. As it is not possible to quit X without crossing an edge with a weight λ, we define it as a **regional minimum lake**.

From a lake zone to the next As any two nodes of X are linked by a path of altitude $< \lambda, \mathrm{diam}(X) = \mu < \lambda$. In a closed ball, diameter and radius are equal, hence X is also a closed ball of radius μ, i.e. a lake zone. According to the preceding lemma, if only one node of X has a flooding level equal to μ, then all its other nodes have a flooding level $\leq \mu$. But as soon one of its nodes has a flooding level $\lambda > \mu$, then all nodes of X have the same flooding level λ. And X is a regional minimum lake as long as its flooding level is lower than the lowest edge ν in its cocycle. As the flooding level within X reaches ν, X remains an open ball $\mathrm{Ball}(p, \nu)$ with a uniform flooding level ν, included in a closed ball $\overline{\mathrm{Ball}}(p, \nu)$ with a flooding level lower or equal than ν outside X.

An Increasing Series of Lakes Containing a Node p. We define the operator $\varepsilon_{ne}p$ which computes the weight of lowest adjacent edge of p ; similarly, $\varepsilon_{Xe}Y$ is the operator which computes the weight of the lowest edge in the cocycle of Y. We now describe the extension of the successive lakes containing a given node p for increasing levels η of flooding.

- for $\eta < \varepsilon_{ne}p$, i.e. a flooding level below $\varepsilon_{ne}p$, the extension of the lake is $X_0 = \{p\}$ and is a regional minimum lake. Hence for $\eta < \varepsilon_{ne}p : X_0 = \{p\}$.
- for $\eta = \varepsilon_{ne}p$, the lake containing p is included in a lake zone $X_1 = B(p, \varepsilon_{ne}p)$. The flood level is equal to η on X_0 and $\leq \eta$ everywhere else on X_1. We have $\mathrm{diam}(X_1) = \varepsilon_{ne}p = \varepsilon_{Xe}X_0$.
- for $\mathrm{diam}(X_1) < \eta < \varepsilon_{Xe}X_1$, the lake is a regional minimum lake with the extension X_1.
- for $\eta = \varepsilon_{Xe}X_1$, the lake containing p is included in a lake zone $X_2 = B(p, \varepsilon_{Xe}X_1)$. The flood level is equal to η on X_1 and $\leq \eta$ everywhere else on X_2. We have $\mathrm{diam}(X_2) = \varepsilon_{Xe}X_1$
- ...
- for $\mathrm{diam}(X_n) < \eta < \varepsilon_{Xe}X_n$, the lake is a regional minimum lake with the extension X_n.
- for $\eta = \varepsilon_{Xe}X_n$, i.e. a flooding level equal to the lowest adjacent edge of X_n, the lake containing p is included in a lake zone $X_{n+1} = B(p, \varepsilon_{Xe}X_n)$. The flood level is equal to η on X_n and $\leq \eta$ everywhere else on X_{n+1}. We have $\mathrm{diam}(X_{n+1}) = \varepsilon_{Xe}X_n$.
- the alternating series of regional minima lakes and lake zones goes on until all nodes of N are flooded.

Dendrogram Structure of the Lake Zone. For $p \in Y$, we define the operator κ_p by $\kappa_p(Y) = B(p, \varepsilon_{Xe}Y)$ and its iteration: $\kappa_p^{(n)}(Y) = \kappa_p\kappa_p^{(n-1)}(Y)$ Starting with the set $X_0 = \{p\}$ we obtain a series of lake zones: $X_0 = \{p\}$, $X_1 = \kappa_p\{p\}, ...,$ $X_n = \kappa_p(X_{n-1}) = \kappa_p^{(n)}\{p\}$.

Obviously, for each node $q \in \kappa_p^{(n)}\{p\}$, there exists a number m such that $\kappa_p^{(n)}\{p\} = \kappa_q^{(m)}\{q\}$.

The sets $\kappa_p^{(n)}\{p\}$ for all n and all nodes p form a hierarchy. Its sets may be organized as a dendrogram. The leaves of the dendrogram are the nodes of G.

Each node $\kappa_p^{(n)}\{p\}$ is linked by an edge with its unique immediate successor $\kappa_p^{(n+1)}\{p\}$ as illustrated in fig.2A.

3 Dominated Floodings

3.1 Lake Level and Lake Extension at a Node p

The preceding section has described how the lake containing a given node is extended as its flooding level increases. Many flooding distributions are physically possible. However there is only one if we consider the highest flooding below a ceiling function ω defined on each node. We consider all floodings of G whose flooding level is lower than the function ω on all its nodes. The supremum of all these floodings also is a valid flooding of G and is the highest flooding of G below ω. We define the ceiling function $\omega(X)$ as the smallest value taken by ω on a node of X.

What will be the level of the flooding and the extension of the lake containing a given node p ? As shown above, the possible lakes containing the node p form an increasing series of nested sets $\kappa^{(n)}\{p\}$, the smallest being $\{p\}$, the largest being the root $\kappa^{(m)}\{p\}$ of the dendrogram.

The operator $\omega(X)$ is decreasing and the operator $\mathrm{diam}(X)$ increasing with X. As the series $\kappa^{(n)}\{p\}$ is increasing with n, we get a series of decreasing values $\omega(\kappa^{(n)}\{p\})$ and a series of increasing values $\mathrm{diam}(\kappa^{(n)}\{p\})$:

- as the set $\{p\}$ has no inside edge, we have $\mathrm{diam}(\kappa^{(0)}\{p\}) = \mathrm{diam}\{p\} = -\infty$. Hence $\omega\{p\} > \mathrm{diam}\{p\} = -\infty$

- if at the root we still have $\omega(\kappa^{(m)}\{p\}) > \mathrm{diam}(\kappa^{(m)}\{p\})$, i.e. the ceiling of p is higher than the root of the dendrogram, then $\tau_p = \mathrm{diam}(\kappa^{(m)}\{p\})$, the lowest flooding value covering the whole domain $\kappa^{(m)}\{p\}$.

- if on the contrary $\omega(\kappa^{(m)}\{p\}) \leq \mathrm{diam}(\kappa^{(m)}\{p\})$, let $k \leq m$ be the smallest index for which $\omega(\kappa^{(k)}\{p\}) \leq \mathrm{diam}(\kappa^{(k)}\{p\})$ (rel. 1). Hence $\omega(\kappa^{(k-1)}\{p\}) > \mathrm{diam}(\kappa^{(k-1)}\{p\})$ (rel. 2), which implies that on $\kappa^{(k-1)}\{p\}$ the flooding level is uniform and higher than $\mathrm{diam}(\kappa^{(k-1)}\{p\})$. On the other hand Rel.1 implies that on $\kappa^{(k)}\{p\}$ the maximal flooding level is $\mathrm{diam}(\kappa^{(k)}\{p\})$. Two possibilities are compatible with both relations:

* if $\omega(\kappa^{(k-1)}\{p\}) \leq \mathrm{diam}(\kappa^{(k)}\{p\})$, then $\tau_p = \tau_{\kappa^{(k-1)}\{p\}} = \omega(\kappa^{(k-1)}\{p\})$
* if $\omega(\kappa^{(k-1)}\{p\}) > \mathrm{diam}(\kappa^{(k)}\{p\})$, then $\tau_p = \tau_{\kappa^{(k-1)}\{p\}} = \mathrm{diam}(\kappa^{(k)}\{p\})$.

3.2 Illustration

Determination of the Flooding Level at the Node c. The ceiling function ω is equal to ∞ on all nodes excepting the nodes $\omega(c) = 6$ and $\omega(h) = 1$. We represent inside a yellow dot the function ω on each node of the dendrogram in fig.2A.

Let us compute the lake level and the extension of the node c. The smallest index for which $\omega(\kappa^{(k)}\{c\}) \leq \mathrm{diam}(\kappa^{(k)}\{c\})$, is $k = 3$, with $\kappa^{(3)}\{c\} = [b, c, d, e, f]$ having a diameter 7, whereas $\omega(\kappa^{(3)}\{c\}) = 6$. For $k = 2$, we get

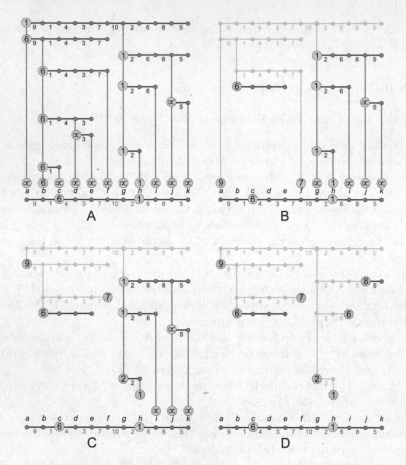

Fig. 2. A: Dendrogram asssociated to the edge weighted graph (red nodes linked by weighted edges). The yellow disks contain the ceiling level of each node of the dendrogram.

B: The lake containing the nodes c also contains the nodes (b, c, d, e) at a flooding level 6.

C: The ancestors of (b, c, d, e) are suppressed and its uncles becom the roots of sub-dendrograms which may be processed separately.

D: Final dendrogram with the flooding levels of the various nodes.

$\kappa^{(2)}\{c\} = [b, c, d, e]$ having a diameter 4, whereas $\omega(\kappa^{(2)}\{c\}) = 6$. According to the preceding analysis the flooding level of $\kappa^{(2)}\{c\} = [b, c, d, e]$ is $\tau_c = \tau_{\kappa^{(2)}\{c\}} = \omega(\kappa^{(2)}\{c\}) = 6$ (see fig.2B).

Pruning the Dendrogram. Fig.2 presents how the upstream of each flooded node is pruned. As the level of $\kappa^{(2)}\{c\}$ is known, the dendrogram may be pruned by discarding all ancestors of $\kappa^{(2)}\{c\}$. For $k > 2$, $\kappa^{(k)}\{c\}$ is an ancestor of c, the flooding level of all its immediate successors which are not ancestors of c, that is, brothers of $\kappa^{(k-1)}\{c\}$ is lower or equal than $\mathrm{diam}(\kappa^{(k)}\{c\})$. The edge linking each brother Y of $\kappa^{(k-1)}\{c\}$ with its father $\kappa^{(k)}\{c\}$ is cut ; like that Y becomes the root of a sub-dendrogram ; as its flooding level is lower or equal than $\mathrm{diam}(\kappa^{(k)}\{c\})$, one sets $\omega(Y) = \omega(Y) \wedge \mathrm{diam}(\kappa^{(k)}\{c\})$. On the same time all ancestors of $\kappa^{(2)}\{c\}$ and the edges linking them are suppressed.

The result of the pruning is illustrated by fig.2B,C. The set $\kappa^{(2)}\{c\} = [b, c, d, e]$ got its flooding level 6 and its upstream is pruned:

- $\kappa^{(3)}\{c\} = [b, c, d, e, f]$ is suppressed and the node $\{f\}$ becomes the root of a sub-dendrogram, with a ceiling value $\omega(\{f\}) = \omega(\{f\}) \wedge \mathrm{diam}(\kappa^{(3)}\{c\}) = 7$. As the sub-dendrogram is reduced to a node, its ceiling value is its flooding value, 7.

- $\kappa^{(4)}\{c\} = [a, b, c, d, e, f]$ is suppressed and the node $\{a\}$ becomes the root of a sub-dendrogram, with a ceiling value $\omega(\{a\}) = \omega(\{a\}) \wedge \mathrm{diam}(\kappa^{(4)}\{c\}) = 9$. As the sub-dendrogram is reduced to a node, its ceiling value is its flooding value, 9.

- $\kappa^{(5)}\{c\}$, the root, is suppressed and the node $[g, h, i, j, k]$ becomes the root of a sub-dendrogram, with a ceiling value $\omega([g, h, i, j, k]) = \omega([g, h, i, j, k]) \wedge \mathrm{diam}(\kappa^{(5)}\{c\}) = 1$.

In summary, as soon a node Y of the dendrogram gets its flooding level, the dendrogram may be pruned, suppressing all ancestors of Y, transforming each uncle Z of Y into the root of a sub-dendrogram, with a ceiling value $\omega(Z) = \omega(Z) \wedge \mathrm{diam}\,\kappa(Z)$.

The final result is obtained by processing each sub-dendrogram separately and is illustrated in fig.2D.

An Algorithm Based on Edge Contractions. The following algorithm constructs the dendrogram and computes the flooding levels by iteratively contracting the edges of each lake. We define $\Lambda(p)$ as a collection of nodes with the same flooding level τ_p as the node p. The result of the algorithm is a list Λ of records of the type $(\Lambda(p), \tau_p)$. The algorithms proceeds by processing the edges still present in the graph in the order of increasing altitudes. Initially $\Lambda = \varnothing$ and for $p \in N : \Lambda(p) = [p]$ and $p = $ "$unflooded$":

As long there are edges to process, let (p, q) be the lowest edge to process:
if $p = $ "$isflooded$" and $q = $ "$isflooded$" take the next edge, else
$X = \overline{\mathrm{Ball}}(p, e_{pq})$; $X_1 = \{x \in X \mid x = $ "$isflooded$"$\}$; $X_2 = \{x \in X \mid \omega(x) \le e_{pq}$ and $x = $ "$unflooded$"$\}$
If $X_1 \cup X_2 = \varnothing : \Lambda(p) = \bigcup_{x \in X} \Lambda(p)$; $\omega(p) = \omega(X)$; contract (o, q) on p
else for each $x \in X/X_1$:

　　$\tau_x = \omega(x) \wedge e_{pq}$
　　$\Lambda = \mathrm{append}[\Lambda, (\Lambda(x), \tau_x)]$
　　contract X on p
　　$p = $ "$isflooded$"

Illustration

- $e_{bc} = 1$; $X_1 \cup X_2 = \varnothing$; $\Lambda(c) = [b, c]$; $\omega(c) = 6$; contract (b, c) on c
- $e_{gh} = 2$; $X_2 = [h]$; $\tau_h = 1$; $\tau_g = 2$; $\Lambda = \mathrm{append}[\Lambda, (h, 1), (g, 2)]$; contract (g, h) on h; $h = $ "$isflooded$"
- $e_{de} = 3$; $X_1 \cup X_2 = \varnothing$; $\Lambda(e) = [d, e]$; $\omega(e) = \infty$; contract (d, e) on e

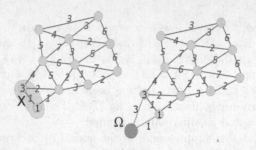

Fig. 3. Adding a dummy node linked to each node x in X by an edge weighted by the offset at x

- $e_{ce} = 4; X_1 \cup X_2 = \varnothing; \Lambda(e) = [b, c, d, e]\,; \omega(e) = 6;$ contract (c, e) on e
- $e_{jk} = 5; X_1 \cup X_2 = \varnothing; \Lambda(e) = [j, k]\,; \omega(k) = \infty;$ contract (j, k) on k
- $e_{hi} = 6; X_1 = [h]\,; \tau_i = 6\ ;\ \Lambda = \text{append}[\Lambda, (i, 6)]\ ;\ $ contract (h, i) on $i;\ i =$ "$isflooded$"
- $e_{ef} = 7; X_1 = [e]\,; \tau_f = 7\ ;\ \Lambda = \text{append}[\Lambda, (f, 7)]\ ;\ $ contract (e, f) on $f;$ $f = $ "$isflooded$"
- $e_{ik} = 8; X_1 = [i]\,; \tau_k = 8\ ;\ \Lambda = \text{append}[\Lambda, (k, 8)]\ ;\ $ contract (i, k) on $k;\ k =$ "$isflooded$"
- $e_{af} = 9; X_1 = [e]\,; \tau_a = 9\ ;\ \Lambda = \text{append}[\Lambda, (a, 9)]\ ;\ $ contract (a, f) on $f;$ $f = $ "$isflooded$"
- $e_{f,k} = 9; p = $ "$isflooded$" and $q = $ "$isflooded$", there is no further edge : end

4   Constrained Highest Floodings on Edge Weighted Graphs as Shortest Distances in an Augmented Graph

4.1   Highest Floodings and Shortest Distances

According to criterion 2, any flooding θ verifies the relation: $\theta_p \leq \theta_q \vee e_{pq}$, for each neighbor q of p. As this relation is to be true for all neighbors of p, we have $\theta_p \leq \bigwedge_{q \text{ neighbor of } p} (\theta_q \vee e_{pq})$ Simultaneously $\theta_p \leq h_p$. So $\theta_p \leq h_p \wedge \bigwedge_{q \text{ neighbor of } p} (\theta_q \vee e_{pq})$ and the highest of them, τ verifies

$$\tau_p = h_p \wedge \bigwedge_{q \text{ neighbor of } p} (\tau_q \vee e_{pq}).$$

If we add to the graph G_e a dummy node Ω with a weight $\tau_\Omega = 0$ linked by a dummy edge (Ω, p) with each node p and holding a weight equal to h_p, we get an augmented graph $\widehat{G_e}$. Relation (6) may be rewritten as $\tau_p = (\tau_\Omega \vee e_{\Omega p}) \wedge \bigwedge_{q \text{ neighbor of } p} (\tau_q \vee e_{qp})$. This formula expresses that the shortest path for the ultrametric flooding distance between Ω and p is $e_{\Omega p} = h_p$ if the path is simply the edge (Ω, p) or is equal to $(\tau_s \vee e_{ps})$ if the path passes through the neighbor s of p, $(\tau_q \vee e_{qp})$ taking its smallest value for $q = s$.

Theorem 1. *The highest flooding of the graph G below a function h defined on the nodes is the shortest ultrametric flooding distance of each node to Ω .*

This theorem permits to use any shortest path algorithm for computing this highest flooding. The simplest recursively applies the relation (6) until stability is reached.

Initialisation: $\tau_p^{(0)} = h_p$

Repeat until $\tau_p^{(m)} = \tau_p^{(m-1)}$: $\tau_p^{(n)} = h_p \wedge \bigwedge_{q \text{ neighbor of } p} \left(\tau_q^{(n-1)} \vee e_{pq} \right)$

Stability is necessarily reached after a number n of iteration as the values of τ decrease and have a lower ceiling equal to 0. As $\tau_p^{(n)} \leq \tau_p^{(n-1)} \leq h_p$, we get an equivalent algorithm with the following sequence: $\tau_p^{(n)} = \tau_p^{(n-1)} \wedge$

$\bigwedge_{q \text{ neighbor of } p} \left(\tau_q^{(n-1)} \vee e_{pq} \right)$.

4.2   The Moore Dijkstra Shortest Path Algorithm [5]

This famous greedy algorithm takes as many steps as there are nodes. At any step, S represents the subset of nodes for which the shortest path is known. For any neighboring node of S, the length of the shortest path for which all edges but the last belong to S constitutes an overestimation of this length. The node with the lowest guess is correctly estimated.

Initialization:

$\quad S = \Omega$; $\overline{S} = N$; for each node p in N : $\tau_p = h_p$

Flooding:

While $\overline{S} \neq \varnothing$ repeat:

$\quad$ Select $j \in \overline{S}$ for which $\tau_j = \min_{i \in \overline{S}} [\tau_i]$

$\quad \overline{S} = \overline{S} \backslash \{j\}$

$\quad$ For any neighbor i of j in $\overline{S}$ do $\tau_i = \min [\tau_i, \tau_j \vee e_{ji}]$

End While

Remark. The dummy node plays no role, nor the nodes with an infinite ceiling value. Without dummy node, the initialisation becomes $S = \varnothing$; for each node p in N verifying $h_p < \infty$, do $\tau_p = h_p$.

The nodes are processed in an increasing order of flooding. If we keep the edges linking each node with the node through which it has been flooded in the algorithm we get a tree. Along each edge of this tree, the level of the flood also is never decreasing.

5   Conclusion

We have given an axiomatic definition of floodings on edge weighted graph. The highest flooding under a ceiling function is a morphological opening of this ceiling function (increasing, anti-extensive and idempotent). The criteria characterizing this flooding permit to express it either as a shortest distance problem on an augmented graph for an ultrametric flooding distance or as a pruning of a dendrogram. The first expression permits to use the shortest path algorithm which is best adapted to each particular problem. The second permits to imagine extremely fast implementations as the dendrogram rapidly splits in sub-dendrograms which may be processed independently.

These results may be transposed on floodings for images [4]. An image f defined on a grid, may be considered as a node weighted graph G; the pixels becoming the nodes, their grey tones the node weights ; the edges connect neighboring pixels/nodes and are not weighted. It may be shown that any flooding τ of an image f is a flooding of the graph G, on which the edges get weights $e_{pq} = f_p \vee f_q$. This results permits to transpose on images all results established on tank networks.

References

1. Marcotegui, B., Beucher, S.: Fast implementation of waterfalls based on graphs. In: ISMM05: Mathematical Morphology and its Applications to Signal Processing, pp. 177–186 (2005)
2. Cousty, J., Bertrand, G., Najman, L., Couprie, M.: Watershed cuts: Minimum spanning forests and the drop of water 1principle. IEEE Transactions on Pattern Analysis and Machine Intelligence 31, 1362–1374 (2009)
3. Meyer, F.: Minimal spanning forests for morphological segmentation. In: ISMM94: Mathematical Morphology and its Applications to Signal Processing, pp. 77–84 (1994)
4. Meyer, F.: Flooding and segmentation. In: Goutsias, J., Vincent, L., Bloomberg, D.S. (eds.) Mathematical Morphology and its Applications to Image and Signal Processing. Computational Imaging and Vision, vol. 18, pp. 189–198 (2002)
5. Moore, E.F.: The shortest path through a maze. In: Proc. Int. Symposium on Theory of Switching, vol. 30, pp. 285–292 (1957)

Graph Matching with Nonnegative Sparse Model

Bo Jiang, Jin Tang, and Bin Luo*

School of Computer Science and Technology,
Anhui University, Hefei 230039, Anhui, China
zeyiabc@163.com, ahhftang@gmail.com, luobin@ahu.edu.cn

Abstract. Graph matching is an essential problem in computer vision and pattern recognition. In this paper, we propose a novel graph matching method based on non-negative sparse model (NSGM). The main feature for our NSGM is that it can generate sparse solution and thus naturally imposes the discrete mapping constraints approximately in the optimization process. In addition, an efficient algorithm was derived to solve NSGM problem. Promising experimental results on both synthetic and real image matching tasks show the effectiveness of the proposed matching method.

Keywords: Graph matching, Sparse model, Nonnegative matrix factorization.

1 Introduction

Many problems of interest in computer vision and pattern recognition can be formulated as a problem of finding consistent correspondences between two sets of features. In computer vision, the problem of establishing correspondences between two sets of features can be effectively solved by attributed relational graph (ARG) matching. The goal of ARG matching (graph matching) is to find a mapping between the two node sets that preserves both unary attributes and binary relationships between nodes as much as possible.

Previous approaches [1–3] have formulated graph matching as an Integer Quadratic Programming (IQP). Since IQP is known to be NP-hard, graph matching is either solved exactly in very restricted setting (bipartite graph matching, for example using the Hungarian method) or approximately. Most of recent literatures on graph matching follow the second way, i.e., developing approximate relaxations to the graph matching problem [4, 2, 5, 6, 3, 1]. Torresani et al.[7] represented graph matching as an energy minimization problem which can be efficiently optimized by dual decomposition. Leordeanu and Hebert [3] proposed a simple and efficient approximate method (spectral matching, SM) to IQP using a spectral relaxation technique. Cour et al.[5] extended SM to spectral matching with affine constraint (SMAC) by incorporating the affine constraints into the spectral relaxation. Comparing with SM, it further encodes the one-to-one matching constraints, therefore can approximate the original IQP problem

---

* Corresponding author.

W.G. Kropatsch et al. (Eds.): GbRPR 2013, LNCS 7877, pp. 41–50, 2013.

more closely. Leordeanu and Hebert [8] proposed an iterative matching method (IPFP) which optimized the IQP in the discrete domain and therefore can satisfy the one-to-one mapping constraints strictly in the optimization process. Zhou et al.[9] proposed a new matching algorithm by exploiting the properties of factorized affinity matrix. Cho et al. [1] interpreted graph matching using the probabilistic framework and proposed a graph matching method based on random walks.

The optimal solution for graph matching problems (IQP) is discrete, binary and thus nonnegative, sparse in nature. To our knowledge, most existing relaxation algorithms do not utilize or emphasize this sparse property. This motivates us to use nonnegative sparse model for graph matching problem. Following this way, we propose a new relaxation algorithm for graph matching problem based on nonnegative sparse model (NSGM). There are two main features for NSGM method. Firstly, NSGM can generate sparse solution and thus can approximately incorporate the mapping constraints (discrete and binary) in optimization process. Secondly, an efficient algorithm can be derived to solve NSGM problem.

The remainder of this paper is organized as follows. In Section 2, we introduce the formulation of ARG matching as an IQP problem. In Section 3, we propose our NSGM graph matching method, and show some benefits of NSGM. In Section 4, we apply our matching method to some matching tasks.

2   Problem Formulation

2.1   Attributed Graph Matching

Assume that two ARGs to be matched are $G^D = (V^D, E^D, A^D, R^D)$ and $G^M = (V^M, E^M, A^M, R^M)$ where V represents a set of nodes, E, edges, and A, unary attributes, R, binary relations. Each node $v_i^D \in V^D$ or edge $e_{ij}^D \in E^D$ has an associated attribute vector $\mathbf{a}_i^D \in A^D$ or $\mathbf{r}_{ij} \in R^D$ [1, 8]. In image feature matching, an attribute vector $\mathbf{a}_i^D \in A^D$ usually describes a local feature descriptor (e.g., SIFT descriptors and shape context), and $\mathbf{r}_{ij} \in R^D$ generally represents the relationship (e.g., geometric distance) between two feature points. The objective of graph matching is to determine the correct correspondences between G^D and G^M. A correspondence mapping is a set C of pairs (or assignments) $(v_i^D, v_{i'}^M)$, where $v_i^D \in V^D$ and $v_{i'}^M \in V^M$. For each assignment $a_i = (v_i^D, v_{i'}^M)$ in C, there is an affinity $\mathbf{W}_{a_i,a_i} = f_a(\mathbf{a}_i^D, \mathbf{a}_{i'}^M)$ that measures how well the node $v_i^D \in V^D$ matches the node $v_{i'}^M \in V^M$. Also, for each pair of assignments (a_i, a_j), where $a_i = (v_i^D, v_{i'}^D)$ and $a_j = (v_j^M, v_{j'}^M)$, there is an affinity $\mathbf{W}_{a_i,a_j} = f_r(\mathbf{r}_{ij}^D, \mathbf{r}_{i'j'}^M)$ that measures how compatible the nodes (v_i^D, v_j^D) in the data graph G^D are with the nodes $(v_{i'}^M, v_{j'}^M)$ in the model graph G^M. Thus, we can use a matrix $\mathbf{W}$ with the diagonal term $\mathbf{W}_{a_i,a_i}$ representing a unary affinity of a correspondence $(v_i^D, v_{i'}^M)$, and the non-diagonal element $\mathbf{W}_{a_i,a_j}$ containing a pair-wise affinity between two assignments $a_i = (v_i^D, v_{i'}^D)$ and $a_j = (v_j^M, v_{j'}^M)$.

We can represent the correspondences by an matching matrix $\mathbf{P}$ where $\mathbf{P}_{ii'} = 1$ implies that the node v_i^D in G^D corresponds to the node $v_{i'}^M$ in G^M, and

$\mathbf{P}_{ii'} = 0$ otherwise. In this paper, we denote $\mathbf{p} \in \{0,1\}^{mn}$ $(m = |V^D|, n = |V^M|)$ as a row-wise vectorized replica of $\mathbf{P}$, i.e., $\mathbf{p} = (\mathbf{P}_{11}...\mathbf{P}_{1n}, ..., \mathbf{P}_{m1}...\mathbf{P}_{mn})$. The graph matching problem can be formulated as an integer quadratic programming (IQP)[8, 5, 1], i.e., finding the indicator vector $\mathbf{p}$ that maximizes the score function as

$$\max_{\mathbf{p}} \mathbf{p}^T \mathbf{W} \mathbf{p} \tag{1}$$

$$s.t. \quad \mathbf{p} \in \{0,1\}^{mn}, \quad \mathbf{Ap} = \mathbf{1} \tag{2}$$

The constraints (Eq.(2)) ensure one-to-one matching between G^D and G^M [8].

2.2   Graph Matching Relaxations

The above IQP problem for graph matching is NP-hard and no efficient algorithm exists. Therefore, lots of approximate algorithms have been proposed to find the solution of this problem [1, 2, 5, 3, 8]. These approaches usually avoid combinatorial searching by approximating the objective function or by relaxing the mapping constraints. In general, a practical approximate solution should have the following two matching properties strongly[4, 5, 8]:

(1) It should maximize the objective function (Eq.(1)) as far as possible;
(2) It should satisfy the constraints (Eq.(2)) as closely as possible.

We call these properties as *Objective Property* and *Constraint Property*, respectively. Usually, the approximate algorithms cannot guarantee that the solution satisfies the constraints strictly, and they obtain the final correspondence solution based on a post-optimization step by using some discretization techniques, which usually lead to weak local optimal solutions for the original IQP problem. Leordeanu and Hebert [3] proposed a spectral technique (SM) to graph matching problem. This method can have strong objective property, i.e., it can find the global maximum of the relaxed problem effectively. However, the method does not hold the constraint property because of the relaxation, namely $\|\mathbf{p}\|_F = 1$. They also recently proposed an iterative matching method (IPFP) [8] which integrates the discretization step and objective function optimization simultaneously. This method optimizes the IQP in the discrete domain and thus can have constraint property strictly. Cour et al. [5] extended SM to spectral matching with affine constraints (SMAC) by incorporating the matching constraints within the relaxation process. Comparing with SM, this method has the constraint property more strongly and therefore can obtain a more effective solution for the graph matching problem.

3   Nonnegative Sparse Graph Matching

The optimal solution for the above IQP graph matching problem (Eq.(1)) is discrete, binary and thus sparse, nonnegative in nature, i.e., there exists small number of positive nonzero elements in the optimal solution. This motivates us to use nonnegative sparse model for graph matching problem. In the following, we first propose our nonnegative sparse graph matching (NSGM) model. Then, an efficient algorithm is introduced to solve it. At last, some benefits of NSGM are demonstrated.

3.1   Relaxation Model

By adding a L_1 norm constraint on the solution $\mathbf{p}$, our NSGM can be formulated as follows:

$$\max_{\mathbf{p}} \mathbf{p}^T \mathbf{W} \mathbf{p} \quad s.t. \quad \|\mathbf{p}\|_1 = 1, \mathbf{p}_i \geq 0. \tag{3}$$

where $\|\mathbf{p}\|_1 = \sum_{i=1}^{mn} |\mathbf{p}_i|$.

The above optimization can be explained as a problem subject to a L_1 norm constraint on the solution. The main feature for this kind of problems is that they can encourage sparse solutions [10, 11, 10–12], i.e., many components of the solution $\mathbf{p}$ are zero. It will be shown that only the relative values between the elements of $\mathbf{p}$ matter, we can fix the L_1 norm of $\mathbf{p}$ to 1.

There are two main features for our NSGM model. Firstly, NSGM can generate sparse solution and thus naturally incorporates the discrete mapping constraints approximately, i.e., it has constraint property approximately. Secondly, an efficient update algorithm can be derived to solve NSGM problem, i.e., it can also have strong objective property. Therefore, our NSGM can approximate the original IQP problem closely and thus can lead to an effective solution for graph matching problem. Our NSGM is most similar to spectral matching method[3], which relaxes the original IQP problem by solving the following problem:

$$\max_{\mathbf{p}} \mathbf{p}^T \mathbf{W} \mathbf{p} \quad s.t. \quad \|\mathbf{p}\|_F = 1, \tag{4}$$

where $\|\mathbf{p}\|_F = \sqrt{\sum_{i=1}^{mn} \mathbf{p}_i^2}$. The main difference between our NSGM and SM is that we impose sparse constraint on the related solution. We demonstrate later that the ability to maintain this constraint leads to a very better solution for graph matching problem.

3.2   Computational Algorithm

In this section, we develop an efficient algorithm to solve our NSGM problem. As discussed in Section 3.1, the optimal solution for NSGM is nonnegative. This motivates us to use a nonnegative matrix factorization (NMF) technique to solve NSGM problem. Since $\mathbf{p}_i \geq 0$, Eq.(3) is equivalent to the following,

$$\max_{\mathbf{p}} \mathbf{p}^T \mathbf{W} \mathbf{p} \quad s.t. \quad \sum_{i=1}^{mn} \mathbf{p}_i = 1, \mathbf{p}_i \geq 0. \tag{5}$$

This problem can be efficiently solved by an iterative algorithm. The algorithm iteratively updates a current solution vector $\mathbf{p}^{(t)}$ as follows [13]:

$$\mathbf{p}_i^{(t+1)} = \mathbf{p}_i^{(t)} \frac{(\mathbf{W}\mathbf{p}^{(t)})_i}{[\mathbf{p}^{(t)}]^T \mathbf{W}\mathbf{p}^{(t)}}. \tag{6}$$

The iteration starts with an initial $\mathbf{p}^{(0)}$ and is repeated until convergence. The optimality and convergence of this update rule can refer to the work [13]. A drawback of Eq. (5) (or Eq. (3)) is not always convex. The final results depend on initializations. Fortunately, there exists a good initialization, i.e., the spectral matching solution [3]. Specifically, we compute the initial $\mathbf{p}^{(0)}$ as $\mathbf{p}_i^{(0)} = (\sum_{i=1}^{mn} \mathbf{q}_i)^{-1}\mathbf{q}_i$ where $\mathbf{q}$ is the principal eigenvector of $\mathbf{W}$. Since $\mathbf{W}$ has nonnegative elements, by Perron-Frobenius theorem, the elements of $\mathbf{q}$ will be in the interval $[0, 1]$[3]. The overall matching algorithm can be summarized as follows:

Step1. Initialization. Compute the principal eigenvector $\mathbf{q}$ of $\mathbf{W}$. Set

$$\mathbf{p}_i^{(0)} = \frac{\mathbf{q}_i}{\sum_{i=1}^{mn} \mathbf{q}_i} \tag{7}$$

Step2. Updating $\mathbf{p}^{(t)} (t = 0, 1, 2...)$ until convergence as follows:

$$\mathbf{p}_i^{(t+1)} = \mathbf{p}_i^{(t)} \frac{(\mathbf{W}\mathbf{p}^{(t)})_i}{[\mathbf{p}^{(t)}]^T \mathbf{W}\mathbf{p}^{(t)}}. \tag{8}$$

Step3. Let $\mathbf{p}^*$ be the convergence solution of Step 2. Compute the final binary matching solution $\tilde{\mathbf{p}}^*$ based on $\mathbf{p}^*$ using Greedy algorithm[3, 5].

3.3   Sparsity and Desirable Matching Properties

The main difference between NSGM and other methods [2, 5, 3] is that a L_1 norm constraint is imposed on the solution $\mathbf{p}$ and thus encourages a sparse solution. Here we show that, by enforcing sparse solution, it can satisfy the constraint property more closely and thus return more effective solution for graph matching problem. This is the key feature for NSGM model. For further illustration, sparsity, objective score and constraint preserving residual (CPR) are first defined. Let $\mathbf{p}^*$ be the convergence solution of the relaxation algorithms. Let $\tilde{\mathbf{p}}^*$ be the final discretization binary solution obtained by performing some discretization processes such as Greedy and Hungarian algorithms[3, 5].

(1) **Sparsity** measures the percentage of non-zero elements in $\mathbf{p}^*$. Firstly, set $\delta = 0.01 \times \texttt{mean}(\mathbf{p}^*)$, then renew $\mathbf{p}_i^* = 0$ if $\mathbf{p}_i^* \leq \delta$, and finally calculate the percentage of nonzero elements in the renewed $\mathbf{p}^*$ as sparsity.
(2) **Objective Score** measures the objective property of the matching algorithm. It is defined as $OS = \tilde{\mathbf{p}}^{*T}\mathbf{W}\tilde{\mathbf{p}}^*$.
(3) **CPR** measures the constraint property. It is defined as $CPR = \frac{1}{m} \min_\alpha \|\tilde{\mathbf{p}}^* - \alpha \mathbf{p}^*\|_F$, where $m = |V^D|$, α is a weighting parameter to compensate the loss of residual due to scaling.

Figure 1(b) shows the solution vector $\mathbf{p}^{(t)}$ of NSGM for the matching between ARGs generated from 2D point sets shown in Figure 1(a). The points and their geometric relationships correspond to the nodes and edges of ARG, respectively. There are no unary attribute and only one binary relation which is the Euclidean

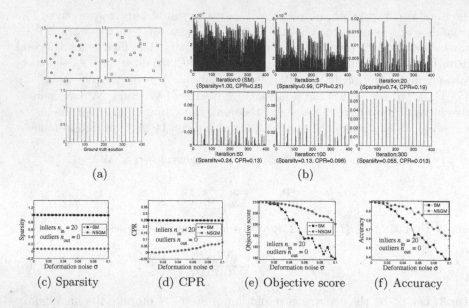

(a)   (b)

(c) Sparsity   (d) CPR   (e) Objective score   (f) Accuracy

Fig. 1. (a) 2D synthetic point sets (top) and ground truth solution (bottom); (b) solution vector $\mathbf{p}^{(t)}$ across the different iterations; (c)-(f) average performance curves for our NSGM vs. SM method on synthetic data matching

distance between two points (see Experimental section in detail). Intuitively, as the iteration increases, $\mathbf{p}^{(t)}$ becomes more and more sparse and therefore approximates the true solution (ground truth) more and more closely. Figure 1(c-f) show the average performance for our NSGM vs. SM method on synthetic data (see Experimental section in detail). Here, the accuracy is measured by the number of detected true matches divided by the total number of ground truths. We can note that:(1) NSGM can generate sparse solution and retain lower CPR value than SM, indicating NSGM can satisfy the matching constraints more closely than SM.(2) NSGM significantly outperforms SM in objective score and accuracy, demonstrating that NSGM can find the final discrete solution for the original IQP problem more optimal than that of SM method. These will be further quantified in the experiments and obviously demonstrate the benefits of our NSGM method.

4 Experiments

In order to evaluate the practicality of our NSGM matching method, we have applied it to the matching tasks including synthetic point matching, feature point matching using CMU image sequence and feature matching on real images. We have used the mapping constraints that one model feature can match at most one data feature and vice-versa. Our method has been compared with other state-of-the-art methods including SM[3], IPFP[8] and SMAC[5].

4.1   Synthetic Point Sets Matching

Our first experiment is based on synthetic 2D point data. Similar to the work [1, 3], we have randomly generated data sets of n_M 2D model points as inlier nodes for G^M. The range of the x-y point coordinates is $\sqrt{n_M/10}$ to enforce an approximate constant density of 10 points over a 1×1 region. We obtain the corresponding n_D nodes in graph G^D by transforming the whole data set with a random rotation and translation and then adding Gaussian noise $N(0, \sigma)$ to the n_M point positions from G^M. The parameter $N(0, \sigma)$ controls the level of position deformation. There are no unary attribute and only one binary relation which is the Euclidean distance between points. The affinity matrix $\mathbf{W}$ is computed by $\mathbf{W}_{ii',jj'} = \exp(\|\mathbf{r}_{ij}^D - \mathbf{r}_{i'j'}^M\|_F^2/\sigma_r)$, where scaling factor σ_r has been set to 0.5 in this experiment and $\mathbf{r}_{ij}^D$ is the Euclidean distance between two points. $\mathbf{W}_{ii',ii'} = 0$. For each deformation noise level σ, we have generated 50 random point sets and then computed the average performances including matching accuracy, objective score, sparse and CPR. Figure 2 shows the comparison results. It is noted that: (1) our NSGM method can generate expected sparse solution and return desirable lower CPR value than other competing methods, indicating that our method has the constraint property more strongly than other methods. Since IPFP puts the discretization step into its optimization process, it can satisfy the mapping constraints strictly (CPR = 0). However, IPFP cannot return the objective score and matching accuracy as high as NSGM method. (2) Comparing with other three methods, our method obviously return the highest matching accuracy and objective score. This clearly demonstrates that our NSGM method can generate more effective solution for graph matching problem.

In addition to the deformation noise, we have also evaluated the effect of our method when outlier nodes (features) exist in both graphs. Here n_{out} outlier feature nodes have been added in both graphs respectively at random positions. Analogously, we have generated 50 random point sets for each outlier level n_{out} and then calculated the average performances. Figure 3 summarizes the results of the different methods as the outlier node number n_{out} varies from 1 to 30. The main feature is that the solution of our method can keep sparse and thus considerably outperforms other competing methods in matching accuracy and objective score. This demonstrates that our method can approximate the constraint

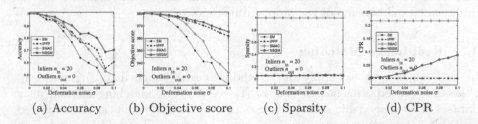

(a) Accuracy   (b) Objective score   (c) Sparsity   (d) CPR

Fig. 2. Comparison results on synthetic point matching when deformation noise exists for the nodes in both graphs

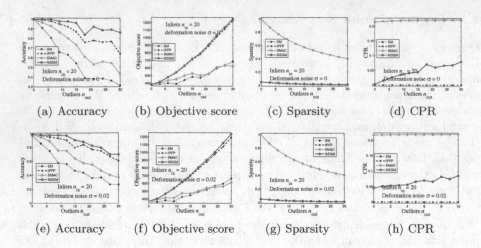

(a) Accuracy   (b) Objective score   (c) Sparsity   (d) CPR

(e) Accuracy   (f) Objective score   (g) Sparsity   (h) CPR

Fig. 3. Comparison results on synthetic point matching when outlier noise and deformation noise exist in both graphs

property closely and thus performs more robustly and effectively for the graphs with outlier nodes.

4.2   Feature Point Matching Across Image Sequence

In this section, we perform feature point matching on CMU house sequence [10, 21, 22], and compare with other methods. For each image, 30 landmark points were manually marked with known correspondences. We have matched all images spaced by 10, 20, 30, $\cdots$ 90 and 100 frames and computed the average accuracy per separation gap. For each image pair, the coordinates of their landmark points have been first normalized to the interval $[0, 1]$, then the affinity matrix has been computed by $\mathbf{W}_{ii',jj'} = \exp(\|\mathbf{r}_{ij}^D - \mathbf{r}_{i'j'}^M\|_F^2/\sigma_r)$, where $\mathbf{r}_{ij}^D$ is the Euclidean distance between two points, σ_r has been set to 0.5. Comparison results are shown in Figure 4. It is noted that NSGM considerably outperforms other three methods as the separation increases. Also, NSGM can generate sparse solution which has desirable constraint property more strongly. These are consistent with the results on the synthetic data experiments and further demonstrate the practicality and benefits of the proposed method.

4.3   Real Image Matching

In this experiment we apply our method to real image matching problem. Following the experimental setting in [1], we test our matching method on a dataset of 30 image pairs containing various images. The results are summarized in Table 1 and some examples are shown in Figure 5. From Table 1, we can note that our NSGM can return higher accuracy and relative score than other methods.

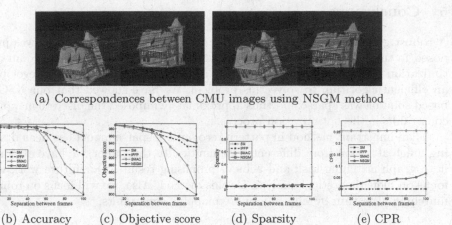

(a) Correspondences between CMU images using NSGM method

(b) Accuracy   (c) Objective score   (d) Sparsity   (e) CPR

Fig. 4. Comparison results on CMU house images

Table 1. Matching performance on the real image dataset (30 pairs)

Methods	NSGM	SMAC	SM
Avg.of accuracy (%)	67.63	52.38	58.66
Avg.of relative score (%)	96.59	75.47	92.56

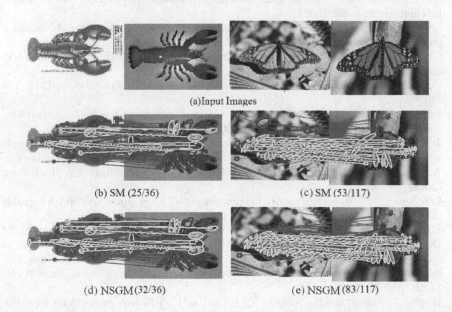

(a)Input Images

(b) SM (25/36)   (c) SM (53/117)

(d) NSGM(32/36)   (e) NSGM(83/117)

Fig. 5. Some results on real image dataset. True matches are represented by yellow lines, and false matches by red lines

5   Conclusions

A robust graph matching method based on nonnegative sparse model was proposed in this paper. We first formulated the graph matching problem to an optimization setting with nonnegative and sparse constraint. Then, we developed an efficient algorithm to solve this problem. We have showed that our NSGM based solution was sparse and thus approximately imposed the discrete mapping constraints in the optimization process.

As an important method in pattern recognition, sparse model has been drawing much attention from different communities. In this paper, we have explored it to graph matching task and achieve promising results. Our future work will focus on more theoretical analysis for this method. Also, we will focus on robust matching algorithm by imposing more mapping constraints.

Acknowledgment. This work is supported by the National Natural Science Foundation of China (No. 61073116, 61211130309).

References

1. Cho, M., Lee, J., Lee, K.M.: Reweighted random walks for graph matching. In: Daniilidis, K., Maragos, P., Paragios, N. (eds.) ECCV 2010, Part V. LNCS, vol. 6315, pp. 492–505. Springer, Heidelberg (2010)
2. Conte, D., Foggia, P., Sansone, C., Vento, M.: Thirty years of graph matching in pattern recognition. International Journal of Pattern Recognition and Artificial Intelligence, 265–298 (2004)
3. Leordeanu, M., Hebert, M.: A spectral technique for correspondence problem using pairwise constraints. In: ICCV, pp. 1482–1489 (2005)
4. Choi, O., Kweon, I.S.: Robust feature point matching by preserving local geometric consistency. CVIU 113, 726–742 (2009)
5. Cour, M., Srinivasan, P., Shi, J.: Balanced graph matching. In: NIPS, pp. 313–320 (2006)
6. Enqvist, O., Josephon, K., Kahl, F.: Optimal correspondences from pairwise constraints. In: ICCV, pp. 1295–1302 (2009)
7. Torresani, L., Kolmogorov, V., Rother, C.: Feature correspondence via graph matching: Models and global optimization. In: Forsyth, D., Torr, P., Zisserman, A. (eds.) ECCV 2008, Part II. LNCS, vol. 5303, pp. 596–609. Springer, Heidelberg (2008)
8. Leordeanu, M., Hebert, M.: An integer projected fixed point method for graph matching and map inference. In: NIPS, pp. 1114–1122 (2009)
9. Zhou, F., Torre, F.D.: Factorized graph matching. In: CVPR, pp. 127–134 (2012)
10. Donoho, D.: Compressed sensing. Technical Report, Stanford University (2006)
11. Donoho, D.: For most large underdetermined systerms of linear equations, the minimal l1-norm solution is also the sparsest solution. In: Comm. Pure Appl. Math., vol. 59 (2006)
12. Duchi, J., Shwartz, S.S., Singer, Y., Chandra, T.: Efficient projections onto the l1-ball for learning in high dimensions. In: ICML (2008)
13. Ding, C., Li, T., Jordan, M.I.: Convex and semi-nonnegative matrix factorization. PAMI 32(1), 45–55 (2010)

TurboTensors for Entropic Image Comparison

Francisco Escolano[1], Edwin R. Hancock[2], Boyan Bonev[1],
and Miguel Angel Lozano[1]

[1] University of Alicante, Spain
{sco,boyan,malozano}@dccia.ua.es
[2] University of York, UK
erh@cs.york.ac.uk

Abstract. In this paper we propose an information-geometric method for comparing superpixel (turbopixel) images. Turbopixels are encoded by tensors and they are referred to as *TurboTensors*. Our methodology has three ingredients. Firstly, we formulate the comparison of the turbopixels topology in terms of the non-rigid alignment of the Isomap embedding of the weighted adjacency matrices; we propose a multi-dimensional information-theoretic dissimilarity measure. Secondly, we formulate the comparison of bags-of-turbopixels through tangent spaces de-projection and multi-dimensional and non-parametric information-theoretic dissimilarity measures. Thirdly, we combine the two latter elements into a flexible energy function whose minimization yields the optimal matching of superpixels images as well as their similarity. In our experiments we show that the proposed method is a useful tool for finding clusters in image sequences. Finally, we show that our approach outperforms state-of-the-art image comparison through non-rigid and affine matching of SURF features.

1 Introduction

Image comparison through matching has been a recurrent topic in computer vision. Two types of existing approaches are: *feature-based* and *segmentation-based*. Feature-based methods rely on the computation of invariant detectors and descriptors followed by either a bag-of-features algorithm usually complemented by matching through RANSAC. Segmentation-based image comparison relies heavily on obtaining a high quality segmentation and then compose a structural representation like a graph or, more recently, a tree. The recent emergence of *superpixels* [1], motivated by the need of label images from a labelled training data [2], provides a methodology for defining over-segmented images so that pixels are grouped in coherent regions in terms of intensity or textural affinity within each of these regions referred to as superpixels. Recently, superpixels have been applied both to video segmentation [3] and object tracking [4]. Concerning image similarity, the Earth Movers Distance (EMD) between superpixels is used in [5] to match two superpixels images. Despite this encouraging methodology which takes into account the topology of the superpixels images to be matched,

W.G. Kropatsch et al. (Eds.): GbRPR 2013, LNCS 7877, pp. 51–60, 2013.

the results obtained show that more progress is needed when dealing with image sequences or videos. Despite their apparent flexibility and relative robustness against image variations, a robust method for quantifying the similarity between superpixels images is a challenging open question. In this paper, we propose an information-geometric methodology for providing a robust similarity measure between superpixels images. Here we use the turbopixels [6] approach. In the section 2 we describe how to characterize each superpixel with a covariance matrix which lies on a Riemannian manifold (tensor space); we also describe how to build a graph of tensors associated to the turbopixels (*turbotensors*) . We also introduce a tangent-space based method for encoding images as bags of turbotensors as well as graphs of turbotensors. Our methodology (section 3) is based on the combination of both structural (embedded graphs of turbotensors) and appearance (bags of turbotensors) dissimilarity measures for the final comparison of turbopixels images. Our experiments, detailed in section 4 shows how powerful and flexible such a combination can be. We also show that the proposed approach improves SURF matching + dissimilarity in terms of retrieval-recall.

2   TurboTensors

2.1   Characterizing TurboPixels

Given an image $\mathcal{I}$, the *TurboPixels* approach [6] produces a planar graph $\mathcal{G} = (\mathcal{V}, \mathcal{E})$ which encodes an over-segmenation of $\mathcal{I}$. The topology of the graph is *relatively stable* (but *not too much stable*) with respect to a different imaging conditions (changes in view point, illumination, and so on) and the main changes occur in the properties of the superpixels. Thus, characterizing the chromatic and textural information contained in superpixels is key to achieve successful image comparison methods based on over-segmentations.

Given a pixel $p \in v_i$, let $\Phi(p) = (f_1(p), \ldots, f_d(p))^T$ be a feature vector including the following region properties: i) the x and y pixel coordinates, ii) the red, green and blue color descriptors, iii) $\mathcal{I}_x(p)$ and $\mathcal{I}_y(p)$, the gradient components,

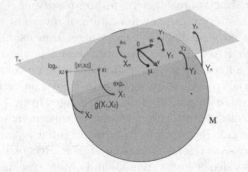

Fig. 1. Riemannian manifold (the sphere) and tangent space T_μ at point μ. Points in the tangent space are the de-projections (log) of their corresponding projections (exp) which lie in the Manifold.

and iv) $\mathcal{I}_{xx}(p)$ and $\mathcal{I}_{yy}(p)$ the Laplacian components. The set of $d \times d$ covariance matrices (tensors) $\mathbf{X}_i = \frac{1}{s-1}\sum_{i=1}^{s}(\Phi(i)-\bar{\Phi})(\Phi(i)-\bar{\Phi})^T$, where $s = |v_i|$ (number of pixels in v_i), lie on a Riemannian manifold $\mathcal{M}$ (see Fig. 1). Such matrices are referred to as *TurboTensors*. For each turbotensor $\mathbf{X} \in \mathcal{M}$ there exists a neighborhood which can be mapped to a given neighborhood in $\mathbb{R}^{d \times d}$. Such a mapping is continuous, bidirectional and one-to-one. As a Riemannian manifold is differentiable, the derivatives at each $\mathbf{X}$ always exist, and such derivatives lie in the so called tangent space $T_{\mathbf{X}}$, which is a vector space in $\mathbb{R}^{d \times d}$. The tangent space at $\mathbf{X}$ is endowed with an inner product $< \mathbf{u}, \mathbf{v} >_{\mathbf{X}} = trace(\mathbf{X}^{-\frac{1}{2}}\mathbf{u}\mathbf{X}^{-1}\mathbf{v}\mathbf{X}^{-\frac{1}{2}})$. The tangent space is also endowed with an exponential map $\exp_{\mathbf{X}} : T_{\mathbf{X}} \to \mathcal{M}$ which maps a tangent vector $\mathbf{u}$ to a point $\mathbf{U} = \exp_{\mathbf{X}}(\mathbf{u}) \in \mathcal{M}$ on the manifold. Such mapping is one-to-one, bidirectional and continuously differentiable and maps u to the point reached by the unique geodesic (minimum-length curve connecting two points in the manifold) from $\mathbf{X}$ to $\mathbf{U}$: $g(\mathbf{X}, \mathbf{U})$. The exponential map is only one-to-one in the neighborhood of $\mathbf{X}$ and this implies that the inverse mapping $\log_{\mathbf{X}} : \mathcal{M} \to T_{\mathbf{X}}$ is uniquely defined in a small neighborhood of $\mathbf{X}$. Therefore, we have the following mappings for going to from the tangent space to the manifold and back (to the tangent space) respectively and the geodesic:

$$\exp_{\mathbf{X}}(\mathbf{u}) = \mathbf{X}^{\frac{1}{2}}\exp(\mathbf{X}^{-\frac{1}{2}}\mathbf{u}\mathbf{X}^{-\frac{1}{2}})\mathbf{X}^{\frac{1}{2}} \, , \, \log_{\mathbf{X}}(\mathbf{U}) = \mathbf{X}^{\frac{1}{2}}\log(\mathbf{X}^{-\frac{1}{2}}\mathbf{U}\mathbf{X}^{-\frac{1}{2}})\mathbf{X}^{\frac{1}{2}} \, , (1)$$

$$g^2(\mathbf{X}, \mathbf{U}) = < \log_{\mathbf{X}}(\mathbf{U}), \log_{\mathbf{X}}(\mathbf{U}) >_{\mathbf{X}} = trace\left(\log^2(\mathbf{X}^{-\frac{1}{2}}\mathbf{U}\mathbf{X}^{-\frac{1}{2}})\right) \, , (2)$$

where we use matrix exponentiation and logarithm.

2.2   Images as Bags of TurboTensors

Given two over-segmented images $\mathcal{I}_X$ and $\mathcal{I}_Y$ and their respective planar graphs $\mathcal{G}_X$ and $\mathcal{G}_Y$, we have two sets of turbopixels $\mathcal{V}_X = \{v_{x_1}, \ldots, v_{x_n}\}$ and $\mathcal{V}_Y = \{v_{y_1}, \ldots, v_{y_m}\}$, with $n = |\mathcal{V}_X|$ and $m = |\mathcal{V}_Y|$ respectively. Following the Riemannian approach and defining a common $\Phi(.)$ $d-$dimensional function for each pixel, each image can be considered as a *bag of turbotensors* of dimension $d \times d$, namely $\mathcal{T}_X = \{\mathbf{X}_1, \ldots, \mathbf{X}_n\}$ and $\mathcal{T}_Y = \{\mathbf{Y}_1, \ldots, \mathbf{Y}_m\}$. In order to compare both bags *from a distributional point of view* (a multi-dimensional generalization of histogramming) it is more convenient to exploit the logarithmic map and work in a tangent space that is suitable for both turbotensor bags. The usual mechanism is to commence by computing the Karcher (weigthed) mean [7] of $\mathcal{Z} = \mathcal{T}_X \cup \mathcal{T}_Y = \{\mathbf{Z_k}\}$ with $k = 1, \ldots, N$ where $N = n + m$. The Karcher mean is defined as $\mu = \arg\min_{\mathbf{Z} \in \mathcal{M}} g^2(\mathbf{Z}_k, \mathbf{Z})$ and it is usually computed after few iterations using the update $\mu^{t+1} = \exp_{\mu^t}(\bar{\mathbf{Z}}^t)$ where $\bar{\mathbf{Z}}^t = \frac{1}{N}\sum_{k=1}^{N}\log_{\mu^t}(\mathbf{Z}_k)$, although more efficient Netwon-based methods have recently been proposed [8]. Since $\log_\mu(\mu) = \mathbf{0}$, we have that μ is the origin of a tangent space where we can de-project the turbotensors both in $\mathcal{T}_X$ and $\mathcal{T}_Y$. For instance, for each $\mathbf{X}_i$ we have

$$vec_\mu(\mathbf{X}_i) = vec_I(\mathbf{u}), \mathbf{u} = \log_\mu(\mathbf{X}_i) \quad vec_I(\mathbf{u}) = (u_{11}\sqrt{2}u_{12} \ldots u_{22}\sqrt{2}u_{23} \ldots u_{dd})^T,$$
$$(3)$$

and similarly for each $\mathbf{Y}_j$. Therefore, we have $d(d+1)/2$-dimensional vectors $vec_\mu(\mathbf{X}_i)$ and $vec_\mu(\mathbf{Y}_j)$ that we can compare through $||vec_\mu(\mathbf{X}_i) - vec_\mu(\mathbf{Y}_i)||^2$. However, the potentially high dimensionality $d \times d$ precludes the adaptation of multi-dimensional point alignment methods like Coherent Point Drift [9] (CPD) to this context. On the other hand, the locality of the tangent space defined at μ and the fact that $||vec_\mu(\mathbf{X}_i) - vec_\mu(\mathbf{Y}_i)||^2 \approx g(\mathbf{X}_i, \mathbf{Y}_j)^2$ allow us to exploit multi-dimensional information theoretic measures for comparing both the de-projected bags of turbotensors and, consequently, the input images $\mathcal{I}_X$ and $\mathcal{I}_Y$. However, as happens in classical bag-of-words, this approach does not consider the rich structural/topological information encoded in the turbopixels graphs.

2.3   Images as TurboTensors Graphs

Let us redefine the turbopixels graphs as attributed graphs: $\mathcal{G}_X = (\mathcal{V}_X, \mathcal{E}_X, \mathcal{W}_X)$ and $\mathcal{G}_Y = (\mathcal{V}_Y, \mathcal{E}_Y, \mathcal{W}_Y)$, where $\mathcal{V}_X$ and $\mathcal{V}_Y$ are the node sets defining the super-pixels, and $\mathcal{E}_X$, $\mathcal{E}_Y$ represent turbopixels common boundaries; $\mathcal{W}_X$ and $\mathcal{W}_X$ are *weighted adjacency matrices* where $\mathcal{W}_X$ is defined as follows

$$\mathcal{W}_X(i,j) = \begin{cases} g(\mathbf{X}_i, \mathbf{X}_j)^2 & \text{if } (i,j) \in \mathcal{E}_X \\ 0 & \text{otherwise} \end{cases}, \tag{4}$$

and similarly for $\mathcal{W}_Y$. The fact that the weigthed adjacency matrices are constructed over geodesics allows us to map the relational structures encoded by $\mathcal{G}_X$ and $\mathcal{G}_Y$ to a Euclidean space using a multi-dimensional scaling technique such as Isomap [10]. Isomap provides quasi iso-metric low-dimensional embeddings for sets of multi-dimensional data using approximate geodesic distances. Such distances (shortest path lengths) are provided by applying the Dijkstra algorithm to a proximity graph. Here we do not need to approximate the geodesic distance between the turbotensors because they lie in a Riemannian manifold. All we have to do is to flatten the manifold whilst imposing structural constraints. To this end we replace the Isomap proximity graph by the weighted adjacency matrices $\mathcal{W}_X$ and $\mathcal{W}_X$. Hence we compare turbotensors through adjacency links not on the basis of their geodesic proximity, but upon the superpixel adjacency. Consequently, after centering $\mathcal{W}_X$ we obtain their centered adjacency matrix $\tau_X = -\mathbf{H}\mathcal{W}_X\mathbf{H}/2$ with $\mathbf{H} = \mathbf{I}_n - \mathbf{1}_n/n$, where $\mathbf{I}_n$ is the $n \times n$ identity matrix and $\mathbf{1}_n$ is the $n \times n$ matrix with all ones entries. Let $\tau_X = \Phi_X \Lambda_X \Phi_X^T$ be the spectral decomposition of τ_X, that is, we have the eigenvectors $\phi_X^{(z)}$ as the columns of Φ_X and the corresponding eigenvalues $\lambda_X^{(z)}$ in the diagonal of Λ_X. Let Φ_X' and Λ_X' the result of re-ordering both eigenvalues and eigenvectors according to the descending eigenvalues. Isomap only considers the d most important positive eigenvalues. Let Φ_{X_d}' be the $n \times d$ matrix obtained by re-ordering Φ_X and discarding rows $d+1\ldots n$. Matrix Λ_{X_d}' is obtained similarly and has dimension $d \times d$. Consequently, the Isomap embedding for the n nodes in $\mathcal{W}_Y$ is encoded by the columns of $\Theta_X = \Lambda_{X_d}'^{1/2} \Phi_{X_d}'^T$ which is a $d \times n$ matrix. The i-th column of Θ_X has the following structure: $\Theta_X^{(i)} = (\sqrt{\lambda_{X_d}'^{(1)}}\phi_{X_d}'^{(1)}(i) \ldots \sqrt{\lambda_{X_d}'^{(d)}}\phi_{X_d}'^{(d)}(i))^T$. This embedding flattens or unrolls the Riemannian manifold so that the embeddings of adjacent turbotensors satisfy $||\Theta_X^{(i)} - \Theta_X^{(j)}||^2 = \psi(g(\mathbf{X}_i, \mathbf{X}_j)^2)_{s.t. e_{ij}=1} \pm \epsilon$, where ϵ is an error term and $\Psi(.)$ is a potential function which corresponds to the identity if and only if $e_{ij} = 1$ and $g(\mathbf{X}_i, \mathbf{X}_j)$ is the true geodesic distance between $\mathbf{X}_i$ and $\mathbf{X}_j$. In uniform areas in the image where turbopixels truly over-segment the image, we find low geodesic distances between adjacent turbopixels and the latter conditions are fulfiled. However, in general we have $||\Theta_X^{(i)} - \Theta_X^{(j)}||^2 \neq g(\mathbf{X}_i, \mathbf{X}_j)^2$ at least at turbopixels defining true low-frequency edges in the image. As a result, we impose the grid-like structure of the turbopixels graph on the embedding and this imposition is stronger at turbopixels defining non-important edges. Therefore, if we apply the latter embedding to $\mathcal{T}_Y$ in order to obtain a $d-$dimensional $\Theta_Y^{(j)}$ and both $\mathcal{W}_X$ and $\mathcal{W}_Y$ encode a similar topology, it is possible to deform the columns in Θ_Y to match the ones in Θ_X. Otherwise, we will have a significant residual error after performing the optimal deformation. In this case, since d is not usually too large it is possible to formulate the problem of matching graphs $\mathcal{G}_X = (\mathcal{V}_X, \mathcal{E}_X, \mathcal{W}_X)$ and $\mathcal{G}_Y = (\mathcal{V}_Y, \mathcal{E}_Y, \mathcal{W}_Y)$ in terms of the non rigid alignment of the multidimensional columns in Θ_X and Θ_Y. Following CPD the non-rigid alignment problem is formulated in terms of a minimization problem which can be solved by a fast EM approach (see details in [9]).

3   Entropic Image Comparison

Given two images $\mathcal{I}_X$ and $\mathcal{I}_Y$ over-segmented by following the turbopixels approach, we quantify their dissimilarity in terms of both their appearance (from the bags of turbotensors de-projected in the tangent space) and in terms of their structure/topology (from the Isomap embeddings). We refer the term encoding appearance dissimilarity as *appear* term and the one encoding embedding dissimilarity as *graph* term.

$$D_{ent}(\mathcal{I}_X, \mathcal{I}_Y) = E_{appear}(vec_\mu(\mathcal{T}_X), vec_\mu(\mathcal{T}_Y)) + \lambda E_{graph}(\Theta_X, \Theta_Y) , \qquad (5)$$

where $vec_\mu(\mathcal{T}_X) = \{vec_\mu(\mathbf{X}_1), \ldots, vec_\mu(\mathbf{X}_n)\}$ (and similarly for $vec_\mu(\mathcal{T}_Y)$) and $\lambda \geq 0$ controls the degree of structural information cosidered. As $vec_\mu(\mathcal{T}_X)$ and $vec_\mu(\mathcal{T}_Y)$ are multi-dimensional distributions which lie in the same tangent space, the term $E_{appear}(.,.)$ can be computed with bypass information-theoretic divergence estimators, that is, estimators which do not need to compute a multi-dimensional histogram and are asymptotically consistent. We apply the same rationale to compute $E_{graph}(.,.)$ but we use a more discriminative bypass divergence, provided that Θ_X and Θ_Y are aligned in advance.

3.1   Divergences in Tangent Spaces

The simplest way of characterizing $\mathcal{T}_X$ is through a generalized Gaussian for tensors [7][11]. Then, after estimating the parameters of two generalized Gaussians from the samples we may compute the symmetrized Kullback-Leibler divergence between the two Gaussians, and even formulate the registration problem in terms of minimizing the latter divergence (see [12] where they extend the approach to more-realistic mixtures of tensors). Such extension is also required in images where turbotensors cannot be assumed to be i.i.d. and thus the central limit theorem is not applicable. In this paper we propose to use more flexible *non-parametric* bypass divergences. One of them is the *Henze and Penrose divergence* [13] between two distributions $\mathcal{F}_X$ and $\mathcal{F}_Y$:

$$D_{HP}(\mathcal{F}_X\|\mathcal{F}_y) = \int \frac{p^2\mathcal{F}_X^2(\mathbf{z}) + q^2\mathcal{F}_Y^2(\mathbf{z})}{p\mathcal{F}_X(\mathbf{z}) + q\mathcal{F}_Y(\mathbf{z})}d\mathbf{z} \ , \tag{6}$$

where $p \in [0,1]$ and $q = 1 - p$. This divergence is the limit of the Friedman-Rafsky run length statistic [14], that in turn is a multi-dimensional generalization based on MSTs (Minimum-Spanning Trees) of the Wald-Wolfowitz test. The Friedman-Rafsky test exploits the fact that the MST relates samples that are close in $\mathbb{R}^{d(d+1)/2}$. Let $vec_\mu(\mathcal{T}_X) = \{vec_\mu(\mathbf{x}_i)\}$ and $vec_\mu(\mathcal{T}_Y) = \{vec_\mu(\mathbf{y}_i)\}$be two sets of samples drawn from $\mathcal{F}_X$ and $\mathcal{F}_Y$, respectively. The steps of the Friedman-Rafsky test are: (a) Build the MST over the samples in $vec_\mu(\mathcal{Z}) = vec_\mu(\mathcal{T}_X) \cup vec_\mu(\mathcal{T}_Y) = \{\mathbf{z}_i\}$; (b) Remove the edges that do not connect a sample from $vec_\mu(\mathcal{T}_X)$ with a sample from $vec_\mu(\mathcal{T}_Y)$; and c) the proportion of non-removed edges converges to $\bar{D}_{HP}(\mathcal{F}_X\|\mathcal{F}_y) = 1 - D_{HP}(\mathcal{F}_X\|\mathcal{F}_y) \in [0,1]$. See an example in Fig. 2.

The second non-parametric (and more efficient) bypass divergence considered in this paper is the *Total Variation kd-Partition Divergence*. Such divergence is based on the kd-partition estimation of a multi-dimensinoal density function introduced in [15]. Let $\mathbf{x}$ be a $d(d+1)/2$-dimensional random variable defined in Ω, and $\mathcal{F}_X$ its pdf. Let $A = \{A_j | j = 1, \ldots, p\}$ be a partition of Ω for which $A_i \cap A_j = \emptyset$ if $i \neq j$ and $\bigcup_j A_j = \Omega$. Then, we can approximate $\mathcal{F}_X(\mathbf{x})$ in each cell as $\hat{F}_{X_{A_j}}(\mathbf{x}) = \frac{n_j}{n\mu(A_j)}$, where: $\mu(A_j)$ is the $d(d+1)/2$-dimensional volume of A_j, n is the number of samples of $\mathbf{x}$, and n_j are the number of these samples inside A_j. The partition is created recursively following the data splitting method of the k-d tree algorithm. At each level, data is split at the median along one axis. Then, data splitting is recursively applied to each subspace until an uniformity

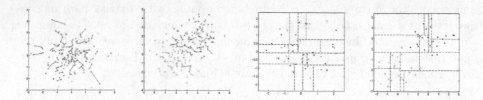

Fig. 2. Left: Henze-Penrose divergences for Gaussians; same mean and covariance (left) with $\bar{D}_{HP}(\mathcal{F}_X\|\mathcal{F}_Y) = 0.5427$ and different means (right) with $\bar{D}_{HP}(f_X\|f_Y) = 0.8191$. Right: same cases for kd-P divergences with $D_{kdP}(\mathcal{F}_X\|\mathcal{F}_Y) = 0.24$ (left) vs $D_{kdP}(\mathcal{F}_X\|\mathcal{F}_Y) = 0.92$ (right).

stop criterion is satisfied. As the distribution of the median of the samples in A_j tends to a normal distribution that can be standardized as:

$$Z_j = \sqrt{n_j}\, 2(med_d(A_j) - min_d(A_j) - max_d(A_j))/(max_d(A_j) - min_d(A_j))$$
(7)

where $med_d(A_j)$, $min_d(A_j)$ and $max_d(A_j)$ are the median, minimum and maximum, respectively, of the samples in cell A_j along dimension d. An improbable value of Z_j, that is, $|Z_j| > 1.96$ (the 95% confidence threshold of a standard normal distribution) indicates significant deviation from uniformity. Non-uniform cells should be divided further. The uniformity test is not applied until there are less than $\sqrt{n}$ data points in each partition, that is, until the level $L_n = \lceil \frac{1}{2} \log_2(n) \rceil$ is reached. Then, if we have two distributions $\mathcal{F}_X$ and $\mathcal{F}_Y$ from which we draw a set $vec_\mu(\mathcal{T}_X)$ of n samples and a set $vec_\mu(\mathcal{T}_X)$ of m samples, respectively. If we apply the partition scheme of the k-d partition algorithm to the set of samples $vec_\mu(\mathcal{Z}) = vec(\mathcal{T}_X) \cup vec_\mu(\mathcal{T}_Y) = \{z_i\}$, the result is a partition A, being $A = \{A_j | j = 1, \ldots, p\}$. For $\mathcal{F}_X$ and $\mathcal{F}_Y$ their respecitve probability at any cell A_j is given by $\mathcal{F}_X(A_j) = \frac{n_j}{n}$ and $\mathcal{F}_X(A_j) = \frac{m_j}{m}$, where n_j is the number of samples of $\mathcal{F}_X$ in cell A_j and m_j is the number of samples of $\mathcal{F}_Y$ in the cell A_j. Since the same partition A is applied to both sample sets, and considering the set of cells A_j a finite alphabet, we can compute the *k-dP total variation divergence* between $\mathcal{F}_X$) and $\mathcal{F}_Y$ as:

$$D_{kdP}(F_X \| F_Y) = \frac{1}{2} \sum_{j=1}^{p} |\mathcal{F}_X(A_j) - \mathcal{F}_X(A_j)| \in [0,1] ,$$
(8)

see an example in Fig. 2.

Considering the two above described divergences we have two alternative definition of the appearance $E_{appear}(vec_\mu(\mathcal{T}_X), vec_\mu(\mathcal{T}_Y)) : \bar{D}_{HP}(\mathcal{F}_X \| \mathcal{F}_y)$ and $D_{kdP}(\mathcal{F}_X \| \mathcal{F}_Y)$.

3.2   Divergences between Embeddings

Let $\mathcal{W}^*$ the optimal non-rigid transformation of Θ_Y to align it with Θ_X. Then, the *normalized-entropy square variation* (NSEV) is [16]

$$D_{NESV}(\Theta_X, \mathcal{W}^*(\Theta_Y)) = (H(\mathcal{W}^*(\Theta_Y)) - H(\Theta_X))^2 / H(\mathcal{W}^*(\Theta_Y)) + H(\Theta_X) ,$$
(9)

where $H(.)$ is the Shannon entropy. What is important in NESV is the fact that it is a multi-dimensional divergence. This implies estimating the Shannon entropy through a bypass approach. In this regard, we exploit the kNN-based bypass estimator proposed by Leonenko [17]. This estimator is both consistent and fast to compute.

In order to define $E_{graph}(\Theta_X, \Theta_Y)$ we have to consider that, in general, $D_{NESV}(\Theta_X, \mathcal{W}^*(\Theta_Y)) \neq D_{NESV}(\mathcal{W}'^*(\Theta_X), \Theta_Y)$, where $\mathcal{W}'$ encodes the optimal way of aligning Θ_X with the static Θ_Y. . Then, it is convenient to symmetrize, as we do with the Kullback-Leibler divergence, to obtain

$$E_{graph}(\Theta_X, \Theta_Y) = D_{NESV}(\Theta_X, \mathcal{W}^*(\Theta_Y)) + D_{NESV}(\mathcal{W}'^*(\Theta_X), \Theta_Y) .$$
(10)

4   Experimental Results

In order to test the proposed dissimilarity measure between over-segmented images we analyze the content of 100 frames of real-time video taken by a wearable device. We have divided the video in 6 fragments attending to both semantical content and observer position: corridor#1,end#1, hall, corridor#2, hall (seen in the opposite sense), and end#2. As both corridors have a similar topology, the combination of non-rigid alignment and NESV provides almost a unique cluster but at the fragments representing transitions between corridors (end of the first corridor, and hall seen backwards). This is due to the fact that the non-rigid alignment of structural constrained Isomap enforces the similarity between images following a similar perspective/vanishing point topology (we present some examples for each zone in Fig. 3).

With a pure topological criterion we can only answer a question of whether a given frame is corridor-like or not. However, in video-analysis applications sometimes we are interesting in answering the following question: in what part of the video is located a given frame? To answer that question it is better to use a pure appearance-based approach. If we analyze the results for E_{appear} for the *Henze-Penrose divergence* , we find that the first corridor is almost perfectly clustered due to the common appearance statistics of its frames; the second corridor less clustered but transitions are well clustered. This is consistend with the fact that the hall and the second corridor suffer from strong illumination effects that make the clustering more local for that kind of divergence. However, such effects are removed by using the *kd-partition total variance divergence* which provides a good clustering of the second corridor and isolates it from the hall. The latter divergence answers the questions: to what corridor belongs a given frame? and, does the frame belongs to the first hall? We combine E_{graph} and E_{appear} with $\lambda = 0.01$ (because the appearance divergences give outcomes in $[0,1]$) and choosing the kd-partition total variance divergence , we are able of enforcing the coherence of the second corridor cluster and simultaneously answering whether a frame belongs to a transition (end of corridor, hall). In Fig. 4-letf we present the retrieval-recall curves for all the latter cases. There is a significant quantitative improving of E_{graph} by E_{appear} and the integration of the two latter elements have a similar result. In all cases we have considered $d = 10$ both for the embedding and the divergences. The average number of turbopixels was 500.

corridor#1   end#1   hall   corridor#2   hall   end#2

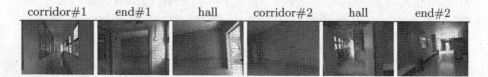

Fig. 3. Examples of frames for each video fragment: corridor#1 (frames #1), end#1 (frame #37), hall (frame #52), corridor#2 (frame #70), hall (backwards) (frames #86) and end#2 (frame #93)

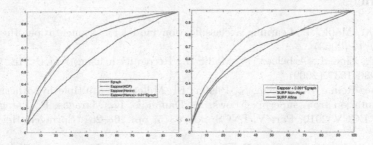

Fig. 4. Retrieval-Recall Curves. Left: Results for E_{graph} vs E_{appear} and also for their combination. Rigth: The complete energy function outperforms SURF matching with CPD both in the non-rigid and the affine versions.

Finally, we compare turbopixels with state-of-the-art feature matching. More precisely, we extract SURF features [18] and perform both non-rigid and affine alignment between images. As a similarity measure we use the NESV divergence [16] considering $d = 64$ (SURF dimensionality). In Fig. 4-right, we show that the turbopixels approach outperforms SURF matching with NESV similarity. The bypass estimator used in the experiments allows to compute high-dimensional entropies (64D) with hundreds of SURF points per image.

5   Conclusions

In this paper we propose an information-geometry approach for image comparison. Our contribution is threefold: (1) we formulate the structural comparison of superpixel-based images from a graph matching perspective which in turn is posed in terms of non-rigid manifold alignment, including also a multi-dimensional information-theoretic dissimilarity measure; (2) we formulate the comparison of bags-of-superpixels through tangent spaces de-projection and multi-dimensional and non-parametric information-theoretic dissimilarity measures; (3) we combine (1) and (2) to outperform SURF matching + SNESV dissimilarity in terms of retrieval-recall. Future work includes the analysis of the role of E_{graph} term in the general case of comparing images taken from different vanishing points (e.g. frontal views, slanted views, and so on) in order to test the usefulness of super-pixels in image retrieval.

Acknowledgements. F. Escolano, B. Bonev and M. Lozano are funded by Project TIN2012-32839 of the Spanish Government and E. R. Hancock is funded by a Royal Society Wolfson Research Merit Award.

References

1. Ren, X., Malik, J.: Learning a classification model for segmentation. In: ICCV, pp. 10–17 (2003)
2. Gu, C., Lim, J., Arbelaez, P., Malik, J.: Recognition using regions. In: CVPR, pp. 1030–1037 (2009)
3. Vazquez-Reina, A., Avidan, S., Pfister, H., Miller, E.: Multiple hypothesis video segmentation from superpixel flows. In: Daniilidis, K., Maragos, P., Paragios, N. (eds.) ECCV 2010, Part V. LNCS, vol. 6315, pp. 268–281. Springer, Heidelberg (2010)
4. Wang, S., Lu, H., Yang, F., Yang, M.H.: Superpixel tracking. In: ICCV (2011)
5. Boltz, S., Nielsen, F., Soatto, S.: Earth mover distance on superpixels. In: ICIP, pp. 4597–4600 (2010)
6. Levinshtein, A., Stere, A., Kutulakos, K., Fleet, D., Dickinson, S., Siddiqi, K.: Turbopixels: Fast superpixels using geometric flows. IEEE Trans. Pattern Anal. Mach. Intell. 31(12), 2290–2297 (2009)
7. Pennec, X., Fillard, P., Ayache, N.: A riemannian framework for tensor computing. International Journal of Computer Vision 66(1), 41–66 (2006)
8. Zhang, F., Hancock, E.: New riemannian techniques for directional and tensorial image data. Pattern Recognition 43(4), 1590–1606 (2010)
9. Myronenko, A., Song, X.B.: Point-set registration: Coherent point drift. EEE Trans. on Pattern Analysis and Machine Intelligence 32(12), 2262–2275 (2010)
10. Tenenbaum, J.B., de Silva, V., Langford, J.C.: A global geometric framework for nonlinear dimensionality reduction. Science 290(5500), 2319–2323 (2000)
11. Pennec, X., Stefanescu, R., Arsigny, V., Fillard, P., Ayache, N.: Riemannian elasticity: A statistical regularization framework for non-linear registration. In: MICCAI, vol. 2, pp. 943–950 (2005)
12. Chiang, M.C., Leow, A., Klunder, A., Dutton, R., Barysheva, M., Rose, S., McMahon, K., de Zubicaray, G., Toga, A., Thompson, P.: Fluid registration of diffusion tensor images using information theory. IEEE Trans. Med. Imaging 27(4), 442–456 (2008)
13. Henze, N., Penrose, M.: On the multi-variate runs test. Annals of Statistics 27, 290–298 (1999)
14. Friedman, J., Rafsky, L.: Mutivariate generalization of the wald-wolfowitz and smirnov two-sample tests. Annals of Statistics 7(4), 697–717 (1979)
15. Stowell, D., Plumbley, M.: Fast multidimensional entropy estimation by k-d partitioning. IEEE Signal Processing Letters 16(6), 537–540 (2009)
16. Escolano, F., Hancock, E., Lozano, M.: Graph matching through entropic manifold alignment. In: CVPR, pp. 2417–2424 (2011)
17. Leonenko, N., Pronzato, L., Savani, V.: A class of renyi information estimators for multidimensional densities. Annals of Statistics 36(5), 2153–2182 (2008)
18. Bay, H., Ess, A., Tuytelaars, T., Gool, L.J.V.: Speeded-up robust features (surf). Computer Vision and Image Understanding 110(3), 346–359 (2008)

Active-Learning Query Strategies Applied to Select a Graph Node Given a Graph Labelling[*]

Xavier Cortés and Francesc Serratosa

Universitat Rovira i Virgili, Departament d'Enginyeria Informàtica i Matemàtiques, Spain
{xavier.cortes,francesc.serratosa}@urv.cat

Abstract. Given two graphs, the aim of graph matching is to find the "best" matching between nodes of one graph and nodes of the other graph. Due to distortions of the data and the complexity of the problem, in some applications, an active and interactive graph algorithm is needed. The active module queries one of the nodes of the graphs and the interactive module receives from an oracle (human or artificial) the node of the other graph that has to be mapped with and considers the new knowledge. We present different active strategies that decide the node to be queried and adapt these strategies to the graph-matching problem.

Keywords: Machine Learning, Active and Interactive Graph Matching, Least Confident, Maximum Entropy, Expected Model Change.

1   Introduction

The key idea behind *active learning* [1], [2] is that a machine learning algorithm can achieve a greater accuracy with fewer classified training examples if it is allowed to choose the data from which it learns. The learner queries some elements and the answerer of the queries decides which classes these elements belong to. The answerer might be another automatic system or a human annotator and in general, it is called an *oracle* since it is assumed its answer is always correct. Active learning is well motivated in many modern machine-learning problems, where unclassified examples may be abundant but finding the class is difficult, time-consuming or expensive to obtain. Active learning scenarios involve evaluating the informativeness of unlabelled instances, thus, the most informative instances are the ones presented to the oracle.

If we put together machine learning and graph matching disciplines, we can define a model in which examples and classes in the machine learning discipline are composed of the set of nodes of one of the graphs and the nodes of the other graphs, respectively. Therefore, in this new framework, we want to learn the labelling between two graphs that is considered to be the best. That is, we wish to find the best labelling between nodes of both graphs but with the minimum necessary help of an oracle.

In this paper, we have analysed several active strategies to select the unlabelled instances and we have adapted them to select a mapping between a pair of nodes given a labelling between graphs. An interactive algorithm for error-tolerant graph

---

[*] This research is supported by the CICYT project DPI 2010-17112.

W.G. Kropatsch et al. (Eds.): GbRPR 2013, LNCS 7877, pp. 61–70, 2013.

matching [3] will query this mapping to the oracle with the aim of increasing the goodness of the labelling. This interactive method is useful in applications that crucial to have a perfect match is crucial but data is very noisy and it is difficult to extract the local parts of objects (medical applications). Conversely, this method is not useful in applications that an unclassified graph has to be compared with a huge number of graphs in a database (fingerprint identification). The human interaction in each graph comparison would increase the run time considerably. Some of the active strategies presented in this paper have been previously presented in [4]. In this paper, they are explained in depth and other ones are included.

The rest of the paper is organised as follows. In section 2 and 3, we present the basic graph-matching notation and summarise the known active strategies. In section 4, we present the adaptation of the active strategies in the graph-matching framework. Finally, in section 5 we show the practical evaluation and we conclude the paper in section 6.

2   Graph Matching

Let g^1 and g^2 be two attributed graphs. We suppose that g^1 and g^2 have the same number of nodes n since they have been enlarged enough to incorporate null nodes. We define nodes in g^1 and g^2 as $v_i^1 \in \Sigma_v^1$ and $v_a^2 \in \Sigma_v^2$ and we define arcs as $e_{ij}^1 \in \Sigma_e^1$ and $e_{ab}^2 \in \Sigma_e^2$, $\forall i, j, a, b \in \{1, \dots, n\}$. Moreover, let f be a bijective labelling between nodes of both graphs. The cost of matching graphs g^1 and g^2, given this isomorphism f, is represented by

$$C_f(g^1, g^2) = \sum_{v_i^1 \in \Sigma_v^1} c_v(v_i^1, v_a^2) + \sum_{e_{ij}^1 \in \Sigma_e^1} c_e(e_{ij}^1, e_{ab}^2) \tag{1}$$

where $f(v_i^1) = v_a^2$ and $f(v_j^1) = v_b^2$. That is, the cost is defined as the addition of the pairwise costs of matching nodes and arcs. These local costs can be represented through two matrices $C_v \in \mathbb{R}^{+^2}$, $C_v[v_i^1, v_a^2] = c_v(v_i^1, v_a^2)$ and $C_e \in \mathbb{R}^{+^4}$, $C_e[v_i^1, v_a^2, v_j^1, v_b^2] = c_e(e_{ij}^1, e_{ab}^2)$ and their definition depends on the application [17], [21]. There are several error-tolerant graph-matching algorithms that return the best isomorphism f between two graphs: Probabilistic relaxation [5], Graduated-Assignment [6], Expectation-Maximisation [7], Bipartite Graph Matching [8] or Graph Matching with Point-Set Correspondence [9], [10]. In fact, the input of these algorithms can be matrices C_v and C_e instead of graphs g^1 and g^2 since matrices capture all the differences between graphs and the minimisation cost is defined through these matrices (1). Considering that the involved graphs have a degree of disturbance and also the exponential complexity of the problem, these algorithms do not return exactly the isomorphism f but a probability matrix related to it (except [8]). We represent this matrix by P where each cell contains $P[v_i^1, v_a^2] = Prob(f(v_i^1) = v_a^2)$. Thus, given the probability matrix P, it is necessary to derive the final labelling f by a discretization process.

In general, if we want to solve the error-tolerant graph-matching problem based on probabilities [5], [6], [7], [9] or [10], given two graphs g^1 and g^2, the general objective function to optimize corresponds to the quadratic assignment problem objective function,

$$C_P(g^1, g^2) = \sum_{v_i^1 \in \Sigma_v^1} \sum_{v_a^2 \in \Sigma_v^2} \sum_{\substack{v_j^1 \in \Sigma_v^1 \\ v_j^1 \neq v_i^1}} \sum_{\substack{v_b^2 \in \Sigma_v^2 \\ v_b^2 \neq v_a^2}} P[v_i^1, v_a^2] P[v_j^1, v_b^2] C_e[v_i^1, v_a^2, v_j^1, v_b^2]$$

$$+ \sum_{v_i^1 \in \Sigma_v^1} \sum_{v_a^2 \in \Sigma_v^2} P[v_i^1, v_a^2] C_v[v_i^1, v_a^2] \tag{2}$$

where P is restricted to

$$\sum_{v_i^1 \in \Sigma_v^1} P[v_i^1, v_a^2] = 1, \forall v_a^2 \in \Sigma_v^2 \quad \text{and} \quad \sum_{v_a^2 \in \Sigma_v^2} P[v_i^1, v_a^2] = 1, \forall v_i^1 \in \Sigma_v^1 \tag{3}$$

3 Query Strategies

This section provides an overview of the query strategies used in different active learning scenarios [1]. We use the following notation. x is an unclassified instance that can be queried, y is one of the classes and θ is a classification model. Finally, the element x_A^* refers to the most informative instance according to some query strategy A. Besides, the conditional probability $P(y|x; \theta)$ represents the posterior class probability of class y given an instance x and a classification model θ.

Uncertainly Sampling. The active learner queries the instances about which it is least certain how to classify. This approach is often straightforward for probabilistic learning models. For instance, when there are only two classes, the sampling strategy simply queries the instance whose posterior probability of a class is nearest ½. For the multiple class case, there are three interesting options.

Least Confident (LC) [11]: This strategy queries the element that its highest probability of belonging to a class is the lower one between all the elements.

$$x_{LC}^* = \text{argmin}_{\forall x} P(y^*|x; \theta) \tag{4}$$

where $y^* = \text{argmax}_{\forall y} P(y|x; \theta)$ is the most likely class labelling.

Margin Sampling (MS) [12]: This strategy aims to incorporate the posterior probability of the second most likely labelled. Intuitively, instances with large margins are easy, since the classifier has little doubt in differentiating between the two most likely class labels. On the contrary, instances with small margins are more ambiguous, thus knowing the true label would help the model to discriminate more effectively between them. If y^1 and y^2 are the first and second most probable class labels under the model θ, respectively, the queried element is,

$$x_{MS}^* = \text{argmin}_{\forall x} \{P(y^1|x; \theta) - P(y^2|x; \theta)\} \tag{5}$$

Maximum Entropy (ME) [11]: This strategy queries the element with maximum Shannon Entropy given the probabilities. The main idea of the method is to query the elements that are more difficult to be classified,

$$x_{ME}^* = \underset{\forall x}{\text{argmax}} \{-\textstyle\sum_{\forall y} P(y|x; \theta) log(P(y|x; \theta))\} \tag{6}$$

Query by Committee [13]: This strategy involves maintaining a committee of models which are all trained on the current labelled set but represent competing hypothesis.

The most informative query is considered to be the instance about which they most disagree. For measuring a level of disagreement, an option is the vote entropy

$$x_{QBC}^* = \underset{\forall x}{\text{argmax}} \left\{ -\sum_{\forall y} \frac{V(y)}{C} log\left(\frac{V(y)}{C}\right) \right\} \tag{7}$$

where $V(y)$ represents the number of votes that a class receives from among the committee member's predictions and C is the number of committees.

Expected Model Change [14]: An active learner queries the instances that would impart the greatest change to the current model if we knew its class. Since probabilistic models are usually trained using a Gradient Ascent technique, the change imparted to the model can be measured by the magnitude of the training gradient. And due to the learning module does not know the true class y of an instance x in advance; we must instead calculate the length as an expectation over the possible classes. Moreover, we assume the resulting magnitude of the training gradient M when the pair $\langle x, y \rangle$ has been added to the model θ is similar to the gradient magnitude of the probability related to $\langle x, y \rangle$. We make this approximation because the gradient magnitude should be nearly zero since the method converged in the previous round training and because we assume the training instances are independent.

$$x_{EMC}^* = \underset{\forall x}{\text{argmax}} \left\{ \sum_{\forall y} P(y|x; \theta) M(\langle x, y \rangle, \theta) \right\} \tag{8}$$

The magnitude of the gradient of C_P (eq. 1) respect variable $P[v_i^1, v_a^2]$ is

$$M(v_i^1, v_a^2) = \sum_{\substack{v_j^1 \in \Sigma_v^1 \\ v_j^1 \neq v_i^1}} \sum_{\substack{v_b^2 \in \Sigma_v^2 \\ v_b^2 \neq v_a^2}} P[v_j^1, v_b^2] C_e[v_i^1, v_a^2, v_j^1, v_b^2] + C_v[v_i^1, v_a^2] \tag{9}$$

Variance Reduction [15]: The aim of this strategy is to query the instance that minimises the learner's future error by minimising its variance. They used the estimated distribution of the model's output to estimate the variance σ_θ^2 of the learner after some new instance x has been labelled to class y.

$$x_{VR}^* = \underset{\forall x}{\text{argmin}} \{\sigma_\theta^2\} \tag{10}$$

Estimated Error Reduction [16]: Similarly to Variance Reduction strategy, the aim of this strategy is to query the instance that minimises the learner's future error but, instead of minimising the variance, the aim is to minimise the expectation of this error E_θ.

$$x_{VR}^* = \underset{\forall x}{\text{argmin}} \{E_\theta\} \tag{11}$$

Density-Weighted Methods [11]: It has been suggested that Uncertainty Sampling and Query by Committee strategies are prone to querying outliers. Although they are the least certain instances, they are not representatives of the other instances in the distribution. Therefore, knowing their label is unlikely to improve accuracy on the data as a whole. The main idea of this strategy is that instances should not only be those that are uncertain, but also those that are representative of the input distribution.

Consider that any strategy can be represented through equations $x_A^* = \underset{\forall x}{\text{argmax}} \{\emptyset_A(x)\}$ or $x_A^* = \underset{\forall x}{\text{argmin}} \{\emptyset_A(x)\}$ where $\emptyset_A(x)$ represents the informativeness of instance x according to some query strategy A. Then, the most informative query is,

$$x^*_{DWM\_A} = \underset{\forall x}{\text{argmax}} \left\{ \frac{\emptyset_A(x)}{W(x)} \right\} \tag{12}$$

or

$$x^*_{DWM\_A} = \underset{\forall x}{\text{argmin}} \{ W(x) \cdot \emptyset_A(x) \} \tag{13}$$

The choice of argmax or argmin depends on the strategy A. The W term weights the informativeness of x by its average distance to all other instances x' in the input distribution; $W(x) = \left(\frac{1}{U} \sum_{\forall x'} distance(x, x') \right)^{\beta}$. Parameter U is the number of instances and parameter β controls the relative importance of the density term.

4   Active Learning Strategies Based on the Probability Matrix

In this section, we present several strategies to select a node v^{1^*} of g^1 that have to be queried to the specialist. The specialist feedback is v^{2^*} which means it believes $f(v^{1^*}) = v^{2^*}$. The pool-based active learning [1] method can be cast directly to our problem since we have access to all the elements to be classified (graph nodes of g^1) and also the predefined classes (graph nodes of g^2). Usually, query elements are selectively drawn from the set of unclassified elements. In our case, an "unclassified element" is a node of g^1 that we don't know its mapping. For this reason, the pool of nodes to be queried is composed of nodes of g^1 that have never been queried before. In the strategies we present, there is a logical function $Q(v_i^1)$ that shows if node v_i^1 has been queried before. This logical function is used y the interactive algorithm [3] to assure a node is not queried several times. Note that in the case that $Q(i) = True$ for all nodes of g^1 then the following strategies returns the empty set. We next show our query strategies applied to graph matching. We have put together the original strategies and the weighted density strategies. In our case, the distance between elements is represented by the cost between two nodes of g^1. Note that we write the cost function c_v instead of the matrix C_v since in this case both parameters of the cost function are nodes of graph g^1.

Uncertainly Sampling. We define three different strategies.
Least Confident (LC): The learner queries the node $v_{LC}^{1^*}$ of g^1 that has not been previously queried and whose maximum probability given the nodes of g^2 is the lower. Node $v_{LC}^{1^*}$ is obtained in two steps. Firstly, we obtain the set of nodes in g^2: $\left\{ v^{2\{1\}}, ..., v^{2\{i\}}, ..., v^{2\{n\}} \right\}$ such that,

$$v^{2\{i\}} = \text{argmax}_{\forall j = \{1,..,n\}} P[v_i^1, v_j^2]; \ \forall i = \{1,..,n\} \tag{14}$$

note that $v^{2\{i\}}$ represents the node selected of g^2 when v_i^1 is considered. For this reason, some of the nodes in this set can appear several times, $v^{2\{i\}} = v^{2\{j\}}; i \neq j$.

And secondly, we select the node in g^1 such that its respective node in the set obtains the minimum probability,

$$v_{LC}^{1^*} = \text{argmin}_{\forall i = \{1,..,n\} | Q(i) = False} P\left[v_i^1, v^{2\{i\}} \right] \tag{15}$$

And the *Weighted Least Confident* (wLC):

$$v_{wLC}^{1^*} = \text{argmin}_{\forall i = \{1,..,n\} | Q(i) = False} \left\{ W(v_i^1) \cdot P\left[v_i^1, v^{2\{i\}} \right] \right\} \tag{15'}$$

Least Confident given the Current Labelling (LCCL): The aim of this strategy is to query the nodes that are matched through the current labelling but they have not been queried before. Therefore, it could be seen as the method tries to minimise the hamming distance between the current labelling and the ideal labelling (the labelling that would have been predicted by the oracle if all the nodes were queried). The learner queries node $v_{LCCL}^{1^*}$ of g^1 that has not been previously queried and it has the minimum probability given the current labelling f. Formally,

$$v_{LCCL}^{1^*} = \text{argmin}_{\forall i=\{1,\dots,n\}|Q(i)=\,False}\, P[v_i^1, f(v_i^1)] \tag{16}$$

And the weighted strategy (wLCCL):

$$v_{wLCCL}^{1^*} = \text{argmin}_{\forall i=\{1,\dots,n\}|Q(i)=\,False}\, \left\{ W(v_i^1) \cdot P\left[v_i^1, v^{2\{i\}}\right] \right\} \tag{16'}$$

Margin Sampling (MS): We define $v^{2\{i\}}$ as the most probable node respect v_i^1 and it is defined as in equation (14). We also define $v^{2'\{i\}}$ as the second most probable node and defined in a similar way than $v^{2\{i\}}$ but without considering this node. Then, the queried element is,

$$v_{MS}^{1^*} = \text{argmin}_{\forall i=\{1,\dots,n\}|Q(i)=\,False}\, \left\{ P\left[v_i^1, v^{2\{i\}}\right] - P\left[v_i^1, v^{2'\{i\}}\right] \right\} \tag{17}$$

And the *weighted Margin Sampling* (wMS):

$$v_{wMS}^{1^*} = \text{argmin}_{\forall i=\{1,\dots,n\}|Q(i)=\,False}\, \left\{ W(v_i^1) \cdot \left(P\left[v_i^1, v^{2\{i\}}\right] - P\left[v_i^1, v^{2'\{i\}}\right] \right) \right\} \tag{17'}$$

Maximum Entropy (ME): The selected node $v_{ME}^{1^*}$ depends on the Shannon Entropy,

$$v_{ME}^{1^*} = \text{argmax}_{\forall i=\{1,\dots,n\}|Q(i)=\,False}\, -\sum_{v_a^2 \in \Sigma_v^2} P[v_i^1, v_a^2] log(P[v_i^1, v_a^2]) \tag{18}$$

And the *weighted Maximum Entropy* (ME):

$$v_{wME}^{1^*} = \text{argmax}_{\forall i=\{1,\dots,n\}|Q(i)=\,False}\, \left\{ -\frac{1}{W(v_i^1)} \cdot \sum_{v_a^2 \in \Sigma_v^2} P[v_i^1, v_a^2] log(P[v_i^1, v_a^2]) \right\} \tag{18'}$$

Query by Committee: The general idea of this strategy would be to use several graph matching algorithms [5], [6], [7], [8], [9], [10] and then use the obtained matchings as models. Due to the huge time consuming that would suppose, we leave the study of how to efficiently implement this method as a future work.

Expected Model Change: We define two strategies.

Maximum Gradient Norm (MGN): The magnitude of the training gradient of variable $P[v_i^1, v_a^2]$ is defined as $M(v_i^1, v_a^2)$ (eq. 9). The learner should query the node $v_{MGN}^{1^*}$ defined through the following equation,

$$v_{MGN}^{1^*} = \underset{\forall i=\{1,\dots,n\} \wedge Q(i)=\,False}{\text{argmax}} \left\{ \sum_{v_a^2 \in \Sigma_v^2} P[v_i^1, v_a^2] M(v_i^1, v_a^2) \right\} \tag{19}$$

And the *weighted Maximum Gradient Norm* (wMGN):

$$v_{wMGN}^{1^*} = \underset{\forall i=\{1,\dots,n\} \wedge Q(i)=\,False}{\text{argmax}} \left\{ \frac{1}{W(v_i^1)} \cdot \sum_{v_a^2 \in \Sigma_v^2} P[v_i^1, v_a^2] M(v_i^1, v_a^2) \right\} \tag{19'}$$

Maximum Probability Change given a Common Labelling (MPC*CL*): The learner should query the instance that if changed its current labelling, would result a maximum increase in its probability,

$$v_{MPCCL}^{1^*} = \underset{\forall i = \{1,...,n\} \wedge Q(i) = False}{\operatorname{argmax}} \left\{ \underset{\forall a = \{1,...,n\}}{\max} \{P[v_i^1, v_a^2]\} - P[v_i^1, f(v_i^1)] \right\} \qquad (20)$$

And the weighted strategy (wMPC*CL*):

$$v_{wMPCCL}^{1^*} = \underset{\forall i = \{1,...,n\} \wedge Q(i) = False}{\operatorname{argmax}} \left\{ \frac{1}{W(v_i^1)} \cdot \left(\underset{\forall a = \{1,...,n\}}{\max} \{P[v_i^1, v_a^2]\} - P[v_i^1, f(v_i^1)] \right) \right\} \quad (20')$$

If $\underset{\forall j = \{1,...,n\}}{\max} \{P[v_i^1, v_a^2]\} > P[v_i^1, f(v_i^1)]$, the current labelling of v_i^1 is not the ideal one, considering only probabilities $P[v_i^1, v_a^2]$, for all $v_a^2 \in \Sigma_v^2$. On the contrary, if $\underset{\forall a = \{1,...,n\}}{\max} \{P[v_i^1, v_a^2]\} = P[v_i^1, f(v_i^1)]$, then, the current labelling is the one that obtains the maximum probability, so, it is the ideal case.

Variance Reduction & Estimated Error Reduction: We do not have implemented these methods. Both methods are based on statistical analysis. In our case, we only have one instance per class. Thus, it seems difficult to be applied in our framework.

5   Practical Evaluation

To experimentally validate our method, we have used some well-known graph databases: Letter high, GREC, COIL-Graph [18]. These databases have different characteristics such as cardinality, diversity, mean number of nodes and so on. Nevertheless, we considered only the (x, y) attribute on the nodes and arcs have no attributes. Several graph matching algorithms and graph-class prototypes have been compared using these datasets [19]. We have not taken into consideration the separation of these graphs into classes in any of the experiments. We have also used House and Hotel datasets [20]. They are composed by a set of reference points associated with different frames of a video. We built graphs as follows. Nodes are the points in the database and edges were generated using a 3-NN method. The first two rows of Table 1 show the main database characteristics. Third and fourth rows of table 1 show the mean and standard deviation of the weighting coefficient $W(v_i^1)$.

Each test set is composed of T elements $\{g^t, g'^t, h^t\}$ that have a pairs of graphs and a matching: $\{\{g^1, g'^1, h^1\} ... \{g^t, g'^t, h^t\} ... \{g^T, g'^T, h^T\}\}$. Graphs g^t, $1 \leq t \leq T$, are

Table 1. Database characteristics

Database	Letter	COIL	GREC	Hotel	House
# Nodes	4.7	11.5	12.8	30	30
# Edges	4.5	11.9	27.1	38.1	39.5
Mean of $W(v_i^1)$	1.5	33.8	222.1	184.6	154.5
St. dev. of $W(v_i^1)$	0.2	8.3	51.5	35.6	33.9
Node noise (σ)	0.75	15	0	10	10
Edge noise (Prob.)	0	0	0.2	0. 04	0.04

the ones in the original databases. T is the number of graphs of these datasets. Graphs g'^t, $1 \le t \le T$, have been randomly generated through distorting g^t. The attribute value of nodes have been modified applying a Gaussian distribution with mean zero and the standard deviation shown in the fourth row of table 1. Finally, some edges have been erased or inserted given the probability shown in the last row of table 1.

Both graphs g^t and g'^t have the same number of nodes. The labelling h^t has been computed together with the generation of the distorted graph g'^t and we assume it is the best labelling between g^t and g'^t. Nevertheless, by construction, we do not guarantee this labelling to obtain the minimum cost.

In this practical evaluation we show the precision of the strategies to select a node that has been incorrectly mapped. We want the active method to suggest a node v^{*t} of g^t that the current matching obtained by the graph-matching algorithm f^t is different from the human's matching h^t. This is because, in this case, the active and interactive methods will help the system to improve its output since the human will try to correct the wrong mapping. We are not interested in the cases that $f^t(v^{*t}) = h^t(v^{*t})$ since the human's interaction is going to be useless. Given an element of the test set $\{g^t, g'^t, h^t\}$ and the obtained matching f^t, the delta function represents the usefulness of the active action as follows,

$$\delta(v^{*t}, f^t, h^t) = \begin{cases} 1 & \text{if} \quad f^t(v^{*t}) \neq h^t(v^{*t}) \\ 0 & \quad otherwise \end{cases} \tag{21}$$

If the active method has been useful then $\delta(w^t, f^t) = 1$. The *Node Precision* of an active strategy is computed as the average usefulness of all the active actions through the elements $\{g^t, g'^t, h^t\}$ of the test set.

$$Node\ Precision = \frac{\sum_{t=1}^T \delta(v^{*t}, f^t, h^t)}{T} \tag{22}$$

Table 2 shows the Node Precision results of the active strategies. LC, LCCL and ME obtain the best results and the ME obtains the highest means. Considering the nature of the databases, less structure have the graphs, better performs MS strategy. Conversely, MGN performs better with more structural information. Finally, MPCCL obtained poor results in all of our tests.

Table 3 shows the Node Precision results of the weighted active strategies and table 4 shows the difference between non-weighted strategies (table 2) and weighted

Table 2. Node Precision of different strategies and random selection. Bold values are higher than 0.75.

	Uncertainly Sampling				Expe. Model Change		
	LC	LCCL	MS	ME	MGN	MPCCL	Random
Letter	0.81	0.81	0.80	0.81	0.56	0.51	0.53
COIL	0.81	0.81	0.78	0.79	0.53	0.55	0.55
GREC	0.79	0.80	0.74	0.77	0.97	0.56	0.34
HOUSE	0.79	0.82	0.77	0.78	1.00	0.77	0.53
HOTEL	0.71	0.77	0.68	0.75	1.00	0.72	0.56
Mean	0.78	0.80	0.75	0.78	0.81	0.62	0.50

strategies (table 3). There are few differences between them and they appear in cases the original strategies obtained poor results. We suppose this is because few nodes are distant from the other ones.

Table 3. Node Precision of different weighted strategies and random selection. Bold values are higher than 0.75.

	Uncertainly Sampling				Expe. Model Change		
	wLC	**wLCCL**	**wMS**	**wME**	**wMGN**	**wMPCCL**	**Random**
Letter	0.83	0.82	0.80	0.81	0.77	0.51	0.53
COIL	0.81	0.81	0.78	0.80	0.72	0.56	0.55
GREC	0.79	0.80	0.74	0.77	0.97	0.56	0.34
HOUSE	0.79	0.82	0.77	0.78	1.00	0.77	0.53
HOTEL	0.71	0.77	0.68	0.77	1.00	0.72	0.56
Mean	0.78	0.80	0.75	0.78	0.89	0.62	0.50

Table 4. Node Precision difference between non-weighted and weighted. Bold values are higher than 0.

	LC	LCCL	MS	ME	MGN	MPCCL	Rand
Letter	0.01	0.01	0.00	0.00	0.21	0.00	0.00
COIL	0.00	0.00	0.00	0.01	0.18	0.00	0.00
GREC	0.00	0.00	0.00	0.00	0.00	0.00	0.00
HOUSE	0.00	0.00	0.00	0.00	0.00	0.00	0.00
HOTEL	0.00	0.00	0.00	0.02	0.00	0.00	0.00

6   Conclusions and Future Work

We have presented different strategies to be applied on an active graph-matching algorithm. These strategies are based on classical active machine learning but they are applied to the case of searching for the best labelling between nodes. Moreover, they are based on the probability matrix between nodes that some sub-optimal algorithms use to iteratively find the best labelling. Experimental validation shows that MGN and LCCL obtain the highest precision for the non-weighted and weighted strategies, respectively.

As a future work, we want to apply these strategies to active and interactive error-tolerant graph matching algorithms. These algorithms improve the obtained labelling between nodes through the feedback of an oracle. Presented strategies decide the node to be queried to the oracle.

References

1. Settles, B.: Active Learning Literature Survey. Computer Science Technical Report 1648, University of Wisconsin-Madison
2. Kotsiantis, S.B.: Supervised Machine Learning: A Review of Classification Techniques. Informatica 31, 249–268 (2007)

3. Serratosa, F., Cortés, X., Solé-Ribalta, A.: Interactive Graph Matching by means of Imposing the Pairwise Costs. In: International Conference on Pattern Recognition, ICPR 2012, Accepted for publication, Tsukuba (2012)
4. Cortés, X., Serratosa, F., Solé-Ribalta, A.: Active Graph Matching based on Pairwise Probabilities between nodes. In: Gimel'farb, G., Hancock, E., Imiya, A., Kuijper, A., Kudo, M., Omachi, S., Windeatt, T., Yamada, K. (eds.) SSPR&SPR 2012. LNCS, vol. 7626, pp. 98–106. Springer, Heidelberg (2012)
5. Fekete, G., Eklundh, J.O., Rosenfeld, A.: Relaxation: Evaluation and Applications. IEEE Transactions on Pattern Analysis and Machine Intelligence 3(4), 459–469 (1981)
6. Gold, S., Rangarajan, A.: A Graduated Assignment Algorithm for Graph Matching. IEEE Transactions on Pattern Analysis and Machine Intelligence 18(4), 377–388 (1996)
7. Luo, B., Hancock, E.R.: Structural graph matching using the EM algorithm and singular value decomposition. IEEE Transactions on Pattern Analysis and Machine Intelligence 23(10), 1120–1136 (2001)
8. Riesen, K., Bunke, H.: Approximate graph edit distance computation by means of bipartite graph matching. Image Vision Comput. 27(7), 950–959 (2009)
9. Sanromà, G., Alquézar, R., Serratosa, F.: A New Graph Matching Method for Point-Set Correspondence using the EM Algorithm and Softassign. Computer Vision and Image Understanding 116(2), 292–304 (2012)
10. Sanromà, G., Alquézar, R., Serratosa, F., Herrera, B.: Smooth Point-set Registration using Neighbouring Constraints. Pattern Recognition Letters, PRL 33, 2029–2037 (2012)
11. Settles, B., Craven, M.: An analysis of active learning strategies for sequence labelling tasks. In: Conference on Empirical Methods in Natural Language Processing, pp. 1069–1078 (2008)
12. Culotta, A., McCallum, A.: Reducing labelling effort for structured prediction tasks. In: National Conference on Artificial Intelligence, pp. 746–751 (2005)
13. Melville, P., Mooney, R.: Diverse ensembles for active learning. In: International Conference on Machine Learning, pp. 584–591 (2004)
14. Settles, B., Craven, M., Ray, S.: Multiple-instance active learning. Advances in Neural Information Processing Systems 20, 1289–1296 (2008)
15. Zhang, T., Oles, F.J.: A probability analysis on the value of unlabelled data for classification problems. In: International Conference on Machine Learning, pp. 1191–1198 (2000)
16. Roy, N., McCallum, A.: Toward optimal active learning through sampling estimation of error reduction. In: International Conference on Machine Learning, pp. 441–448 (2001)
17. Sanfeliu, A., Fu, K.: A distance measure between attributed relational graphs for pattern recognition. IEEE Trans. on Sys. Man and Cybern 13, 353–362 (1983)
18. Riesen, K., Bunke, H.: IAM graph database repository for graph based pattern recognition and machine learning. In: da Vitoria Lobo, N., Kasparis, T., Roli, F., Kwok, J.T., Georgiopoulos, M., Anagnostopoulos, G.C., Loog, M. (eds.) SSPR & SPR 2008. LNCS, vol. 5342, pp. 287–297. Springer, Heidelberg (2008)
19. Serratosa, F., Cortés, X., Solé-Ribalta, A.: Component Retrieval based on a Database of Graphs for Hand-Written Electronic-Scheme Digitalisation. Expert Systems With Applications, ESWA 40, 2493–2502 (2013)
20. Caetano, T.S., Caelli, T., Schuurmans, D., Barone, D.A.C.: Graphical Models and Point Pattern Matching. IEEE Trans. Pattern Analysis and Machine Intelligence 28(10), 1646–1663 (2006)
21. Solé, A., Serratosa, F., Sanfeliu, A.: On the Graph Edit Distance cost: Properties and Applications. International Journal of Pattern Recognition and Artificial Intelligence, IJPRAI 26(5) (2012)

GMTE: A Tool for Graph Transformation and Exact/Inexact Graph Matching

Mohamed Amine Hannachi[1,2], Ismael Bouassida Rodriguez[1,2,3],
Khalil Drira[1,2], and Saul Eduardo Pomares Hernandez[1,4]

[1] CNRS, LAAS, 7 avenue du colonel Roche, F-31400 Toulouse, France
[2] Univ de Toulouse, LAAS, F-31400 Toulouse, France
[3] ReDCAD, University of Sfax, B.P. 1173, 3038 Sfax, Tunisia
[4] Computer Science Department, Instituto Nacional de Astrofísica, Óptica y
Electrónica, C.P. 72840, Tonantzintla, Puebla, Mexico
{hannachi,bouassida,khalil,sepomare}@laas.fr

Abstract. Multi-labelled graphs are a powerful and versatile tool for
modelling real applications in diverse domains such as communication
networks, social networks, and autonomic systems, among others. Due
to dynamic nature of such kind of systems the structure of entities is
continuously changing along the time, this because, it is possible that
new entities join the system, some of them leave it or simply because
the entities relations change. Here is where graph transformation takes
an important role in order to model systems with dynamic and/or evo-
lutive configurations. Graph transformation consists of two main tasks:
graph matching and graph rewriting. At present, few graph transformation tools support multi-labelled graphs. To our knowledge, there is
no tool that support inexact graph matching for the purpose of graph
transformation. Also, the main problem of these tools lies on the limited expressiveness of rewriting rules used, that negatively reduces the
range of application scenarios to be modelling and/or negatively increase
the number of rewriting rules to be used. In this paper, we present the
tool GMTE Graph Matching and Transformation Engine. GMTE handles directed and multi-labelled graphs. In addition, to the exact graph
matching, GMTE handles the inexact graph matching. The approach
of rewriting rules used by GMTE combines Single PushOut rewriting
rules with edNCE grammar. This combination enriches and extends the
expressiveness of the graph rewriting rules. In addition, for the graph
matching, GMTE uses a conditional rule schemata that supports complex comparison functions over labels. To our knowledge, GMTE is the
first graph transformation tool that offers such capabilities.

1 Introduction

Graphs are a powerful and natural way of modelling complex systems on an
intuitive level. Graph-based modelling is applied in diverse domain such as communication networks, social networks, autonomic systems, data representing,
entity relationship and UML diagrams, and visualization of software architectures. Due to the dynamic nature of such systems, graph transformation concept

W.G. Kropatsch et al. (Eds.): GbRPR 2013, LNCS 7877, pp. 71–80, 2013.
© Springer-Verlag Berlin Heidelberg 2013

takes an important role in order to model dynamic behavior by describing the evolution of graph based structures.

The research area of graph transformation dates back to the seventies, but development of software tools that support this formalism only begun twenty years later. These tools have made graph transformation more and more popular and widely used as a modelling paradigm. Currently, numerous graph transformation tools are under development, covering a wide spectrum of applications such as bioinformatics, process management, object-oriented modeling, architectural design, reengineering, distributed systems, etc. PROGRES [1], AGG [2], GROOVE [3] are well-known tools, which support general purpose graph transformation. For specific purpose we could find VMTS [4], GreAT [5], ATOM3 [6] for model transformation.

In this paper, we expose our general graph matching and transformation engine $GMTE$[1] handling exact and inexact graph matching with expressive graphs and rewriting rules. In section 2, we present the preliminaries concept. In Section 3, we introduce exact graph matching, extensions of model graph definition, vertex matching and consistent valuation building. Graph updating process where rewriting rules consider variable labels, both positive and negative application conditions, connection instructions, modification instruction and conditional rule schemata with label calculation, is presented in this section. Section 4 deals with inexact graph matching based on the graph edit distance and bipartite matching. Comparison to a reference tool is performed in Section 5. The concluding remarks and perspectives are discussed in Section 6.

2   Preliminaries

In this section, we establish the fundamental definitions used in this paper and give the formal problem statement. This paper investigates the subgraph matching and transformation for directed and multi-labeled graphs.

Definition 1. *A multi-labeled graph G is defined as a 6-tuple $G = V, E, L_V, D_{L_V}, L_E, D_{L_E})$, where V is the set of vertices $E \subseteq V \times V$ is the set of edges. D_{L_V} and D_{L_E} are the definition domains of vertex labels and edge labels. $L_V : V \to D_{L_V}$ is the function assigning labels to vertices and $L_E : E \to D_{L_E}$ is the function assigning labels to edges.*

The main idea of graph transformation is the rule-based modification of graphs. The foundation of a rule is a pair of graphs (L,R), called the left-hand side L and the right-hand side R. Applying the rule $p = (L, R)$ means finding a match of L in the source graph and replacing L by R, leading to the target graph of the graph transformation. The problem of finding a match of L is treated by *graph isomorphism.*

Definition 2. *Two graphs $G_1 = (V_1, E_1)$ and $G_2 = (V_2, E_2)$ are isomorphic if there exists a mapping $M \subseteq V_1 \times V_2$ such that for all pairs of vertices $v_i, v_j \in V_1$,*

---

[1] Graph Matching and Transformation Engine ($GMTE$), available at
   http://homepages.laas.fr/khalil/GMTE/

$(v_i, v_j) \in E_1$ *if and only if* $(M(v_i), M(v_j)) \in E_2$. *M is, in this case, a graph isomorphism between G_1 and G_2. A subgraph isomorphism is an isomorphism between G_1 and a subgraph G' of G_2.*

In contrast to the exact graph matching, the inexact (or approximate, error-tolerance) graph matching allows nodes or edges mismatch or both. *Graph edit distance* is one of the most commonly used and well-known approach that defines similarity between graphs. The distance between two graphs is measured by applying a sequence of edit operations (i.e. node and edge insertion, deletion, or substitution) in order to transform one graph into the other. For each edit operation, a cost is assigned. The cost of an edit series is the sum of the individual edit operations. The graph edit distance is the minimum cost necessary for transforming one graph to another.

Definition 3. *Let $G_1 = (V_1, E_1)$ and $G_2 = (V_2, E_2)$ be two graphs. The graph edit distance of G_1 and G_2 defined by*

$$d(G_1, G_2) = \min_{(e_1 \ldots e_k) \in P(G_1, G_2)} \sum_{i=0}^{k} C(e_i)$$

where $P(G_1, G_2)$ denotes the set of edit paths transforming G_1 into G_2, and C denotes the edit cost function.

3   Graph Matching and Transformation Engine

In this section we deal with concepts and theory that our tool is built on them. The *GMTE* encompasses two main processes: the first one is called the pattern matching process and the second one is the graph updating process.

3.1   Exact Matching Process

The matching process between two graphs $G_1 = (V_1, E_1)$ and $G_2 = (V_2, E_2)$ consists of finding a mapping M which assigns nodes from G_1 to nodes from G_2, taking into account some predefined constraints. The mapping M is a set of node pairs. Each pair represents the mapping of a node from G_1 with node from G_2 if and only if the mapping is a bijection, preserve adjacency (if two nodes are adjacent in the first graph, their image by the bijection should be adjacent), so M is called isomorphism.

The graph matching algorithm implemented within the *GMTE*, is the one defined in [7]. The algorithm considers a Breadth-first search approach similar to that introduced by the algorithm described by Messmer and Bunke in [8]. The choice to rely on such approach is motivated by the fact that this algorithm is highly effective in cases where matching involves several similar graphs.

For *GMTE* the definition of the left-hand side L is extended to allow the use of variable attributes for nodes and edges labels.

Definition 4. *Let $S_X = X_1, ..., X_i$ be a set of variables. A model graph MG is defined as a 6-tuple $MG = (V, E, L_V, D_{L_V}, L_E, D_{L_E})$, where V, E, L_V and L_E have the same descriptions as given in Definition 1. $D_{L_V} = [(L_1 \cup S_X^1) \cup ... \cup (L_n \cup S_X^n)]$ is the definition domain of vertex labels, with $S_X^1, ..., S_X^n$ are subsets of S_X. $D_{L_E} = [(E_1 \cup S'^1_X) \cup ... \cup (E_n \cup S'^n_X)]$ is the definition domain of edge labels, with $S'^1_X, ..., S'^n_X$ are subsets of S_X.*

Considering the extension previously introduced, we establish some new notions related to label and vertex matching and to valuation merging. Two labels are considered matchable if and only if they are either constant and having same type or if the labels from the model graph is variable it must have the same type as the constant label from the input graph. Based on this notion, two nodes or edges are matchable if and only if they have same number of labels, their labels are, two by two, matchable in respect of their occurrence order, and the results of all parameter matching are consistent. Two valuations Val_1 and Val_2 are consistent if and only if, for every pair $(x_1, value_1)$ belonging to Val_1 and every pair $(x_2, value_2)$ belonging to Val_2, x_1 and x_2 are two different variables (syntactically) or represent the same variable and associate it with the same value $x_1 = x_2$ and $value_1 = value_2$. Typically, a vertex $v_1(x, y, x, 3)$ is not matchable with $v_2(1, 1, 2, 3)$ but is matchable with $v_3 = (1, 2, 1, 3)$ and gives in this case the valuation set $\{(x, 1), (y, 2)\}$ as a result.

3.2 Graph Updating Process

Dealing with graph transformation two major technical problems arise: how to delete L from G and how to connect R with the remaining part of it. To cope with these problems $GMTE$ combines two approaches. The first one is the simple pushout SPO [9], where a graph transformation rule is (L, K, R). L and R are the left and right hand side graphs of the rule and K is a common subgraph of L and R that will be preserved after the rule application. Also K has a second role which consists of the part that the added nodes will be connected to. The removal of $L \setminus K$ raises the problem of dangling edges (edges without starting or ending node). This approach implies that dangling edges are deleted once $L \setminus K$ is removed.

According to [10] there is another powerful mechanism which is based on *connection instructions* who enrich the formalism of how to connect R. This approach, called *Node Controlled Embedding*, allows connecting nodes from the right-hand side graph to the neighbours of removed nodes from the left-hand side graph.

The $GMTE$ rewriting techniques uses the last extension of the NCE which is the $edNCE$ used for directed and edge labelled graph. For $edNCE$ grammars a connection instruction is of the form $(m, \mu, p/q, x, d, d')$ with obvious meaning: a q-labelled edge should be established between x (node from R) and every μ-labelled node of host graph G that is a p-neighbour of m (node from L), with $d, d' \in \{in, out\}$, where d is the old edge direction and d' the new one. The host graph is multi-labelled so μ, p and q are sets of labels. To determine the applicability of a transformation rule, the approaches presented above are based on the existence of an instance of the sub-graph L. In some cases, it is necessary to express additional conditions to specify conditions relating to the absence of an occurrence in the host graph. Such conditions or restrictions are called negative application conditions (NAC) [11].

3.3   Conditional Rule Schemata

The conditional rule schemata introduced in [12] extend graph transformation with operations on labels. Within the *GMTE* we have two types of operations. Operations on nodes and edges labels of the *Add zone* are called *functions*. When operations are used within nodes and edges labels of the *Delete zone* and *Invariant zone* they are called *conditions*.

Rule graphs used by the *GMTE* are multi-labelled so conditions and functions could be used on any label of node and edge with the respect to the following two conditions:

– Conditions are Boolean expression built in an arithmetic expressions, used as label of nodes and edges within the *Delete zone* and *Invariant zone*. Functions used only on the *Add zone*.
– Var (*Addzone*) $\subseteq Var$ (*Delete zone* $+$ *Invariant zone*).

where $Var(Y)$ is set of all variables in zone Y. Constraint label formalism is added in order to extend the expressiveness of the matching and the transformation. We used the *muParser* [2] which is an extensible high performance math expression parser library written in $C++$. The main objective of this library is to provide a fast and easy way of parsing mathematical expressions.

3.4   Rule and Application Condition

In *GMTE*, the basic representation of a rule is a single graph combining all of the following four zones:

– *Invariant zone*: a subgraph that needs to be present in the input graph in order for the rule to be applicable, this pattern is preserved after the rule application;
– *Absent zone*: a subgraph that must be absent in the input graph to allow the application of the rule;
– *Delete zone*: a subgraph that needs to be present in the input graph in order to be deleted after the rule application;
– *Add zone*: a subgraph that will be added after the rule application.

As we can see, the presence of the *Invariant zone* and *Delete zone* form the positive application condition, and the absence of the *Absent zone* forms the

---

[2] A fast math parser library Version 2.2.0 (*muParser*), available at `http://muparser.beltoforion.de/`

negative application condition. The rule within the *GMTE* could be extended to meet a much more powerful transformation mechanism assured by the connection instructions as it was defined in the previous Section.

Two main rule application approaches are implemented within the *GMTE*. The first one is to simply apply all rules listed in a rule file to a given input graph. The second one is to recursively apply all rules listed in a rule file to a given input graph, and to all graphs generated by such applications.

4   Inexact Matching

In this section we introduce our approach for inexact graph matching. Graph edit distance is one of the most flexible graph similarity measures. Our approach in-line with [13]. In [13] the authors proposed to compute graph edit distance based on bipartite graph matching by means of the Linear Sum Assignment Problem. Their algorithm performs quite efficiently, but it is limited in that it is often applicable for matching two graphs with equal number of nodes. In case of subgraph matching, the authors expand the cost matrix to get a square matrix. Therefore, this expansion increases computation time. As we can see in the sequel, our approach tackles efficiently this problem, a) using a modified version of the *LSAP* algorithm and b) possibility to directly work on a rectangular cost matrix.

4.1   Node/Edge Edit Distance

To compute the distance between two nodes or edges, we use a modified version of the the Heterogeneous Euclidean Overlap Metric (HEOM) [14] which handles numeric and string labels. But first, we will start by defining a metric to measure the distance between two labels. Given two labels l_i, l_i' (i is the index of the label within the node or the edge attributes) the distance is measured by $labelDistance(l_i, l_i')$ defined as follow:

$$labelDistance(l_i, l_i') = \begin{cases} 1 & \text{if } l_i \text{ or } l_i' \text{ is missing,} \\ \frac{ed(l_i, l_i')}{\max(|l_i|, |l_i'|)} & \text{if } l_i \text{ or } l_i' \text{ are strings,} \\ \frac{|l_i - l_i'|}{1 + |l_i - l_i'|} & \text{if } l_i \text{ or } l_i' \text{ are numerics.} \end{cases}$$

The string edit distance of s and t, denoted by $ed(s, t)$ is the minimal atomic string operations (character insertion, deletion or substitution) needed to transform s into t. when labels are strings, we could see that if they are equal then the distance is 0, and if they are totally different, the distance is 1. In case labels are numeric, the distance is 0 if they have same value and the distance is $\simeq 1$ if $|l_i - l_i'| \to \infty$. We define node/edge distance as follow:

$$\delta(n, n') = \sqrt{\sum_{i=0}^{\max(|n|, |n'|)} (labelDistance(l_i, l_i'))^2}$$

$|n|$ denotes the number of labels within a node or an edge. n and n' could be either two nodes or two edges.

4.2 Cost Matrix

Let $G_1 = (V_1, E_1)$ and $G_2 = (V_2, E_2)$ be two multi-labelled graph, where $\mid V_1 \mid = n$ and $\mid V_2 \mid = m$. The cost matrix defined as $C = (c_{ij})_{n \times m}$, where each element $c_{ij} \geq 0$ correspond to the cost of assigning the ith of V_1 to the jth element of V_2. To enhance the matching, information about edges distance could be added to the cost matrix. The technique is somehow similar to [15], except for the cost matrix is rectangular to reduce computation time. For each c_{ij} (assignment cost of node u_i to the node v_j) an adjacency edge cost matrix is generated. The cost resulting from the minimum-cost edge assignment for all edges connected to u_i and v_j is added to c_{ij}.

$$C_{ij} = c_{ij} + \min\{\sum cost(e_{u_i}, e_{v_j})\}$$

where $\min\{\sum cost(e_{u_i}, e_{v_j})\}$ is computed by the algorithm using the adjacency-edge cost matrix of node u_i and v_j. The problem is then to determine the minimum cost of assigning node from G to node from G'.

4.3 Bipartite Graph Matching

Standard graph matching procedures assign nodes and edges of one graph to another using some kind of search tree and trying to minimize the global edit cost. Let n and m be the number of nodes in G and G'. We have $\frac{n!}{m!}$ possibilities of assigning nodes from G to G'. The time complexity of such brute a force algorithm is $O(n^m)$. However, according to [16] the process of assigning nodes can be solved as Linear Sum Assignment Problem $LSAP$. For the previously defined cost matrix the problem is how to match each row to a different column in such a way that the sum of the assignment is minimal. In other words, we want to select n elements of C, so that there is exactly one element in each row and one in each column, and the sum of the corresponding sum is minimal.

4.4 Assignment Algorithm

In [17] the paper considers the classic linear assignment problem with a min-sum objective function, and the most efficient and easily available codes for its solution. Also it gives a survey describing the different approaches in the literature, presenting their implementations, and pointing out similarities and differences. Then it selects eight codes and introduces a wide set of dense instances containing both randomly generated and benchmark problems. According to [17], the modified versions of the Volgenant Jonker algorithm [18] is one of the fastest to solve dense linear assignment problem instances. In [18] authors have made a significant modification that led to speed up the algorithm. This modification is based on a selection procedure that selects a number of small cost elements from the cost matrix.

The *GMTE* adopt the modified versions of the Volgenant Jonker algorithms [18]: for square problem $LAPMOD$ and the non-square problem $LAPMODrow$.

The *LAPMODrow* is a special version of the core oriented *LAP* algorithm, used where the cost matrix is constructed with less rows than columns. For the bipartite matching, *LAPMOD* is used when graphs have equal number of nodes and *LAPMODrow* is used when the nodes number in one graph is less than the other.

5   Comparison with Other Tools

In this section, we compare *GMTE* with other graph transformation tools. The comparison is summarized in Table 1 and covers different criteria. These criteria have been presented in different papers like [3,19]. *GMTE* is categorized as a general purpose tool, however, it can be used through the expressiveness of his rule for model transformations as well.

The third criterion considers advanced rule features, these features increase the expressiveness of rules. *GMTE* and GROOVE support parallel rule application. This rule will be applied to all subgraphs that satisfy the application conditions. Also, *GMTE* like the other tools support standard case application (application on a precise or random matching). *GMTE* supports recursive rules application. GROOVE, PROGRES, Fujaba and GrGen [20] support using regular expressions on edge labels, *GMTE* supports label calculation known as label condition and functions on node and edge labels. Moreover, *GMTE* combines the *SPO* and *edNCE* to support a reach formalism for gluing and embedding technique through the use of connection instructions. Most of the presented tools in this section support either the SPO or the DPO approach, which reduce the expressiveness of the rule.

Table 1. Comparison between *GMTE* and other tools

Tool	Purpose	Typing	Advanced rule features	Editing
AGG	General purpose	Required	-	Graphical
PROGRES	General purpose	Required	Set nodes Star rules Regular expression	Graphical
GReAT	Model transformation	Required	Match condition Recursive pattern	Graphical
GrGen	Multi-purpose	Required	Regular expressions	Textual
VIATRA2	Model transformation	Required	Recursive patterns	Textual
GROOVE	General purpose	Optional	Regular expressions Quantification	Graphical
VMTS	Model transformation	Required	Quantification	Textual
ATOM3	Model transformation	Required	Triple Graph Grammar	Graphical
GMTE	General purpose	Optional	Parallel application, Label calculation, Connection instructions Codification instructions	Textual

The final criterion is whether a tool provides a graphical user interface for editing graphs and rules or is text-based only. *GMTE* can reads the rule graph and the host graph description from input XML files. The standard used is GraphML [21], which is an XML-based file format for graphs. Its main features include supporting directed, undirected, and mixed graphs, hypergraphs, hierarchical graphs, graphical representations, references to external data, application-specific attribute data, and light-weight parsers.

6 Conclusion

In this paper we addressed the problem of tools for graph matching and graph transformation. We presented a tool capable of performing matching and transformation for multi-labelled graphs. Also, we enhanced the rewriting system by extending the expressiveness of rules. To our knowledge, *GMTE* is the first tool that implements the *edNCE* approach and combines it with *SPO* approach, in order to get a rich formalism for both gluing and connecting technique. As well, we improved the formalism through the use of conditional rule schemata which are rule schemata equipped with a Boolean term built on arithmetic expressions. This allows to control rule applications by comparing values of labels. Also GMTE support inexact graph and subgraph matching, by efficiently computing the graph edit distance based on bipartite matching by means of faster and adaptive version the Volgenant-Jonker assignment algorithm.

For other part, we are working on extending our tool with the use of graph transformation system with time. Also, we are looking to implement a faster algorithm for inexact matching in order to reduce computation time. In order to show the efficiency of our approach, we plan to improve autonomic approaches like [22] with graph capabilities for reconfiguration consistency checking purpose.

Acknowledgment. This research is supported by the ITEA2 A2NETS (Autonomic Services in M2M Networks) project[3].

References

1. Schürr, A., Winter, A., Zündorf, A.: The PROGRES approach: Language and environment. In: Ehrig, H., Engels, G., Kreowski, H.J., Rozenberg, G. (eds.) Handbook of Graph Grammars and Computing by Graph Transformation: Applications, Languages, and Tools, vol. 3, pp. 487–550. World Scientific (1999)
2. Taentzer, G.: AGG: A tool environment for algebraic graph transformation. In: Münch, M., Nagl, M. (eds.) AGTIVE 1999. LNCS, vol. 1779, pp. 481–488. Springer, Heidelberg (2000)
3. Ghamarian, A.H., de Mol, M., Rensink, A., Zambon, E., Zimakova, M.: Modelling and analysis using groove. STTT 14(1), 15–40 (2012)
4. Lengyel, L., Levendovszky, T., Charaf, H.: Constraint validation support in visual model transformation systems. Acta Cybern. 17(2), 339–357 (2005)

---

[3] https://a2nets.erve.vtt.fi/

5. Balasubramanian, D., Narayanan, A., van Buskirk, C.P., Karsai, G.: The graph rewriting and transformation language: Great. ECEASST 1 (2006)
6. de Lara, J., Vangheluwe, H.: AToM^3: A tool for multi-formalism and meta-modelling. In: Kutsche, R.-D., Weber, H. (eds.) FASE 2002. LNCS, vol. 2306, pp. 174–188. Springer, Heidelberg (2002)
7. Guennoun, K., Drira, K., Diaz, M.: A proved component-oriented approach for managins dynamic software architectures. In: Proc. 7th IASTED International Conference on Software Engineering and Application, Marina Del Rey, CA, USA (2004)
8. Messmer, B.T., Bunke, H.: Efficient subgraph isomorphism detection: A decomposition approach. IEEE Trans. Knowl. Data Eng. 12(2), 307–323 (2000)
9. Ehrig, H., Korff, M., Löwe, M.: Tutorial introduction to the algebraic approach of graph grammars based on double and single pushouts. In: Ehrig, H., Kreowski, H.-J., Rozenberg, G. (eds.) Graph Grammars 1990. LNCS, vol. 532, pp. 24–37. Springer, Heidelberg (1991)
10. Rozenberg, G. (ed.): Handbook of graph grammars and computing by graph transformation: volume I. foundations. World Scientific Publishing Co., Inc., River Edge (1997)
11. Habel, A., Heckel, R., Taentzer, G.: Graph grammars with negative application conditions. Fundamenta Informaticae 26, 287–313 (1995)
12. Plump, D., Steinert, S.: Towards graph programs for graph algorithms. In: Ehrig, H., Engels, G., Parisi-Presicce, F., Rozenberg, G. (eds.) ICGT 2004. LNCS, vol. 3256, pp. 128–143. Springer, Heidelberg (2004)
13. Fankhauser, S., Riesen, K., Bunke, H.: Speeding up graph edit distance computation through fast bipartite matching. In: Jiang, X., Ferrer, M., Torsello, A. (eds.) GbRPR 2011. LNCS, vol. 6658, pp. 102–111. Springer, Heidelberg (2011)
14. Wilson, D.R., Martinez, T.R.: Improved heterogeneous distance functions. J. Artif. Int. Res. 6(1), 1–34 (1997)
15. Riesen, K., Bunke, H.: Approximate graph edit distance computation by means of bipartite graph matching. Image Vision Comput. 27(7), 950–959 (2009)
16. Riesen, K., Neuhaus, M., Bunke, H.: Bipartite Graph Matching for Computing the Edit Distance of Graphs, pp. 1–12 (2007)
17. Dell'Amico, M., Toth, P.: Algorithms and codes for dense assignment problems: the state of the art. Discrete Appl. Math. 100(1-2), 17–48 (2000)
18. Volgenant, A.: Linear and semi-assignment problems: A core oriented approach. Computers & OR 23(10), 917–932 (1996)
19. Fuss, C., Mosler, C., Ranger, U., Schultchen, E.: The jury is still out: A comparison of agg, fujaba, and progres. ECEASST 6 (2007)
20. Geiß, R., Batz, G.V., Grund, D., Hack, S., Szalkowski, A.: GrGen: A fast SPO-based graph rewriting tool. In: Corradini, A., Ehrig, H., Montanari, U., Ribeiro, L., Rozenberg, G. (eds.) ICGT 2006. LNCS, vol. 4178, pp. 383–397. Springer, Heidelberg (2006)
21. Brandes, U., Eiglsperger, M., Herman, I., Himsolt, M., Marshall, M.S.: GraphML progress report. In: Mutzel, P., Jünger, M., Leipert, S. (eds.) GD 2001. LNCS, vol. 2265, p. 501. Springer, Heidelberg (2002)
22. Ben-Halima, R., Jmaiel, M., Drira, K.: A QoS-oriented reconfigurable middleware for self-healing Web services. In: Proceedings of the IEEE International Conference on Web Services (ICWS 2008), Beijing, China. IEEE Computer Society Press (2008)

A Comparison of Explicit and Implicit Graph Embedding Methods for Pattern Recognition

Donatello Conte[1], Jean-Yves Ramel[2], Nicolas Sidère[2],
Muhammad Muzzamil Luqman[3], Benoît Gaüzère[4], Jaume Gibert[4],
Luc Brun[4], and Mario Vento[1]

[1] Università di Salerno
Via Ponte Don Melillo, 1, 84084 Fisciano(SA) Italy
{dconte,mvento}@unisa.it
[2] Université François Rabelais de Tours, Laboratoire Informatique (EA6300)
64 Avenue Jean Portalis, 37200 Tours France
{ramel,sidere}@univ-tours.fr
[3] L3i Laboratory, University of La Rochelle
17042 La Rochelle France
muhammad_muzzamil.luqman@univ-lr.fr
[4] GREYC UMR CNRS 6072
14000 Caen France
{benoit.gauzere,jaume.gibert,luc.brun}@ensicaen.fr

Abstract. In recent years graph embedding has emerged as a promising solution for enabling the expressive, convenient, powerful but computational expensive graph based representations to benefit from mature, less expensive and efficient state of the art machine learning models of statistical pattern recognition. In this paper we present a comparison of two implicit and three explicit state of the art graph embedding methodologies. Our preliminary experimentation on different chemoinformatics datasets illustrates that the two implicit and three explicit graph embedding approaches obtain competitive performance for the problem of graph classification.

1 Introduction

Two important challenges related to graphs concern the structural pattern recognition field: first of all, graph based methods like graph matching are computationally demanding hence restricting the application of such methods. Secondly, despite numerous theoretical results on graphs, the graph space has no strong algebraic properties. It is for example not a group nor a vector space. Such a lack of mathematical properties on the graph's space does not allow to readily combine structural and statistical pattern recognition methods.

Graph embedding methods map either explicitly or implicitly graphs into high dimensional spaces hence allowing to perform the basic mathematical computations required by various statistical pattern recognition techniques. Graph embedding methods appear thus as an interesting solution to address graph clustering and classification problems.

W.G. Kropatsch et al. (Eds.): GbRPR 2013, LNCS 7877, pp. 81–90, 2013.
© Springer-Verlag Berlin Heidelberg 2013

The graph embedding methods are formally categorized as implicit graph embedding or explicit graph embedding. The implicit graph embedding methods are based on graph kernels. A graph kernel is a function that can be thought of as a dot product in some implicitly existing vector space. Instead of mapping graphs from graph space to vector space and then computing their dot product, the value of the kernel function is evaluated in graph space. Such an implicit embedding satisfies all properties of a dot product. Since it does not explicitly map a graph to a point in vector space, a strict limitation of implicit graph embedding is that it does not permit all operations that could be defined on vector spaces. Further reading on state of the art methods of graph kernels and implicit graph embedding could be found in [2].

On the other hand, explicit graph embedding methods explicitly embed an input graph into a feature vector and thus enable the use of all the methodologies and techniques devised for vector spaces. The vectors obtained by an explicit graph embedding method can also be employed in a standard dot product for defining an implicit graph embedding function between two graphs. An important property of explicit graph embedding is that graphs of different size and order need to be embedded into a feature vector of determined size. The selection of the axis of this feature vector requires thus a fine analysis of the analysed dataset in order to selected features representative of the set while remaining sufficiently generic to describe any input graph. We refer the interested reader to [16] for further reading on classical explicit graph embedding techniques.

Similarly to the previous study described in [6], in this paper we propose a comparison between two implicit graph embedding methods based on graph kernels ([1,5]) and three methods of explicit graph embedding with comparable behavior ([7,12,17]). The difference between these techniques will be illustrated on classification problems using chemoinformatic datasets, such as those from IAM [14], the predictive toxicology challenge (PTC) dataset [20] and the MAO dataset from GREYC [5]. In Section 2 and Section 3 we describe the respective graph kernels and explicit graph embedding methods. Section 4 details the experimental evaluation alongwith a discussion on the classification performance of these methods. Finally, the paper concludes in Section 5.

2   Classification by Graph Kernels Methods

2.1   Method 1: Laplacian Graph Kernel

The graph edit distance between two graphs is defined as the minimal number of vertices and edge removal/addition/relabeling required to transform one graph into an other [13]. Unfortunately, even though the edit distance defines a metric under weak conditions, this distance is not definite negative. Consequently, kernels directly based on the edit distance are not definite positive and hence do not correspond to a valid kernel.

Let us consider a set of input graphs $\{G_1, \ldots, G_n\}$ defining our graph test database. Given a kernel k, the gram matrix K associated to this database is an $n \times n$ matrix defined by $K_{i,j} = k(G_i, G_j)$. As denoted by Steinke [19], the inverse of K (or its pseudo inverse if K is not invertible) may be considered as a regularization operator on the set of vector of dimension n. Conversely, the inverse (or pseudo inverse) of any definite positive regularization operator may be considered as a kernel.

From this point view, designing a "good" graph kernel comes up to define for each dataset a Gram matrix K whose associated norm penalizes the mapping of different values to similar graphs. One scheme to design a kernel consists thus to first build a definite positive regularization operator and then to take its inverse (or its pseudo inverse) to obtain a kernel. Let us consider the Laplacian operator defined as follows: given the set of graphs, $\{G_1, \ldots, G_n\}$, we first consider the $n \times n$ adjacency matrix $W_{i,j} = e^{-\frac{d(G_i, G_j)}{\sigma}}$ where σ is a tuning variable and $d(\cdot, \cdot)$ denotes the edit distance [15]. The normalized Laplacian of $\{G_1, \ldots, G_n\}$ is then defined as $L = I - D^{-\frac{1}{2}} W D^{-\frac{1}{2}}$ where D is a diagonal matrix defined by $D_{i,i} = \sum_{j=1}^{n} W_{i,j}$.

Well known results from spectral graph theory ([3]) establish that L is a symmetric, semi definite positive matrix whose eigenvalues belongs to the interval $[0, 2]$. Unfortunately, the Laplacian is also well know for being non invertible since the eigenvector vector $\mathbf{1} = (1, \ldots, 1)$ is associated to the eigenvalue 0. The only semi definite property of the Laplacian matrix forbids a direct inversion of this matrix. Moreover, the pseudo inverse of the Laplacian induces numerical instabilities which does not lead to a reliable kernel. Therefore, following Smola [18], we rather regularize the spectrum of L. The regularized version of L, denoted as $\tilde{L}$, is defined as $\tilde{L} = I + \lambda L$, where λ is a regularisation coefficient. The regularized laplacian $\tilde{L}$ is invertible and its inverse $K = \tilde{L}^{-1}$ is taken as a kernel. Using a classification or regression scheme, such a kernel leads to map to close values graphs having a small edit distance (and thus a strong similarity).

The implicit embedding induced by the Graph Laplacian kernel is not fixed by some a priori rules but is deduced from a regularization of the matrix of pairwise distances between objects. From this point of view, this implicit embedding is close from the explicit embedding proposed by Jouili and Tabbone [10] which additionally requires a dimensionality reduction step.

2.2   Method 2: Treelet Kernel

Treelet kernel [5] is a graph kernel based on a bag of non linear patterns which computes an explicit distribution of each pattern within a graph. This method explicitly enumerates the set of treelets included within a graph. The set of treelets, denoted $\mathcal{T}$, is defined as the 14 trees having a size lower than or equals to 6 nodes. Thanks to the limited number of different patterns encoding treelets, an efficient enumeration of the number of occurrences of each labeled pattern within a graph can be computed by algorithm defined in [5]. Labeling information included within treelets is encoded by a canonical key which is defined such as if treelets have a same structure, their canonical key is similar if and only if the two treelets are isomorphic. Each treelet being uniquely identified by the index of its pattern and its canonical key, any graph G can be associated to a vector

$f(G)$ which encodes the number of occurrences of each treelet $t \in \mathcal{T}$ by $f_t(G)$. Note that this vector representation may be of very high dimension since it may encode all possible treelets according to all possible nodes and edges labellings defined for a graph family. In chemoinformatics, such a vector representation may have a dimension higher than 4.25×10^9 [5] which forbids its explicit vector embedding. Treelet kernel between graphs is defined as a sum of sub kernels between common treelets of both graphs:

$$K_{\mathcal{T}}(G, G') = \sum_{t \in \mathcal{T}(G) \cap \mathcal{T}(G')} k(f_t(G), f_t(G')) \tag{1}$$

where $\mathcal{T}(G)$ encodes the set of treelets included within G and $k(.,.)$ defines any positive definite kernel between real numbers such as linear, Gaussian or polynomial kernel. Each sub kernel $k(.,.)$ encodes the similarity of the number of occurrences for each treelet t common to both graphs to be compared.

In order to improve the accuracy of treelet kernel, each treelet can be weighted according to a prediction task:

$$K_{\mathcal{T}}(G, G') = \sum_{t \in \mathcal{T}(G) \cap \mathcal{T}(G')} w(t) k(f_t(G), f_t(G')) \tag{2}$$

As described in [4], each weight $w(t) \in \mathbb{R}^+$ can be computed in a sparse and optimal way for a given training set by using multiple kernel learning (MKL). Using sparsity promotes the selection of relevant treelets according to the prediction task and $w(t)$ can thus be understood as a measure of the importance of treelet t for the prediction task.

3   Classification by Graph Embedding Methods

3.1   Method 1: Topological Embedding

One of the main challenges of graph embedding is to preserve topological information provided by the graph representation after transformation into a feature vector. The topological embedding method proposed in [17] provides an interesting answer to this problem by using a generic lexicon of topological structures that could be enumerated in graphs during the computation of the vectorial signature of the graphs. However, this lexicon must be comprehensive enough to ensure discrimination from a graph to another. They have therefore decided to take as a baseline the non-isomorphic graphs network presented in [9]. The network presents all graphs composed of n edges up to N (where N is the maximum number of edges). Thereafter, the term pattern will refer to a subgraph element of the non-isomorphic graph network.

For example, Figure 1 shows the non-isomorphic graph network until the fourth rank giving a lexicon of 11 patterns.

The vectorial representation of a graph topology will be built by counting the occurrences of each pattern of the lexicon. In other way, each element of the

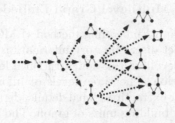

Fig. 1. The non-isomorphic graph network used to embed the topology

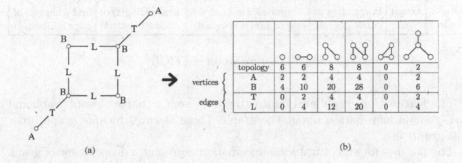

	$\circ$	$\circ\!\!-\!\!\circ$				
topology	6	6	8	8	0	2
A	2	2	4	4	0	2
B	4	10	20	28	0	6
T	0	2	4	4	0	2
L	0	4	12	20	0	4

Fig. 2. Matrix (b) corresponding to vectorial signature of graph presented in (a)

vector is the frequency of apparition of a pattern, which represents a descriptor of a part of the graph. Thus, the topology of the graph is embedded in the vectorial representation. This vectorial representation needs now to be enriched by encapsulating the information provided by labels that can be associated to the edges and vertices. As each of these labels can be composed with several attributes, the inclusion of this information can be problematic regarding the nature (numerical) and the number of attributes constituting a label. Two ways are proposed to by-pass these problems :

1. the first method consists of discretizing numerical attributes to obtain symbolic attributes. Then a combination of all these symbolic attributes can be realized to list all possible labels.
2. the second method is to perform a clustering in the label space using attributes as feature vectors. This results in some new classes of labels where their number can be controlled.

The construction of the vectorial representation can then be performed by filling all the cells of the matrix generated with the frequency of each pattern and each label for this pattern. The matrix (see Fig. 2b) presents an example of the proposed vectorial representation for the graph represented on Fig. 2a. More details about this topological embedding method can be found in [17].

3.2   Method 2: Fuzzy Multilevel Graph Embedding (FMGE)

The Fuzzy Multilevel Graph Embedding method (FMGE) performs multilevel analysis of graph to extract discriminatory information of three different levels. These include the graph level information, structural level information and the elementary level information. The three levels of information represent three different views of graph for extracting global details, details on topology of graph and details on elementary building units of graph. The feature vector of FMGE is named Fuzzy Structural Multilevel Feature Vector - FSMFV (see Fig. 3).

Graph order	Graph size	Embedding of node degree	Embedding(s) of subgraph(s) homogenity	Embedding(s) of node attribute(s)	Embedding(s) of edge attribute(s)

Fig. 3. Feature vector of FMGE

The features for graph level information represent a coarse view of graph and give general information about the graph. These features include graph order and graph size.

The features for structural level information represent a deeper view of graph and are extracted from the node degrees and subgraph homogeneity in graph. Subgraph homogeneity is represented by computing resemblance attributes for the nodes and edges of graph. The resemblance attributes for an edge is computed from the attributes on its neighboring nodes. The resemblance for a numeric attribute (a) is computed as a ratio of this attribute's values on neighboring nodes of an edge (a_1 and a_2) (see Eq. 3). Whereas the resemblance for a symbolic attribute (b) is computed as a ratio of this attribute's values on neighboring nodes of an edge (b_1 and b_2) (see Eq. 4).

$$resemblance(a_1, a_2) = min(|a_1|, |a_2|)/max(|a_1|, |a_2|) \tag{3}$$

$$resemblance(b_1, b_2) = \begin{vmatrix} 1 & b_1 = b_2 \\ 0 & otherwise \end{vmatrix} \tag{4}$$

The third level of information is extracted by penetrating into further depth and more granular view of graph and employing details of the elementary building blocks of graph. These features represent the information extracted from the node and edge attributes. The node degree, numeric resemblance attributes, numeric node attributes and numeric edge attributes are embedded by fuzzy histograms whereas the symbolic resemblance attributes, symbolic node attributes and symbolic edge attributes are embedded by crisp histograms. FMGE learns the intervals, for constructing these histograms, during an unsupervised learning phase and employs the learned intervals during graph embedding phase [12].

The feature vector obtained by FMGE is based on histogram encoding of the multilevel information extracted from graph. The number of features in the vector is directly dependent on the number of bins employed for constructing these histograms. The use of high dimensional histograms is explicitly built into the method as it enables FMGE to provide a more robust encoding of information and enables it to generalize to unseen graphs. However, the feature vector can become sparse and confuse between classes of graphs. In order to reduce the size of FMGE feature vector and to remove the unimportant features for a given graph dataset, we select the subset of top-ranked features on the basis of ranks obtained through the Relief algorithm [11].

3.3   Method 3: Attribute Statistics Based Embedding

The attribute statistics based embedding of graphs is a simple and efficient way of expressing the labelling information stored in nodes and edges of graphs in a rather naive feature vector. It basically consists in computing frequencies of appearences of very simple subgraph structures such as nodes with certain labels or node-edge-node structures with specific label sequences. Formally, consider a set of graphs $\mathcal{G} = \{g_1, \ldots, g_N\}$, with $g_i = (V_i, E_i, \mu_i, \nu_i)$ being the ith graph in the set with labelling alphabet L_{V_i} for the nodes and L_{E_i} for the edges. We assume that all graphs in $\mathcal{G}$ have the same labelling alphabets, this is $L_{V_i} = L_{V_j}$ and $L_{E_i} = L_{E_j}$ for all $i, j \in \{1, \ldots, N\}$. We do not assume, however, that each node and edge label occurs in each graph. Let $L_V = \{\alpha_1, \ldots, \alpha_p\}$ and $L_E = \{\omega_1, \ldots, \omega_q\}$ be the common labelling alphabets.

For each graph $g = (V, E, \mu, \nu) \in \mathcal{G}$, we define p unary features measuring the number of times each label in L_V appears in the graph, this is

$$U_i = \#(\alpha_i, g) = |\{v \in V \mid \alpha_i = \mu(v)\}|. \tag{5}$$

Binary features for edges are defined by computing how many times each possible sequence of node-edge-node labels appears in the graph. In particular,

$$\begin{aligned} B_{ij}^k &= \#([\alpha_i \leftrightarrow \alpha_j]_{\omega_k}, g) \\ &= |\{e = (u, v) \in E \mid \alpha_i = \mu(u) \wedge \alpha_j = \mu(v) \wedge \omega_k = \nu(e)\}|. \end{aligned} \tag{6}$$

Note that, since graphs are undirected, these features are symmetric, this is, $B_{ij}^k = B_{ji}^k$ for all $i, j \in \{1, \ldots, p\}$. We can then just consider half of them and always assume that $i \leq j$. This results in defining $\frac{1}{2} \cdot q \cdot p \cdot (p+1)$ binary features.

The final embedding configuration is the ensemble of all this features. Note than another interpretation of these features is their relation with random walks. In particular the random walk graph kernel implicilty computes the number of random walks of any length in each graph. In the attribute statistics based embedding case, one just considers walks of length 0 (node labels appearences) and walks of length 1 (node-edge-node label sequences). Although much simpler and local, the fact that these features are explicitly built makes them interesting and flexible enough to provide robust results in several classification problems. Distance correlation with edit distance or their extension to continuous attributed graphs [8,7] have also been shown in the literature.

4   Experimental Results

This section deals with the experimentation aiming at evaluate and compare the implicit and explicit embedding approaches. In particular, we consider classification tasks applied to chemoinformatics datasets.

4.1   The Considered Application and the Dataset

We have conducted experiments on four datasets of molecules. Molecules are easily converted into graphs by representing atoms as nodes and the covalent bonds as edges. Nodes are labeled with chemical symbols and edges by the valence of the linkage:

AIDS. This dataset consists of two classes (*active, inactive*) of 2000 graphs representing molecules with activity against HIV or not.

Mutagenicity. This dataset is divided in two classes regarding the mutagenicity (one of the numerous adverse properties of a compound that hampers its potential to become a marketable drug) of 4337 molecules.

Predictive Toxicology Challenge (PTC). This dataset deals with the predicting of the outcome of biological tests for the carcinogenicity of chemicals using information related to chemical structure only (*positive or negative*) on four catgories of animals : female rats (FR), male rats (MR), female mice (FM), male mice (MM) with about 240 graphs per set.

Monoamine Oxidase Dataset (MAO). This problem is defined on a set of 68 molecules divided into two classes: the molecules that inhibit the monoamine oxidase (antidepressant drugs) and those that do not.

These datasets are issued from public repositories. *AIDS* and *Mutagenicity* come from the IAM database repository[1], while *PTC* and *MAO* are both available in the GREYC's Chemistry databank[2]. Classification accuracy is measured by following the classification scheme designed by the datasets authors ([14,5]). For *AIDS* or *Mutagenicity*, a validation subset is used to optimize an SVM and the classification accuracy is obtained on an independant test subset. We used a k-fold cross-validation approach for PTC ($k=68$) and MAO ($k=10$) to parameterize an SVM and obtain the classification mean rates.

4.2   Results and Comparison

Table 1 shows the classification rates achieved by the 6 methods. Taking into account the four datasets, all implicit and explicit methods seems to be competitive and comparable. Of course, depending on the data, some variations can appear but these variations are small and they rather not be a criterion for choosing one method over another. Thus, other considerations should be taken into account. In particular, computational complexity or parameterization dependancy should be evaluated for all these approaches.

---

[1] http://www.iam.unibe.ch/fki/databases/iam-graph-database
[2] https://brun101.users.greyc.fr/CHEMISTRY/index.html

Table 1. Classification results for different methods and datasets

	Mutagenicity	AIDS	MAO	PTC			
				FM	MM	FR	MR
Laplacian kernel	70.2	92.6	90.0	59.2	55.2	57.7	60.9
Treelet kernel	77.1	99.1	91.2	58.7	61.9	60.4	60.8
Treelet kernel with MKL	77.6	99.7	94.1	64.2	64.6	71.2	64.8
Topological Embedding	77.2	99.4	91.2	65.9	67.5	68.7	63.7
FMGE	76.5	99.0	92.1	63.9	66.3	60.0	59.9
Attribute Statistics	76.5	99.6	90.6	64.8	63.1	67.9	59.7

On top of these things, an important remark one should be aware of is the fact that most of the discussed methodologies might present some restrictions in order to be evaluated in other pattern recognition problems. For instance, explicit embeddings may be favored over implicit ones whenever the explicit vector representation is required by some algorithms which require more than the dot product. Indeed, graph kernels are limited to kernel methods such as SVM. On the other hand, though, implicit ones are usually defined between two graphs whereas most of explicit methods require the whole dataset to compute the graph embedding. So, implicit methods may be favored other explicit one whenever the access to the whole dataset is limited.

5   Conclusions

Graph embedding for pattern recognition is a recent emerging trend to enable the pattern recognition community to benefit from the representative power of graph based structural approaches of pattern recognition and the computational power of machine learning models of statistical pattern recognition approaches. We have outlined two graph kernel based implicit graph embedding methods and three explicit graph embedding methods. Our initial experimentation on different chemoinformatic databases for the problem of graph classification illustrates that all the methods under consideration obtain competitive performance in terms of classification rates. Our future research goals are to take forward this study on the comparison of implicit and explicit graph embedding methods for revealing the strengths of these methods in terms of learning abilities, automatic parameter optimization, computational complexity and other interesting criteria.

References

1. Brun, L., Conte, D., Foggia, P., Vento, M.: A graph-kernel method for re-identification. In: Kamel, M., Campilho, A. (eds.) ICIAR 2011, Part I. LNCS, vol. 6753, pp. 173–182. Springer, Heidelberg (2011)
2. Bunke, H., Riesen, K.: Recent advances in graph-based pattern recognition with applications in document analysis. Pattern Recognition 44(5), 1057–1067 (2011)

3. Chung, F.R.K.: Spectral Graph Theory (CBMS Regional Conference Series in Mathematics, No. 92). American Mathematical Society (February 1997)
4. Gaüzère, B., Brun, L., Villemin, D.: Graph kernels based on relevant patterns and cycle information for chemoinformatics. In: Proceedings of ICPR 2012. IAPR, pp. 1775–1778. IEEE (November 2012)
5. Gaüzére, B., Brun, L., Villemin, D.: Two new graphs kernels in chemoinformatics. Pattern Recognition Letters 33(15), 2038–2047 (2012)
6. Gaüzère, B., Hasegawa, M., Brun, L., Tabbone, S.: Implicit and explicit graph embedding: Comparison of both approaches on chemoinformatics applications. In: Gimel'farb, G., Hancock, E., Imiya, A., Kuijper, A., Kudo, M., Omachi, S., Windeatt, T., Yamada, K. (eds.) SSPR&SPR 2012. LNCS, vol. 7626, pp. 510–518. Springer, Heidelberg (2012)
7. Gibert, J., Valveny, E., Bunke, H.: Graph embedding in vector spaces by node attribute statistics. Pattern Recognition 45(9), 3072–3083 (2012)
8. Gibert, J., Valveny, E., Bunke, H., Fornés, A.: On the correlation of graph edit distance and L_1 distance in the attribute statistics embedding space. In: Gimel'farb, G., Hancock, E., Imiya, A., Kuijper, A., Kudo, M., Omachi, S., Windeatt, T., Yamada, K. (eds.) SSPR&SPR 2012. LNCS, vol. 7626, pp. 135–143. Springer, Heidelberg (2012)
9. Jaromczyk, J., Toussaint, G.: Relative neighborhood graphs and their relatives. In: Proceedings of the IEEE (1992)
10. Jouili, S., Tabbone, S.: Graph embedding using constant shift embedding. In: Ünay, D., Çataltepe, Z., Aksoy, S. (eds.) ICPR 2010. LNCS, vol. 6388, pp. 83–92. Springer, Heidelberg (2010)
11. Luqman, M.M., Ramel, J.Y., Lladós, J.: Improving Fuzzy Multilevel Graph Embedding through Feature Selection Technique. In: Gimel'farb, G., Hancock, E., Imiya, A., Kuijper, A., Kudo, M., Omachi, S., Windeatt, T., Yamada, K. (eds.) SSPR & SPR 2012. LNCS, vol. 7626, pp. 243–253. Springer, Heidelberg (2012)
12. Luqman, M.M., Ramel, J.Y., Lladós, J., Brouard, T.: Fuzzy multilevel graph embedding. Pattern Recognition 46(2), 551–565 (2013)
13. Neuhaus, M., Bunke, H.: Bridging the Gap Between Graph Edit Distance and Kernel Machines. World Scientific Publishing Co., Inc., River Edge (2007)
14. Riesen, K., Bunke, H.: Iam graph database repository for graph based pattern recognition and machine learning. In: da Vitoria Lobo, N., Kasparis, T., Roli, F., Kwok, J.T., Georgiopoulos, M., Anagnostopoulos, G.C., Loog, M. (eds.) SSPR & SPR 2008. LNCS, vol. 5342, pp. 287–297. Springer, Heidelberg (2008)
15. Riesen, K., Bunke, H.: Approximate graph edit distance computation by means of bipartite graph matching. Image Vision Computing 27(7), 950–959 (2009)
16. Riesen, K., Bunke, H.: Graph classification based on vector space embedding. IJPRAI 23(6), 1053–1081 (2009)
17. Sidere, N., Héroux, P., Ramel, J.Y.: Vector representation of graphs: Application to the classification of symbols and letters. In: ICDAR, pp. 681–685. IEEE Computer Society (2009)
18. Smola, A.J., Kondor, R.I.: Kernels and regularization on graphs. In: XV Annual Conference on Learning Theory, pp. 144–158 (2003)
19. Steinke, F., Schlkopf, B.: Kernels, regularization and differential equations. Pattern Recognition 41(11), 3271–3286 (2008)
20. Toivonen, H., Srinivasan, A., King, R., Kramer, S., Helma, C.: Statistical evaluation of the predictive toxicology challenge 2000–2001. Bioinformatics 19(10), 1183–1193 (2003)

Adjunctions on the Lattice of Dendrograms

Fernand Meyer

CMM-Centre de Morphologie Mathématique,
Mathématiques et Systèmes, MINES ParisTech, France
fernand.meyer@mines-paristech.fr

Abstract. Dendrograms are used in hierarchical classification. They also are useful structures in image processing, for segmentation or filtering purposes. The structure of a hierarchy is univocally expressed by a ultrametric ecart. The hierarchies form a complete lattice on which two adjunctions will be defined.

1 Introduction

Hierarchies are the classical structures for representing a taxonomy. The most famous taxonomy is the Linnaean system. Each genus is the union of all species it contains, which in turn is the union of animals it contains.

As hierarchies are nested partitions of a domain, they are also encountered in image segmentation. Multiple segmentations of increasing coarseness are produced. Each level of the hierarchy contains a partition of the image and from level to level only fusions of regions take place [4].

Partitions are thus the simplest hierarchies, with only one level. The algebraic structure of partitions has been studies by Heijmans, Serra and Ronse [2], [11], [7]. Often one is not interested in partitioning the total domain of an image, but one wants to get the masks of some objects of interest. These masks are disjoint sets but do not partition the domain ; they constitute a partial partition as introduced by Ch. Ronse [5].

A series of nested partitions, where each coarser partition is obtained by merging regions of finer partitions, constitutes a hierarchy. We have a partial hierarchy or dendrogram, if the lowest level contains particular interest zones ; in higher levels, some preexisting regions become larger, eventually merge and others appear. Consider a topographic surface which is flooded such that all lakes have the same altitude. For each flooding level, the lakes form a partial partition. From one level to a higher level, the extension of a lake may grow, new lakes appear, existing lakes merge. The corresponding partial hierarchy is called min-tree [8] and is often used for image filtering.

The paper is organized as follows. A first part gives an axiomatic definition of dendrograms and hierarchies. The second derives an ultrametric half distance derived from a stratification index. An order relation between dendrograms organizes the hierarchies as a complete lattice. Finally, two adjunctions are defined on dendrograms. Combining erosion and dilation in an adjunction produces openings and closings, from which the classical morphological filters may then be derived [3].

W.G. Kropatsch et al. (Eds.): GbRPR 2013, LNCS 7877, pp. 91–100, 2013.

2   Dendrograms and Hierarchies

The axiomatic definition of dendrograms and hierarchies is due to Benzecri [1]. Let E be a domain with a finite number of elements are called points (for instance the pixels of an image) and $\mathcal{P}(E)$ the family of subsets of E. Let $\mathcal{X}$ be a subset of $\mathcal{P}(E)$, on which we consider an arbitrary order or preorder relation relation $\prec$ (in the present work $\prec$ is the inclusion relation $\subset$ between sets) The union of all sets belonging to $\mathcal{X}$ is called support of $\mathcal{X}$: $\mathrm{supp}(\mathcal{X})$. The subsets of $\mathcal{X}$ may be structured into:

* the summits : $\mathrm{Sum}(\mathcal{X}) = \{A \in \mathcal{X} \mid \forall B \in \mathcal{X} : A \prec B \Rightarrow A = B\}$
* the leaves : $\mathrm{Leav}(\mathcal{X}) = \{A \in \mathcal{X} \mid \forall B \in \mathcal{X} : B \prec A \Rightarrow A = B\}$
* the nodes : $\mathrm{Nod}(\mathcal{X}) = \mathcal{X} - \mathrm{Leav}(\mathcal{X})$
* the predecessors : $\mathrm{Pred}(A) = \{B \in \mathcal{X} \mid A \prec B\}$
* the immediate predecessors :
   $\mathrm{ImPred}(A) = \{B \in \mathcal{X} \mid \{U \mid U \in \mathcal{X}, A \prec U \text{ and } U \prec B\} = (A, B)\}$
* the successors : $\mathrm{Succ}(A) = \{B \in \mathcal{X} \mid B \prec A\}$

Fig.5, at the end of the document, presents a dendrogram in which the letters represent subsets of $\mathcal{P}(E)$; and $A \to B$ means that $B \prec A$. Then A is the summit ; B is the predecessor of (D, E, F, H, I) and the immediate predecessor of (D, E, F) ; (H, I, E, F, J, G) are leaves ; (J, G) are successors of A and C, and immediate successors of C.

2.1   Dendrograms

We now structure $\mathcal{X}$ as a tree or a dendrogram (also called "partial hierarchy")

Dendrograms : $\mathcal{X}$ is a dendrogram if and only if the set $\mathrm{Pred}(A)$ of the predecessors of A, with the order relation induced by $\prec$ is a total order. The maximal element of this family is a summit, which is the unique summit containing A.

Proposition 1. *The following properties are equivalent:*
1)$\mathcal{X}$ is a dendrogram
2) $U, V, A \in \mathcal{X} : A \subset U$ and $A \subset V \Rightarrow U \subset V$ or $V \subset U$
3) $U, V \in \mathcal{X} : U \not\subseteq V$ and $V \not\subseteq U \Rightarrow U \cap V = \varnothing$
4) Any element $A \in \mathcal{X} - \mathrm{Sum}(\mathcal{X})$ possesses a unique immediate predecessor (valid as we suppose E and $\mathcal{X}$ finite)

Proposition 2. *A family $(A_i)_{i \in I}$ of sets in $\mathcal{X}$ with a non empty intersection is completely ordered for $\subset$.*

A dendrogram is said to be connected if it possesses a unique summit. Finite dendrograms are classically represented as a tree : each element $A \in \mathcal{X}$ is a node of the tree, and is linked by an edge with its unique immediate predecessor.

Consider a dendrogram Π verifying : $A \in \mathrm{supp}(\Pi) \Rightarrow \mathrm{Pred}(A) = A$. Such a dendrogram has only one hierarchical level is called *partial partition* (partial partitions have been introduced by C.Ronse in [5]). If $\mathrm{supp}(\Pi) = E$, then it is called partition.

2.2   Hierarchies

Definition 1. *We call hierarchy $\mathcal{H}$ a dendrogram verifying $\bigcup \mathrm{Leav}(\mathcal{H}) = \mathrm{supp}(\mathcal{H})$*

Proposition 3. *A dendrogram $\mathcal{X}$ is a hierarchy if and only any element A of $\mathcal{X}$ is the union of all other elements of $\mathcal{X}$ contained in A:*
$\forall A \in \mathcal{X} : \bigcup \{B \in \mathcal{X} \mid B \subset A \; ; B \neq A\} = \{A, \emptyset\}$

For understanding the difference between dendrograms and hierarchies, we give an example of each used all along the paper.

Hierarchy : A prototype of hierarchy is a series of nested partitions, where each coarser partition is obtained by merging regions of finer partitions. The leaves are the regions of the finest partition ; their union constitutes the support of the hierarchy.

Dendrogram : A prototype of a dendrogram is constituted by the lake distribution of a topographic surface. Each lake is included in all lakes with a higher level. The leaves are the lakes when they just cover the regional minima. The union of all lakes is larger than the union of the leaves. For each flooding level, the lakes form a partial partition. From one level to a higher level, the extension of a lake may grow, new lakes appear, existing lakes merge.

2.3   Stratification Index and Partial Ultrametric Distances (PUD)

Consider a dendrogram or hierarchy $\mathcal{X}$; $\mathcal{X}$ is a stratified hierarchy, if it is equipped with an index function st from $\mathcal{X}$ into the interval $[0, L]$ of $\mathbb{R}$ which is strictly increasing with the inclusion order:
$\forall A, B \in \mathcal{X} : A \subset B$ and $B \neq A \Rightarrow \mathrm{st}(A) < \mathrm{st}(B)$.
As E is finite, the number of distinct stratification levels is finite. We suppose that for all $A \in \mathcal{X} : \mathrm{st}(A) < L$ and set $\mathrm{st}(\varnothing) = L$.

Each dendrogram $\mathcal{X}$ with a stratification index st induces on the points $p, q \in E$ a partial ultrametric distance $\chi(p, q)$. If no set of $\mathcal{X}$ contains both p and q, then $\chi(p, q) = L$. Otherwise, the family $(A_i)_{i \in I}$ of sets of $\mathcal{X}$ containing both p and q has a non empty intersection, and as established above, is completely ordered for $\subset$. Thus it possesse a smallest element A and $\chi(p, q) = \mathrm{st}(A)$. In particular $\chi(p, p)$ is the stratification index of the smallest set $\mathcal{X}$ of containing p ; if no set of $\mathcal{X}$ contains p, then $\chi(p, p) = \mathrm{st}(\varnothing) = L$; such a point is called "alien" of $\mathcal{X}$.

Properties : χ has the following properties:
$\forall p, q \in E : \chi(p, q) = \chi(q, p)$
$\forall p, q, r \in E : \chi(p, q) \leq \max \{\chi(p, r), \chi(r, q)\}$

Remark: χ is not a distance but an ecart as $\chi(p, q) = 0$ does not necessarily imply $p = q$. We call it partial ultrametric distance.

Properties of the Balls of a Partial Ultrametric Distance

Closed balls, defined as $\mathrm{Ball}(p,\rho) = \{q \in E \mid \chi(p,q) \leq \rho\}$ have strange properties:

* *Each element of a closed ball* $\mathrm{Ball}(p,\rho) = \{q \in E \mid \chi(p,q) \leq \rho\}$ *is centre of this ball.*
* *Two closed balls* $\mathrm{Ball}(p,\rho)$ *and* $\mathrm{Ball}(q,\rho)$ *with the same radius are either disjoint or identical.*
* *The radius of a ball is equal to its diameter.* If A is a set of the dendrogram, p an arbitrary point of A, then A is the ball of center p and of radius equal to the diameter of A (the maximal distance between two points of A.

Inversely, the closed balls of a partial ultrametric distance χ form a dendrogram $\mathcal{X}$.

2.4   Partial Partitions by Thresholding Partial Hierarchies

Consider a partial hierarchy $\mathcal{X}$ with its associated PUD χ. By thresholding the PUD at level λ one obtains a partial binary ultrametric half distance (PBUD):

$$\pi_\lambda(x,y) = \begin{array}{l} 1 \text{ if } \chi(x,y) > \lambda \\ 0 \text{ if } \chi(x,y) \leq \lambda \end{array} \text{ associated to a partial partition } \Pi_\lambda.$$

Aliens and Singletons

Aliens and singletons of partial partitions. We define aliens and singletons of a partial partition π:

* Singletons are characterized by: $\forall p, q \in E$, $p \neq q,: \pi(p,q) = 1$ and $\pi(p,p) = 0$.
* The support of π is the set of points p verifying : $\pi(p,p) = 0$
* Aliens, which are points outside the support are characterized by: $\forall p \in E$: $\pi(p,p) = 1$ implying $\forall q \in E : \pi(p,p) \leq \pi(p,q) \vee \pi(q,r)$ so that $\pi(p,q) = 1$

Aliens and singletons of dendrograms. Consider a PUD χ and its thresholds π_λ at level λ. For increasing values of λ, the partial partitions π_λ obtained by thresholding χ have increasing supports $\mathrm{supp}_\lambda(\chi) = \{p \in E : \chi(p,p) \leq \lambda\}$. Let $p \in E$ be a point verifying $\chi(p,p) = \lambda$ and the partition π_μ obtained by thresholding χ at the level μ. And consider $\nu = \bigwedge_{q \neq p} \chi(p,q)$. Since $\chi(p,p) \leq \chi(p,q) \vee \chi(q,p)$, we have $\nu \geq \lambda$. The status of p will vary in the partitions π_ν for increasing levels ν :

* $\mu < \lambda : \pi_\mu(p,p) = 1$ and p does not belong to the support of π_μ and is an alien
* $\lambda \leq \mu < \nu : \pi_\mu(p,p) = 0$ but for $q \neq p$, $\pi_\mu(p,q) = 1$ and p is a singleton
* $\nu \leq \mu : \pi_\mu(p,p) = 0$ and for there exists $q \neq p$ such that $\pi_\mu(p,q) = 0$ and p is a regular node.

3   The Lattice of Hierarchies

3.1   Order Relation between Hierarchies and Partial Hierarchies

Let $\mathcal{A}$ and $\mathcal{B}$ be two dendrograms with their associated PUD : $\chi_{\mathcal{A}}$ and $\chi_{\mathcal{B}}$. The following relation defines an order relation between the hierarchies: $B \leq A \Leftrightarrow$ $\forall p, q \in E \quad \chi_{\mathcal{A}}(p,q) \leq \chi_{\mathcal{B}}(p,q)$.

It follows that $\forall p \in E : \mathrm{Ball}_{\mathcal{B}}(p, \rho) \subset \mathrm{Ball}_{\mathcal{A}}(p, \rho)$. We say that the hierarchy $\mathcal{A}$ is coarser than the hierarchy $\mathcal{B}$ and that the hierarchy $\mathcal{B}$ is finer than the hierarchy $\mathcal{B}$.

For each $p \notin \mathrm{supp}(\mathcal{A}) : \chi_{\mathcal{A}}(p,p) = L$ which implies that $\chi_B(p,p) = L$, so that $p \notin \mathrm{supp}(\mathcal{B})$.

The smallest dendrogram has an empty support and contains only aliens, i.e. points p verifying $\forall q \in E$, $\chi(p,q) = L$.

The smallest hierarchy has E as support and contains only singletons $\forall p \neq q \in E$, $\chi(p,q) = L$, and $\forall p \in E$, $\chi(p,q) = 0$. The largest hierarchy is E itself, whose PUD verifies: $\forall p, q \in E : \chi(p,q) = 0$

To binary PUDs $\chi_{\mathcal{A}}$ and $\chi_{\mathcal{B}}$ correspond partitions and partial partitions. Their closed balls verify : $\mathrm{Ball}_{\mathcal{B}}(p,0) \subset \mathrm{Ball}_{\mathcal{A}}(p,0)$, the aliens remaining outside the balls. Hence the tiles of the finer partition $\mathcal{B}$ are included in the tiles of the coarser partition $\mathcal{A}$ which is coherent with the usual definition of the order between partitions.

3.2   The Lattice of Dendrograms

Consider a family of dendrograms $(\mathcal{A}_i)_{i \in I}$, the PUD of the dendrogram $\mathcal{A}_i$ being χ_i.

Infimum of Hierarchies. The infimum $\wedge \mathcal{A}_i$ is the largest dendrogram which is smaller than each $\mathcal{A}_i$ and its PUD is the smallest verifying for all elements of the family $\chi_{\wedge \mathcal{A}_i} \geq \chi_i$. As the supremum of PUDs is a PUD, we have $\chi_{\wedge \mathcal{A}_i} = \bigvee_i \chi_i$.

And $\mathrm{supp}_\lambda \chi_{\wedge \mathcal{A}_i} = \bigwedge_i \mathrm{supp}_\lambda \mathcal{A}_i$.

Supremum of Dendrograms. The supremum $\vee \mathcal{A}_i$ is the smalles dendrogram which is larger than each $\mathcal{A}_i$ and its PUD is the largest verifying for all elements of the family $\chi_{\vee \mathcal{A}_i} \leq \chi_i$. Unfortunately $\bigwedge_i \chi_i$ is not a PUD and $\chi_{\vee \mathcal{A}_i} = \widetilde{\bigwedge_i \chi_i}$ is the largest partial ultrametric distance which is lower than $\bigwedge_i \chi_i$. It exists as the set of ultrametric distances lower than $\bigwedge_i \chi_i$ is not empty and contains the largest dendrogram whose PUD verifies $\forall p, q \in E$, $\chi(p,q) = 0$. As this family is closed by supremum it has a largest element.

Illustration

Fig.1 presents two hierarchies HA and HB through their nested partitions. The supremum and infimum of both hierarchies also are represented. The infimum

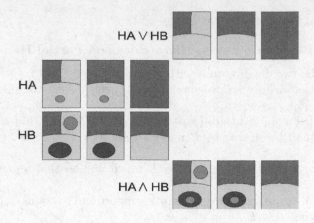

Fig. 1. Two hierarchies HA and HB and their derived supremum and infimum

takes for each threshold the intersection of the corresponding partitions, obtained through intersection of the tiles. The supremum is obtained by keeping only the boundaries existing in each component.

4   Adjunctions on Partial Hierarchies

4.1   Erosion and Dilations by a Structuring Element of Binary Sets

Adjunctions are the mother of morphology. Consider a lattice, A, B two arbitrary elements of the lattice ; the operator δ and ε iff $\delta A < B \Leftrightarrow A < \varepsilon B$ [9]. Then δ and ε are increasing operators, δ a dilation and ε an erosion, $\varepsilon\delta$ a closing and $\delta\varepsilon$ an opening.

Consider the classical erosion and dilation of binary sets by a structuring element. Chosing a point O in the domain E, we associate to each point x the vector $\overrightarrow{Ox}$. Inversely we associate to each affine vector $\overrightarrow{Ox}$ its extremity x. We write $x + b$ for the extremity of the vector $\overrightarrow{Ox} + \overrightarrow{Ob}$ and define $X_b = \bigcup_{x \in X} x + b$, the set X translated by the vector $\overrightarrow{Ob}$ (X_{-b} for the translation $\overrightarrow{bO}$). We have the following equivalence: $B_p \subset X \Leftrightarrow \forall b \in B : p + b \in X \Leftrightarrow \forall b \in B : p \in X_{-b}$ from which we derive two classical and equivalent formulations of the erosion of a set X by a structuring element $B : X \ominus B = \{p \in X \mid B_p \subset X\} = \bigwedge_{b \in B} X_{-b}$. It appears that each of these formulations, which are equivalent for sets lead to two distinct adjunctions in the case of partial hierarchies.

4.2   A First Adjunction Based on the Supremum and Infimum of Translated PUD

Consider a dendrogram $\mathcal{X}$ and its PUD χ. If we translate all elements of $\mathcal{X}$ by a translation $\overrightarrow{bO}$ we get a new hierarchy $\mathcal{X}_{-b}$ with a PUD χ_b defined by

$\chi_b(p,q) = \chi(p-b, q-b)$. To the eroded hierarchy $\mathcal{X} \ominus B = \bigwedge_{b \in B} \mathcal{X}_{-b}$ corresponds the PUD $\bigvee_{b \in B} \chi_{-b}$ defined by $\bigvee_{b \in B} \chi_{-b}(p,q) = \bigvee_{b \in B} \chi(p-b, q-b \mid b \in B)$. The adjunction dilation is then $\mathcal{X} \oplus B = \bigvee_{b \in B} \mathcal{X}_b$ with the associated PUD $\overbrace{\bigwedge_{b \in B} \chi_b}$. For showing that the first $\mathcal{X} \ominus B = \bigwedge_{b \in B} \mathcal{X}_b$ is an erosion and the second $\mathcal{X} \oplus B = \bigvee_{b \in B} \mathcal{X}_b$ a dilation, we have to show that they form an adjunction: for any two hierarchies $\mathcal{X}, \mathcal{Y} \in \mathcal{X}(E) : \mathcal{X} \oplus B < \mathcal{Y} \Leftrightarrow \mathcal{X} < \mathcal{Y} \ominus B$.

We will prove the adjunction through the PUD χ and ζ associated to the hierarchies $\mathcal{X}$ and $\mathcal{Y}$: $\mathcal{X} < \mathcal{Y} \ominus B \Leftrightarrow \chi > \bigvee_{b \in B} \zeta_{-b} \Leftrightarrow \forall b \in B : \chi > \zeta_{-b} \Leftrightarrow \forall b \in B :$

$$\chi_b > \zeta \Leftrightarrow \overbrace{\bigwedge_{b \in B} \chi_b} > \zeta$$

Remains to establish : $\overbrace{\bigwedge_{b \in B} \chi_b} > \zeta \Leftrightarrow \overbrace{\bigwedge_{b \in B} \chi_b} > \zeta$:

* $\overbrace{\bigwedge_{b \in B} \chi_b} > \zeta \Rightarrow \bigwedge_{b \in B} \chi_b > \zeta$ since $\overbrace{\bigwedge_{b \in B} \chi_b}$ is the largest ultrametric ecart below $\bigwedge_{b \in B} \chi_b$

* Suppose now $\bigwedge_{b \in B} \chi_b > \zeta$. Since ζ is an ultrametric ecart below $\bigwedge_{b \in B} \chi_b$, it is smaller or equal to the largest ultrametric ecart below $\bigwedge_{b \in B} \chi_b$, that is $\overbrace{\bigwedge_{b \in B} \chi_b}$

This completes the proof :

$$\mathcal{X} < \mathcal{Y} \ominus B \Leftrightarrow \chi > \bigvee_{b \in B} \zeta_{-b} \Leftrightarrow \overbrace{\bigwedge_{b \in B} \chi_b} > \zeta \Leftrightarrow \overbrace{\bigwedge_{b \in B} \chi_b} > \zeta \Leftrightarrow \mathcal{X} \oplus B < \mathcal{Y}$$

The expression of the PUD is

$$\chi \ominus B(p,q) = \left[\bigvee_{b \in B} \chi_{-b} \right](p,q) = \bigvee \{ \chi(p-b, q-b) \mid b \in B \}$$

$$\chi \oplus B(p,q) = \left[\overbrace{\bigwedge_{b \in B} \chi_b} \right](p,q) = \overline{\bigwedge \{ \chi(p+b, q+b) \mid b \in B \}}$$

If there exists a $b \in B$, such that $\chi(p-b, p-b) = \lambda$, then $\chi \ominus B(p,p) \geq \lambda$; in other words, aliens of E for χ are dilated by the structuring element B.

We illustrate in figures 2 and 3 the erosion and the opening of a one dimensional hierarchy by a structuring element made of three pixels.

4.3 Adjunction on Hierarchies and Partial Hierarchies, Defined on a Tile by Tile Basis

Adjunctions on Partial Partition. The second formulation of the erosion for sets $\{ p \in X \mid B_p \subset X \}$ will now be adapted to a partial partition with its PUD χ. Two points p and q belong to the same tile of the partition eroded by a structuring element B, if they are centers of disks entirely included in the same tile of the initial partition (see fig.4), which is the case if and only if all pairs

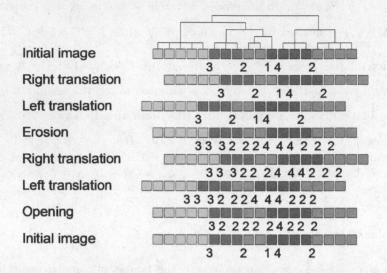

Fig. 2. Erosion and opening by a segment of 3 pixels: intermediate steps

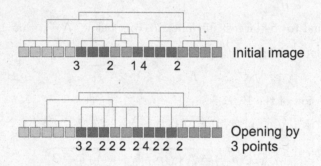

Fig. 3. Dendrogram of an initial image and its opening by a segment of 3 points

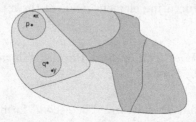

Fig. 4. The points p and q belong to the same tile of the partition eroded by a disk, as they are centers of disks entirely included in the same tile of the initial partition

$x, y \in B_p \cup B_q$ belong to the same tile of the partition, i.e. $\chi(x,y) = 0$. Hence the PUD $\varepsilon\chi$ of the eroded hierarchy is $\delta\chi(p,q) = \bigvee \{\chi(x,y) \mid x,y \in B_p \cup B_q\}$.

Consider now a point p such that B_p is not included in any tile of the partition χ. For each B_p, there exists $s,t \in B_p$ such that $\chi(s,t) = 1$, hence $\delta\chi(p,p) = \bigvee \{\chi(x,y) \mid x,y \in B_p \cup B_p\} = 1$, showing that p is an alien in the eroded partition χ. On the other hand if there exists a tile of the partition containing B_p and the erosion of this tile is reduced to a singleton p, then $\delta\chi(p,p) = \bigvee \{\chi(x,y) \mid x,y \in B_p \cup B_p\} = 0$. In other terms, this erosion of partial partitions adjusts the support of the partial partition by including aliens when necessary, and is identical with the erosion defined by Ronse in [7] (but distinct from the adjunction defined by J.Serra for partitions [10], where each tile of a partition is eroded and dilated separately, empty spaces being filled with singletons).

Adjunctions on Dendrograms. The expression established for partial partitions is still valid for arbitrary dendrograms:
$\delta\chi(p,q) = \bigvee \{\chi(x,y) \mid x,y \in B_p \cup B_q\}$.
It may be reformulated as the supremum of three terms:
$\delta\chi(p,q) = \bigvee \{\chi(x,y) \mid x,y \in B_p \cup B_q\} = \bigvee \{\chi(p + b_1, p + b_2) \mid b_1, b_2 \in B\}$
$\vee \bigvee \{\chi(p + b_1, q + b_2) \mid b_1, b_2 \in B\} \vee \bigvee \{\chi(q + b_1, q + b_2) \mid b_1, b_2 \in B\}$.
The first and last terms are dominated by the central term. Indeed, for each couple $b_1, b_2 \in B : \chi(p + b_1, p + b_2) \leq \chi(p + b_1, q + b_2) \vee \chi(q + b_2, p + b_2)$.

We obtain like that a simpler expression for this dilation : $\delta\chi = \bigvee \{\chi(p + b_1, q + b_2) \mid b_1, b_2 \in B\}$, PUD of the erosion of the hierarchy $\varepsilon\mathcal{X}$. The adjunct dilation $\delta\mathcal{X}(x,y)$ of the dendrogram is defined by the erosion of its PUD
$\varepsilon\chi(x,y) = \bigwedge \{\chi(x - b_1, y - b_2) \mid b_1, b_2 \in B\}$.
The couple $(\varepsilon\mathcal{X}, \delta\mathcal{X})$ forms an adjunction for the partial hierarchies.

4.4 Ordering the Adjunctions on Partial Hierarchies or Partitions

Both adjunctions established above are ordered as:

- $\bigvee \{\chi(p + b, q + b) \mid b \in B\} \leq \bigvee \{\chi(p + b_1, q + b_2) \mid b_1, b_2 \in B\}$, showing that the partial hierarchy $\varepsilon\mathcal{X}$ is coarser than the partial hierarchy $\mathcal{X} \ominus B$

- $\bigwedge \{\chi(p - b_1, q - b_2) \mid b_1, b_2 \in B\} \leq \bigwedge \{\chi(p - b, q - b) \mid b \in B\}$ showing that the partial hierarchy $\delta\mathcal{X}$ is finer than the partial hierarchy $\mathcal{X} \oplus B$

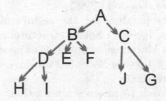

Fig. 5. A dendrogram

If the origin belongs to the structuring element we have the following order relations between the partial hierarchies $\varepsilon \mathcal{X} \leq \mathcal{X} \ominus B \leq \mathcal{X} \leq \mathcal{X} \oplus B \leq \delta \mathcal{X}$.

5   Conclusion

We have established two adjunctions for partial hierarchies, also valid for hierarchies, partitions and partial partitions through the associated PUD. The adjustment of the supports is treated automatically thanks to the introduction of aliens. Aliens and singletons have distinct definitions and distinct fates in the transformations. The morphological corpus can now be derived from these adjunctions. Iterating erosions or dilations increase their sizes. An erosion followed by its adjunct dilation produces an opening γ, a dilation followed by its adjunct erosion a closing φ. The classical filters $\gamma\varphi$, $\varphi\gamma$, $\gamma\varphi\gamma$, $\varphi\gamma\varphi$ may then be derived. Alternate sequential filters may be obtained by concatenating alternatively openings and closings of increasing sizes. One may also imagine geodesic dilations of hierarchies, by iterating an elementary geodesic dilation of a hierarchy under a hierarchy : $\delta \mathcal{X} \wedge \mathcal{Y}$ and $\varepsilon \mathcal{X} \vee \mathcal{Y}$.

References

1. Benzécri, J.P.: L'analyse des données 1. La taxinomie, ch. 3, pp. 119–153. Dunod (1973)
2. Heijmans, H.: Connected morphological operators for binary images. Computer Vision and Image Understanding 73(1), 99–120 (1999)
3. Matheron, G.: Filters and lattices. In: Serra, J. (ed.) Mathematical Morphology Volume II: theoretical advances. Academic Press, London (1988)
4. Meyer, F.: An overview of morphological segmentation. International Journal of Pattern Recognition and Artificial Intelligence 17(7), 1089–1118 (2001)
5. Ronse, C.: Partial partitions, partial connections and connective segmentation. J. Math. Imaging Vis. 32(2), 97–125 (2008)
6. Ronse, C.: Reconstructing masks from markers in non-distributive lattices. Appl. Algebra Eng. Commun. Comput. 19, 51–85 (2008)
7. Ronse, C.: Adjunctions on the lattices of partitions and of partial partitions. Appl. Algebra Eng., Commun. Comput. 21(5), 343–396 (2010)
8. Salembier, P., Garrido, L.: Connected operators based on region-tree pruning. In: Goutsias, J., Vincent, L., Bloomberg, D.S. (eds.) Mathematical Morphology and its Applications to Image and Signal Processing, vol. 18, pp. 169–178. Springer, US (2002)
9. Serra, J. (ed.): Image Analysis and Mathematical Morphology. II: Theoretical Advances. Academic Press, London (1988)
10. Serra, J.: Morphological operators for the segmentation of colour images. In: Bilodeau, M.S.M., Meyer, F. (eds.) Space, structure, and randomness. Contributions in honor of Georges Matheron in the fields of geostatistics, random sets, and mathematical morphology. Lecture Notes in Statistics, vol. 183, pp. 223–255, xviii, 395 p. Springer, New York (2005)
11. Serra, J.: A lattice approach to image segmentation. J. Math. Imaging Vis. 24, 83–130 (2006)

A Continuous-Time Quantum Walk Kernel
for Unattributed Graphs

Luca Rossi[1], Andrea Torsello[1], and Edwin R. Hancock[2]

[1] Department of Environmental Science, Informatics and Statistics,
Ca' Foscari University of Venice, Italy
{lurossi,torsello}@dsi.unive.it
[2] Department of Computer Science, University of York, YO10 5GH, UK
edwin.hancock@york.ac.uk

Abstract. Kernel methods provide a way to apply a wide range of learning techniques to complex and structured data by shifting the representational problem from one of finding an embedding of the data to that of defining a positive semidefinite kernel. In this paper, we propose a novel kernel on unattributed graphs where the structure is characterized through the evolution of a continuous-time quantum walk. More precisely, given a pair of graphs, we create a derived structure whose degree of symmetry is maximum when the original graphs are isomorphic. With this new graph to hand, we compute the density operators of the quantum systems representing the evolutions of two suitably defined quantum walks. Finally, we define the kernel between the two original graphs as the quantum Jensen-Shannon divergence between these two density operators. The experimental evaluation shows the effectiveness of the proposed approach.

Keywords: Graph Kernels, Graph Classification, Continuous-Time Quantum Walk, Quantum Jensen-Shannon Divergence.

1 Introduction

Graph-based representations have become increasingly popular due to their ability to characterize in a natural way a large number of systems which are best described in terms of their structure. Concrete examples include the use of graphs to represent shapes [1], metabolic networks [2], protein structure [3], and road maps [4]. Unfortunately, our ability to analyse this wealth of data is severely limited by the restrictions posed by standard pattern recognition techniques, which usually require the graphs to be first embedded into a vectorial space, a procedure which is far from being trivial. The reason for this is that there is no canonical ordering for the nodes in a graph and a correspondence order must be established before analysis can commence. Moreover, even if a correspondence order can be established, graphs do not necessarily map to vectors of fixed length, as the number of nodes and edges can vary.

Kernel methods [5], whose best known example is furnished by support vector machines (SVMs) [6], provide a neat way to shift the problem from that of

W.G. Kropatsch et al. (Eds.): GbRPR 2013, LNCS 7877, pp. 101–110, 2013.

finding an embedding to that of defining a positive semidefinite kernel, via the well-known kernel trick. In fact, once we define a positive semidefinite kernel $k : X \times X \to \mathbb{R}$ on a set X, then we know that there exists a map $\phi : X \to H$ into a Hilbert space H, such that $k(x,y) = \phi(x)^{\top} \phi(y)$ for all $x, y \in X$. Thus, any algorithm that can be formulated in terms of scalar products of the $\phi(x)$s can be applied to a set of data on which we have defined our kernel. As a consequence, we are now faced with the problem of defining a positive semidefinite kernel on graphs rather than computing an embedding. However, due to the rich expressiveness of graphs, also this task has proven to be difficult.

Many different graph kernels have been proposed in the literature [7–9]. Graph kernels are generally instances of the family of R-convolution kernels introduced by Haussler [10]. The fundamental idea is that of defining a kernel between two discrete objects by decomposing them and comparing some simpler substructures. For example, Gärtner et al. [7] propose to count the number of common random walks between two graphs, while Borgwardt and Kriegel [8] measure the similarity based on the shortest paths in the graphs. Shervashidze et al. [9], on the other hand, count the number of graphlets, i.e. subgraphs with k nodes. Note that these kernels can be defined both on unattributed and attributed graphs, although we will restrict our analysis to the simpler case of unattributed graphs, while the more general case will be the focus of future work. Another interesting approach is that of Bai and Hancock [11], where the authors investigate the possibility of defining a graph kernel based on the Jensen-Shannon kernel.

In this paper, we introduce a novel kernel on unattributed graphs where we probe the graph structure through the evolution of a continuous-time quantum walk [12, 13]. In particular, we are taking advantage of the fact that the interference effects introduced by the quantum walk seem to be enhanced by the presence of symmetrical motifs in the graph [14, 15]. To this end, we define a walk onto a new structure that is maximally symmetric when the original graphs are isomorphic. Finally, to define the kernel we make use of the quantum Jensen-Shannon divergence, a measure which has recently been introduced as a means to compute the distance between quantum states [16, 17].

The remainder of this paper is organized as follows: Section 2 provides an essential introduction to the basic terminology required to understand the proposed quantum mechanical framework. With these notions to hand, we introduce our graph kernel in Section 3. Section 4 illustrates the experimental results, while the conclusions are presented in Section 5.

2   Quantum Mechanical Background

Quantum walks are the quantum analogue of classical random walks [13]. In this paper we consider only continuous-time quantum walks, as first introduced by Farhi and Gutmann in [12]. Given a graph $G = (V, E)$, the state space of the continuous-time quantum walk defined on G is the set of the vertices V of the graph. Unlike the classical case, where the evolution of the walk is governed by a stochastic matrix (i.e. a matrix whose columns sum to unity), in the quantum case the dynamics of the walker is governed by a complex unitary matrix i.e., a matrix that multiplied by its conjugate transpose yields the identity matrix. Hence, the evolution of the quantum walk is reversible, which implies that quantum walks are non-ergodic and do not possess a limiting distribution. Using Dirac notation, we denote the basis state corresponding to the walk being at vertex $u \in V$ as $|u\rangle$. A general state of the walk is a complex linear combination of the basis states, such that the state of the walk at time t is defined as

$$|\psi_t\rangle = \sum_{u \in V} \alpha_u(t) |u\rangle \tag{1}$$

where the amplitude $\alpha_u(t) \in \mathbb{C}$ and $|\psi_t\rangle \in \mathbb{C}^{|V|}$ are both complex.

At each point in time the probability of the walker being at a particular vertex of the graph is given by the square of the norm of the amplitude of the relative state. More formally, let X^t be a random variable giving the location of the walker at time t. Then the probability of the walker being at the vertex u at time t is given by

$$\Pr(X^t = u) = \alpha_u(t)\alpha_u^*(t) \tag{2}$$

where $\alpha_u^*(t)$ is the complex conjugate of $\alpha_u(t)$. Moreover $\alpha_u(t)\alpha_u^*(t) \in [0,1]$, for all $u \in V$, $t \in \mathbb{R}^+$, and in a closed system $\sum_{u \in V} \alpha_u(t)\alpha_u^*(t) = 1$.

Recall that the adjacency matrix of the graph G has elements

$$A_{uv} = \begin{cases} 1 \text{ if } (u,v) \in E \\ 0 \text{ otherwise} \end{cases} \tag{3}$$

The evolution of the walk is governed by Schrödinger equation, where we take the Hamiltonian of the system to be the graph adjacency matrix, which yields

$$\frac{d}{dt}|\psi_t\rangle = -iA|\psi_t\rangle \tag{4}$$

Given an initial state $|\psi_0\rangle$, we can solve Equation 4 to determine the state vector at time t

$$|\psi_t\rangle = e^{-iAt}|\psi_0\rangle = \Phi e^{-i\Lambda t}\Phi^\top |\psi_0\rangle , \tag{5}$$

where $A = \Phi \Lambda \Phi^T$ is the spectral decomposition of the adjacency matrix.

2.1 Quantum Jensen-Shannon Divergence

A pure state is defined as a state that can be described by a ket vector $|\psi_i\rangle$. Consider a quantum system that can be in a number of states $|\psi_i\rangle$ each with probability p_i. The system is said to be in the ensemble of pure states $\{|\psi_i\rangle, p_i\}$. The density operator (or density matrix) of such a system is defined as

$$\rho = \sum_i p_i |\psi_i\rangle \langle \psi_i| \tag{6}$$

The Von Neumann entropy [18] of a density operator ρ is

$$H_N(\rho) = -Tr(\rho \log \rho) = -\sum_j \lambda_j \log \lambda_j, \tag{7}$$

where the λ_js are the eigenvalues of ρ. With the Von Neumann entropy to hand, we can define the quantum Jensen-Shannon divergence between two density operators ρ and σ as

$$D_{JS}(\rho, \sigma) = H_N\left(\frac{\rho + \sigma}{2}\right) - \frac{1}{2}H_N(\rho) - \frac{1}{2}H_N(\sigma) \tag{8}$$

This quantity is always well defined, symmetric and negative definite [19]. It can also be shown that $D_{JS}(\rho, \sigma)$ is bounded, i.e., $0 \leq D_{JS}(\rho, \sigma) \leq 1$. Let $\rho = \sum_i p_i \rho_i$ be a mixture of quantum states ρ_i, with $p_i \in \mathbb{R}^+$ such that $\sum_i p_i = 1$, then we can prove that

$$H_N(\sum_i p_i \rho_i) \leq H_S(p_i) + \sum_i p_i H_N(\rho_i) \tag{9}$$

where the equality is attained if and only if the states ρ_i have support on orthogonal subspaces. By setting $p_1 = p_2 = 0.5$, we see that

$$D_{JS}(\rho, \sigma) = H_N\left(\frac{\rho + \sigma}{2}\right) - \frac{1}{2}H_N(\rho) - \frac{1}{2}H_N(\sigma) \leq 1 \tag{10}$$

Hence D_{JS} is always less or equal than 1, and the equality is attained only if ρ and σ have support on orthogonal subspaces.

3   QJSD Kernel

Given two graphs $G_1(V_1, E_1)$ and $G_2(V_2, E_2)$ we build a new graph $\mathcal{G} = (\mathcal{V}, \mathcal{E})$ where $\mathcal{V} = V_1 \cup V_2$, $\mathcal{E} = E_1 \cup E_2 \cup E_{12}$, and $(u, v) \in E_{12}$ only if $u \in V_1$ and $v \in V_2$. With this new structure to hand, we define two continuous-time quantum walks $|\psi_t^-\rangle = \sum_{u \in V} \psi_{0u}^- |u\rangle$ and $|\psi_t^+\rangle = \sum_{u \in V} \psi_{0u}^+ |u\rangle$ on $\mathcal{G}$ with starting states

$$\psi_{0u}^- = \begin{cases} +\frac{d_u}{C} & \text{if } u \in G_1 \\ -\frac{d_u}{C} & \text{if } u \in G_2 \end{cases} \qquad \psi_{0u}^+ = \begin{cases} +\frac{d_u}{C} & \text{if } u \in G_1 \\ +\frac{d_u}{C} & \text{if } u \in G_2 \end{cases} \tag{11}$$

where d_u is the degree of the node u and C is the normalisation constant such that the probabilities sum to one.

We let the two quantum walks evolve until a time T and we define the average density operators ρ_T and σ_T over this time as

$$\rho_T = \frac{1}{T}\int_0^T |\psi_t^-\rangle \langle \psi_t^-| \, dt \qquad \sigma_T = \frac{1}{T}\int_0^T |\psi_t^+\rangle \langle \psi_t^+| \, dt \tag{12}$$

In other words, we defined two mixed systems with equal probability of being in any of the pure states defined by the quantum walks evolutions.

Then, given two unattributed graphs G_1 and G_2, we define the quantum Jensen-Shannon kernel $k_T(G_1, G_2)$ between them as

$$k_T(G_1, G_2) = D_{JS}(\rho_T, \sigma_T) \tag{13}$$

where ρ_T and σ_T are the density operators defined as in Eq. 12. Note that this kernel is parametrised by the time T. As it is not clear how we should set this parameter, in this paper we propose to let $T \to \infty$. However, in Section 4 we will show that a proper choice of T can yield an increased average accuracy in an SVM classification task.

We now proceed to show some interesting properties of our kernel. First, however, we need to prove the following

Lemma 1. *If G_1 and G_2 are two isomorphic graphs, then ρ_T and σ_T have support on orthogonal subspaces.*

Proof. We need to prove that

$$(\rho_T)^\dagger \sigma_T = \frac{1}{T^2} \int_0^T \rho_{t_1} \, \mathrm{d}t_1 \int_0^T \sigma_{t_2} \, \mathrm{d}t_2 = \mathbf{0} \tag{14}$$

where $\mathbf{0}$ is the matrix of all zeros, $\rho_t = |\psi_t^-\rangle\langle\psi_t^-|$ and $\sigma_t = |\psi_t^+\rangle\langle\psi_t^+|$. Note that if $\rho_{t_1}^\dagger \sigma_{t_2} = \mathbf{0}$ for every t_1 and t_2, then $(\rho_T)^\dagger \sigma_T = \mathbf{0}$. We now prove that if G_1 is isomorphic to G_2 then $\langle\psi_{t_1}^- | \psi_{t_2}^+\rangle = 0$ for every t_1 and t_2.

Let $U = e^{-iAt}$ be the unitary evolution operator of the quantum walk. If $t_1 = t_2 = t$, then $\langle\psi_0^- | (U^t)^\dagger U^t |\psi_0^+\rangle = 0$ since $(U^t)^\dagger U^t$ is the identity matrix and the initial states are orthogonal by construction. On the other hand, if $t_1 \neq t_2$, we have $\langle\psi_0^- | U^{\Delta t} |\psi_0^+\rangle = 0$ where $\Delta t = t_2 - t_1$. To conclude the proof we rewrite the previous equation as

$$\langle\psi_0^- | U^{\Delta t} |\psi_0^+\rangle = \sum_k \psi_{k0}^+ \sum_l \psi_{l0}^+ U_{lk}^{\Delta t}$$

$$= \sum_{k_1} \psi_{k_1 0}^+ \sum_l \psi_{l0}^+ U_{lk_1}^{\Delta t} - \sum_{k_2} \psi_{k_2 0}^+ \sum_l \psi_{l0}^+ U_{lk_2}^{\Delta t}$$

$$= \sum_l \psi_{l0}^+ \left(\sum_{k_1} \psi_{k_1 0}^+ U_{lk_1}^{\Delta t} - \sum_{k_2} \psi_{k_2 0}^+ U_{lk_2}^{\Delta t} \right) = 0 \tag{15}$$

where the indices l, k, k_1 and k_2 run over the nodes of $\mathcal{G}, G_1$ and G_2 respectively. To see that Eq. 15 holds, note that U is a symmetric matrix and it is invariant to graph symmetries, i.e., if u and v are symmetric then $U_{uu}^{\Delta t} = U_{vv}^{\Delta t}$, and that if G_1 and G_2 are isomorphic, then $k_1 = k_2$ and $\psi_{1:k_1 0}^+ = \psi_{k_1+1:k_2 0}^+$.

Corollary 1. *Given a pair of graphs G_1 and G_2, the kernel satisfies the following properties: 1) $0 \leq k_T(G_1, G_2) \leq 1$ and 2) if G_1 and G_2 are isomorphic, then $k_T(G_1, G_2) = 1$.*

Proof. The first property is trivially proved by noting that, according to Eq. 13, the kernel between G_1 and G_2 is defined as the quantum Jensen-Shannon divergence between two density operators, and then recalling that the value of quantum Jensen-Shannon divergence is bounded to lie between 0 and 1.

The second property follows again from Eq. 13 and Theorem 1. It is sufficient to note that the quantum Jensen-Shannon divergence reaches its maximum value if and only if the density operators have support on orthogonal spaces.

Unfortunately we cannot prove that our kernel is positive semidefinite, but both empirical evidence and the fact that the Jensen-Shannon Divergence is negative semidefinite on pure quantum states [19] while our graph similarity is maximal on orthogonal states suggest that it might be.

3.1   Kernel Computation

We conclude this section with a few remarks on the computational complexity of our kernel. Recall that $|\psi_t\rangle = e^{-iAt} |\psi_0\rangle$, then we rewrite Eq. 12 as

$$\rho_T = \frac{1}{T} \int_0^T e^{-iAt} |\psi_0\rangle \langle \psi_0| e^{iAt} \, dt \qquad (16)$$

Since $e^{-iAt} = \Phi e^{-i\Lambda t}\Phi^\top$, we can rewrite the previous equation in terms of the spectral decomposition of the adjacency matrix,

$$\rho_T = \frac{1}{T} \int_0^T \Phi e^{-i\Lambda t}\Phi^\top |\psi_0\rangle \langle \psi_0| \Phi e^{i\Lambda t}\Phi^\top \, dt \qquad (17)$$

The (r, c) element of ρ_T can be computed as

$$\rho_T(r,c) = \frac{1}{T} \int_0^T \left(\sum_k \sum_l \phi_{rk} e^{-i\lambda_k t} \phi_{lk} \psi_{0l}^- \right) \left(\sum_m \sum_n \psi_{0m}^\dagger \phi_{mn} e^{i\lambda_n t} \phi_{cn} \right) dt \qquad (18)$$

Let $\bar{\psi}_k = \sum_l \phi_{lk}\psi_{0l}$ and $\bar{\psi}_n = \sum_m \phi_{mn}\psi_{0n}^\dagger$, then

$$\rho_T(r,c) = \frac{1}{T} \int_0^T \left(\sum_k \phi_{rk} e^{-i\lambda_k t} \bar{\psi}_k \sum_n \phi_{cn} e^{i\lambda_n t} \bar{\psi}_n \right) dt \qquad (19)$$

which can be finally rewritten as

$$\rho_T(r,c) = \sum_k \sum_n \phi_{rk} \phi_{cn} \bar{\psi}_k \bar{\psi}_n \frac{1}{T} \int_0^T e^{i(\lambda_n - \lambda_k)t} \, dt \qquad (20)$$

If we let $T \to \infty$, Eq. 20 further simplifies to

$$\rho_T(r,c) = \sum_{\lambda_k \in \tilde{\Lambda}} \sum_m \sum_n \phi(\lambda_k)_{r,m} \phi(\lambda_k)_{c,n} \bar{\psi}_m \bar{\psi}_n \qquad (21)$$

where $\tilde{\Lambda}$ is the set of unique eigenvalues of A and $\phi(\lambda_k)$ is the matrix whose columns are the eigenvectors associated with λ_k. As a consequence, we see that the complexity of computing the QJSD kernel is upper bounded by that of computing the eigendecomposition of $\mathcal{G}$, i.e. $O(|\mathcal{V}|^3)$.

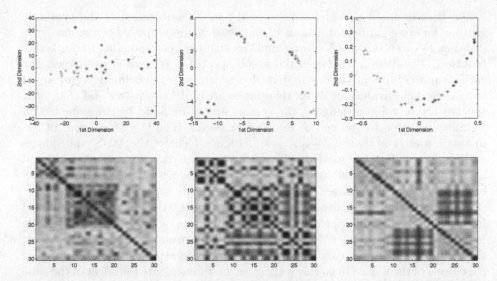

Fig. 1. Two-dimensional MDS embeddings of the synthetic data (top row) on different distance matrices (bottom row). From left to right, the distance is computed as the edit distance between the graphs, the distance between the graph spectra and the distance associated with the QJSD kernel.

4   Experimental Results

In this section, we evaluate the performance of our kernel and we compare it with a number of well-known alternative graph kernels, namely the classic random walk kernel [7], the shortest-path kernel [8] and a set of graphlet kernels [9]. We test different variants of the graphlet kernel, where we vary the graphlet sizes $k \in \{3, 4\}$ and the type of graphlets (all possible size k graphlets vs only those which are fully connected).

The experiments are performed on three different standard dataset, namely MUTAG, Enzymes and PPI. Table 1 reports some statistics about these datasets. MUTAG is a dataset of 188 mutagenic aromatic and heteroaromatic compounds labeled according to whether or not they have a mutagenic effect on the Gram-negative bacterium *Salmonella typhimurium*. Enzymes is a dataset of graphs representing protein tertiary structures that consists of 600 enzymes from the BRENDA enzyme database. Finally, the PPI dataset consists of protein-protein interaction (PPIs) networks related to histidine kinase from two different groups: 40 PPIs from *Acidovorax avenae* and 46 PPIs from *Acidobacteria*. To these three datasets, we add a fourth set of 30 synthetically generated graphs, 10 for each class. The graphs belonging to each class were sampled from a generative model with size 12,14 and 16 respectively. Details about the generative model can be found in [20].

We first evaluate the Multidimensional Scaling embedding of the synthetic graphs for three different distance matrices, namely the edit distance, the distance between the graph spectra and the distance corresponding to our kernel function. The distance between the graph spectra is computed as follows. For each graph G with adjacency matrix A, we compute the column vector s_G of the ordered eigenvalues of A. As the graphs are of different sizes and thus their spectra are of different lengths, the vectors are all made to be the same length by padding zeros to the end of the shorter vector. The (i, j)th element of the distance matrix is then $d_{ij} = ||s_i - s_i||$. Figure 1 shows the MDS embeddings and the graph distance matrices. It is clear that the distance matrix associated with our kernel has a well-defined block structure which is reflected in the MDS embedding, where the three classes seem to be easily separable.

A second experiment uses a binary C-SVM to test the efficacy of our kernel for classification. We perform 10-fold cross validation, where for each sample we independently tune the value of C, the SVM regularizer constant, by considering the training data from that sample. The process is averaged over 100 random partitions of the data. Given this setting, we first investigate the effect of the time parameter in the classification accuracy. Fig. 2 shows the value of the average accuracy ($\pm$ standard error) on the synthetic dataset as the time parameter T varies. Here the red horizontal line shows the mean accuracy for $T \to \infty$. The plot shows that the choice of the time greatly influences the performance of our kernel, as we can clearly see that the average accuracy reaches a maximum before stabilizing around the asymptotic value. This should be compared with the average accuracy that we achieve for $T \to \infty$, which, although not optimal, is not too far from the maximum. however, a more detailed study of the time parameter is beyond the scope of this paper and will thus be the subject of future work.

Finally, Table 2 reports the average classification accuracies ($\pm$ standard error) of the different kernels. As we can see, the proposed kernel achieves the best result on three out of four datasets. The poor accuracy on the Enzymes dataset is likely to be linked to the presence of disjoint graphs, as this will affect the way in which the walk spreads through the graph. Note, however, that this is a particularly hard dataset where the structures of the graphs provide limited information about the underlying class structure. In fact, all kernels based only on graph structure perform only marginally better than random guess, and node and edge attributes need to be taken into account too.

Table 1. Statistics on the graph datasets

datasets	# graphs	# classes	avg # nodes	disjoint
Synth	30	3 (10 each)	13.77	N
MUTAG	188	2 (125 vs. 63)	17.93	N
Enzymes	600	6 (100 each)	32.63	Y
PPI	86	2 (40 vs. 46)	109.60	N

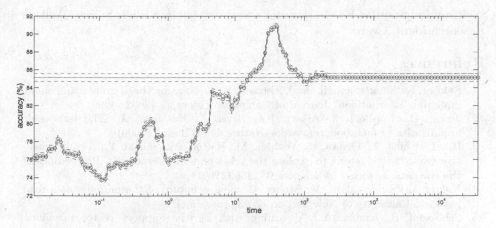

Fig. 2. The mean accuracy ($\pm$ standard error) of the QJSD kernel as the time parameter T varies. The red horizontal line shows the mean accuracy for $T \rightarrow \infty$.

Table 2. Classification accuracy ($\pm$ standard error) on unattributed graph datasets. QJSD is the proposed kernel, SP is the shortest-path kernel [8], RW is the random walk kernel [7], while G_k (CG_k) denotes the graphlet kernel computed using all graphlets (all the connected graphlets, respectively) of size k [9].

Kernel	Synth	MUTAG	Enzymes	PPI
QJSD	**85.20 $\pm$ 0.47**	**86.55 $\pm$ 0.15**	24.20 $\pm$ 0.38	**78.43 $\pm$ 0.30**
SP	74.90 $\pm$ 0.33	85.02 $\pm$ 0.17	**28.55 $\pm$ 0.42**	66.14 $\pm$ 0.40
RW	78.53 $\pm$ 0.43	77.87 $\pm$ 0.21	22.15 $\pm$ 0.37	69.70 $\pm$ 0.30
G_3	79.33 $\pm$ 0.39	82.04 $\pm$ 0.14	24.87 $\pm$ 0.22	51.95 $\pm$ 0.44
G_4	83.60 $\pm$ 0.48	81.89 $\pm$ 0.13	28.60 $\pm$ 0.21	73.14 $\pm$ 0.37
CG_3	56.57 $\pm$ 0.47	66.43 $\pm$ 0.08	19.92 $\pm$ 0.27	52.89 $\pm$ 0.50
CG_4	81.57 $\pm$ 0.54	69.08 $\pm$ 0.15	23.05 $\pm$ 0.06	61.56 $\pm$ 0.41

5   Conclusions

In this paper, we have introduced a novel kernel on unattributed graphs where we probe the graph structure using the time evolution of a continuous-time quantum walk. More precisely, given a pair of graphs we computed the quantum Jensen-Shannon divergence between the evolutions of two quantum walks on a suitably defined union of the original graphs. With the quantum Jensen-Shannon divergence to hand, we established our graph kernel. We performed an extensive experimental evaluation and we demonstrated the effectiveness of the proposed approach. Future work will focus on incorporating node and edge labels information, as well as studying the role of the time parameter more in depth.

Acknowledgments. Edwin Hancock was supported by a Royal Society Wolfson Research Merit Award.

References

1. Siddiqi, K., Shokoufandeh, A., Dickinson, S., Zucker, S.: Shock graphs and shape matching. International Journal of Computer Vision 35, 13–32 (1999)
2. Jeong, H., Tombor, B., Albert, R., Oltvai, Z., Barabási, A.: The large-scale organization of metabolic networks. Nature 407, 651–654 (2000)
3. Ito, T., Chiba, T., Ozawa, R., Yoshida, M., Hattori, M., Sakaki, Y.: A comprehensive two-hybrid analysis to explore the yeast protein interactome. Proceedings of the National Academy of Sciences 98, 4569 (2001)
4. Kalapala, V., Sanwalani, V., Moore, C.: The structure of the united states road network. University of New Mexico (2003) (preprint)
5. Schölkopf, B., Smola, A.J.: Learning with kernels: Support vector machines, regularization, optimization, and beyond. MIT Press (2001)
6. Vapnik, V.: Statistical learning theory (1998)
7. Gärtner, T., Flach, P.A., Wrobel, S.: On graph kernels: Hardness results and efficient alternatives. In: Schölkopf, B., Warmuth, M.K. (eds.) COLT/Kernel 2003. LNCS (LNAI), vol. 2777, pp. 129–143. Springer, Heidelberg (2003)
8. Borgwardt, K., Kriegel, H.: Shortest-path kernels on graphs. In: Fifth IEEE International Conference on Data Mining, p. 8. IEEE (2005)
9. Shervashidze, N., Vishwanathan, S., Petri, T., Mehlhorn, K., Borgwardt, K.: Efficient graphlet kernels for large graph comparison. In: Proceedings of the International Workshop on Artificial Intelligence and Statistics. Society for Artificial Intelligence and Statistics (2009)
10. Haussler, D.: Convolution kernels on discrete structures. Technical report, UC Santa Cruz (1999)
11. Bai, L., Hancock, E.: Graph kernels from the Jensen-Shannon divergence. Journal of Mathematical Imaging and Vision, 1–10 (2012)
12. Farhi, E., Gutmann, S.: Quantum computation and decision trees. Physical Review A 58, 915 (1998)
13. Kempe, J.: Quantum random walks: an introductory overview. Contemporary Physics 44, 307–327 (2003)
14. Emms, D., Wilson, R., Hancock, E.: Graph embedding using quantum commute times. Graph-Based Representations in Pattern Recognition, 371–382 (2007)
15. Rossi, L., Torsello, A., Hancock, E.R.: Approximate axial symmetries from continuous time quantum walks. In: Gimel'farb, G., Hancock, E., Imiya, A., Kuijper, A., Kudo, M., Omachi, S., Windeatt, T., Yamada, K. (eds.) SSPR & SPR 2012. LNCS, vol. 7626, pp. 144–152. Springer, Heidelberg (2012)
16. Majtey, A., Lamberti, P., Prato, D.: Jensen-Shannon divergence as a measure of distinguishability between mixed quantum states. Physical Review A 72, 052310 (2005)
17. Lamberti, P., Majtey, A., Borras, A., Casas, M., Plastino, A.: Metric character of the quantum Jensen-Shannon divergence. Physical Review A 77, 052311 (2008)
18. Nielsen, M., Chuang, I.: Quantum computation and quantum information. Cambridge University Press (2010)
19. Briët, J., Harremoës, P.: Properties of classical and quantum jensen-shannon divergence. Physical review A 79, 52311 (2009)
20. Torsello, A., Rossi, L.: Supervised learning of graph structure. Similarity-Based Pattern Recognition, 117–132 (2011)

Relevant Cycle Hypergraph Representation for Molecules

Benoît Gaüzère[1], Luc Brun[1], and Didier Villemin[2]

[1] GREYC UMR CNRS 6072, Caen, France
[2] LCMT UMR CNRS 6507, Caen, France
{benoit.gauzere,didier.villemin,luc.brun}@ensicaen.fr

Abstract. Chemoinformatics aims to predict molecule's properties through informational methods. Some methods base their prediction model on the comparison of molecular graphs. Considering such a molecular representation, graph kernels provide a nice framework which allows to combine machine learning techniques with graph theory. Despite the fact that molecular graph encodes all structural information of a molecule, it does not explicitly encode cyclic information. In this paper, we propose a new molecular representation based on a hypergraph which explicitly encodes both cyclic and acyclic information into one molecular representation called relevant cycle hypergraph. In addition, we propose a similarity measure in order to compare relevant cycle hypergraphs and use this molecular representation in a chemoinformatics prediction problem.

Keywords: Graph Kernel, Chemoinformatics, Relevant Cycles.

1   Introduction

Chemoinformatics consists in predicting molecule's properties from their similarity. Most of existing methods, called fingerprint methods, encode molecules as collections of chemical descriptors and deduce similarity between molecules from the similarity of their collections of descriptors. Another approach consists in using the molecular graph $G = (V, E, \mu, \nu)$ representation associated to a molecule. Unlabeled graph (V, E) encodes molecular structural information while labelling function μ maps each vertex to an atom's label corresponding to its chemical element and labelling function ν characterizes each edge by the valency (single, double, triple or aromatic) of the corresponding atomic bond which connects two atoms. Hydrogen atoms are implicitly encoded into molecular graph representation using the valency of atoms.

Considering molecular graph representation, similarity between molecules can be deduced from the similarity of their molecular graphs. Graph kernels can be understood as symmetric graph similarity measures. Using a semi definite positive kernel, the value $k(G, G')$, where G and G' encode two graphs, corresponds to a scalar product between two vectors $\psi(G)$ and $\psi(G')$ in an Hilbert space. Graph kernels thus provide a natural connection between structural and statistical pattern recognition fields.

W.G. Kropatsch et al. (Eds.): GbRPR 2013, LNCS 7877, pp. 111–120, 2013.

A large family of graph kernels defined in chemoinformatics is based on bag of patterns. These methods extract a bag of patterns from graphs and deduce similarity between graphs from similarity between their bags. Most of existing graph kernels based on bags of patterns are defined on linear patterns [8]. Such methods are generally limited by the lack of expressiveness of linear patterns to encode structural information of graphs. In order to encode more structural information, some methods are defined on non-linear patterns. For example, tree-pattern kernel [9] is based on an implicit enumeration of tree-patterns, i.e. trees where a vertex can appear more than once. Another approach, called treelet kernel [4], computes an explicit enumeration of a limited set of subtrees which allows to perform an a-posteriori feature weighting step [5]. Others graph kernels aim to transform a molecular graph into a set of chemical relevant groups [3] or a set of cycles [7,6] but these methods do not define a valid kernel or do not allow to encode relationships between cyclic and acyclic parts of a molecule.

In this paper, we propose to define a new molecular representation encoded by an hypergraph which aims to encode adjacency relationships between cyclic and acyclic parts of a molecule. After a presentation of existing methods to encode molecular cyclic information in Section 2, we define in Section 3 our new molecular representation. In addition, we propose in Section 4 a method to apply treelet kernel on this new molecular representation. This method allows us to use our new molecular representation to predict molecule's properties. Section 5 shows results obtained by our contribution to a chemoinformatics problem.

2   Encoding Cyclic Information

Most of existing graph kernels based on bags of patterns applied to chemoin-formatics are based on the molecular graph representation (Section 1). Whereas this representation allows to encode most of the structural information of a given molecule, it does not explicitly encode some special combinations of atoms, such as cycles, which may have a particular influence on molecule's properties. In order to highlight such particular groups of atoms, Frölich et al. [3] have proposed to encode a molecule by a set of predefined subgraphs composing the associated molecule. These predefined subgraphs correspond to chemical relevant groups of atoms and are generally defined by cycles or connected atom groups. Then, similarity between molecules is deduced by an optimal matching between two sets of relevant groups. Unfortunately, the kernel defined from this optimal assignment may lead to a non positive definite kernel [12], hence restricting the application field of this kernel.

Some other approaches aim to encode a molecule by a subset of its cycles. A first approach, proposed by Horváth [7], consists in computing the set of simple cycles of a molecule. Then, similarity between two molecules is defined as a sum of two kernels encoding respectively the cyclic and acyclic similarities between both molecules. Similarity between cycles is defined by the number of common simple cycles and similarity between acyclic parts by a tree-pattern kernel [9]. An extension of this method only computes the set of relevant cycles [6], as defined by Vismara [13], of the molecular graph hence providing a better computational efficiency. Whereas this approach provides an explicit encoding of cyclic information, the cyclic system is encoded by a set of cycles which does not encode relationships between cycles.

In order to encode additional information, Gaüzère et al. [5] have proposed to encode the set of relevant cycles and their adjacency relationships within the relevant cycle graph. The similarity between molecules can then be deduced by combining a kernel on relevant cycle graphs which encodes cyclic system similarity and a kernel on molecular graphs which encodes the similarity of molecules based on atom's relationships. Despite the fact that this approach leads to good results on experiments involving cyclic molecules, this representation, as the one of Horváth [6], separates cyclic and acyclic information by defining two different molecular representations. Then, global similarity between molecules is computed using two distinct similarity measures, each of them being applied on one representation. This separation induces a loss of adjacency relationships between cyclic and acyclic parts of molecules. In the following, we propose a new molecular representation which aims to merge cyclic and acyclic information into one molecular representation and hence encodes adjacency relationships between cyclic and acyclic parts.

3   Encoding Topological Relationships between Cyclic and Acyclic Parts

In order to encode adjacency relations between cyclic and acyclic parts of a molecule, we propose to define a molecular representation which aims to represent a set of atoms encoding a cycle as a single vertex. For any graph G, a simple cycle is defined as a subgraph $C = (V', E', \mu, \nu)$ of $G = (V, E, \mu, \nu)$ where each vertex $v \in V'$ has a degree equal to 2. Each cycle $C \subseteq G$ can be represented as a vector $C \in \{0, 1\}^{|E|}$ where C_i equals 1 if i is an edge of C and 0 otherwise. Using this vector representation, the set of vectors encoding cycles of G defines a vector space [13]. Given this vector space, the union of all bases of minimum length defines the set of relevant cycles, denoted $\mathcal{C_R}$. The length of a base is defined as the sum of lengths of its cycles.

Adjacency relationships between relevant cycles can be encoded by the relevant cycle graph [5]. This graph is defined as $G_C = (\mathcal{C_R}, E_{\mathcal{C_R}}, \mu_{\mathcal{C_R}}, \nu_{\mathcal{C_R}})$ where each vertex $c \in \mathcal{C_R}$ corresponds to a relevant cycle. Each vertex c is associated to the set of vertices $V(c)$ corresponding to the set of atoms included within c and the set of edges $E(c)$ corresponding to the set of atomic bonds forming cycle c. By extension, $E(\mathcal{C_R})$ denotes the set of atomic bonds belonging to a relevant cycle of $\mathcal{C_R}$. An edge (c_1, c_2) is in $E_{\mathcal{C_R}}$ if $V(c_1) \cap V(c_2) \neq \emptyset$, i.e. if c_1 and c_2 share at least one vertex of the molecular graph (Figure 1). The labelling function $\mu_{\mathcal{C_R}}(c)$ is defined as a canonical code of the cyclic sequence of vertex and edge labels defining c. In the same way, the label function $\nu_{\mathcal{C_R}}(e)$ of an edge $e = (c, c')$ is defined as a canonical code of the path common to c and c'. Despite the fact that this relevant cycle graph encodes adjacency information

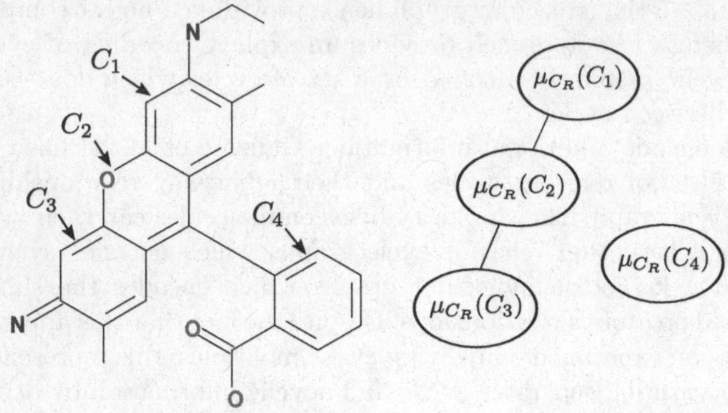

(a) Molecular graph.   (b) Relevant cycle graph.

Fig. 1. A cyclic molecular graph and its relevant cycle graph representation

of the molecular cyclic system, all adjacency information involving vertices and edges of a molecular graph which are not included within a cycle is missing. For example, acyclic parts connected to C_1, C_3 and C_4 and connection between C_2 and C_4 in Figure 1(a) are not encoded within the associated relevant cycle graph representation (Figure 1(b)).

Therefore, in order to add this information into our molecular representation, a first approach consists in adding missing vertices and edges to our relevant cycle graph. Unfortunately, such an approach can not handle the case where an atom is connected to two distinct relevant cycles. As shown in Figure 2(a), the atom labeled O is connected by an unique edge to two distinct cycles in the molecular graph representation. This adjacency relationship can not be encoded by a simple graph where an edge connects only two vertices. Therefore, in order to handle such relationships, we propose to define a new hypergraph representation of the molecular graph.

A directed hypergraph [1,2] $H = (V, E)$ is defined as a set of vertices V and a set $E = E^e \cup E^h$ encoding the union of a set of edges $E^e \subset V \times V$ and a set of hyperedges $E^h \subset \mathcal{P}(V) \times \mathcal{P}(V)$ where $\mathcal{P}(V)$ denotes the set of all subsets of V. An ordered hyperedge $e = (s_u, s_v)$ with $s_u = \{u_1, \ldots, u_i\}$ and $s_v = \{v_1, \ldots, v_j\}$ defines an adjacency relation between sets $\{u_1, \ldots, u_i\}$ and $\{v_1, \ldots, v_j\}$, as illustrated in Figure 2(b). In the following, we assume that if $\exists e = (s_1, s_2) \in E$ then $\exists e' = (s_2, s_1) \in E$ and e and e' are considered as a same unique hyperedge. Such a definition allows us to represent relationships between an acyclic atom and a set of cycles, each cycle being encoded as a vertex.

A molecular graph $G = (V, E, \mu, \nu)$ can now be encoded as a relevant cycle hypergraph $H_{RC}(G) = (V_{RC}, E_{RC})$. Within relevant cycle graph representation, the set of vertices $C_{\mathcal{R}}$ encodes the set of atoms $V(C_{\mathcal{R}})$ and the set of atomic bonds $E(C_{\mathcal{R}})$ which belong to a cycle. Considering such a representation, missing molecular graph information corresponds to atoms and atomic bonds not included within a cycle. These sets are respectively defined by the complement

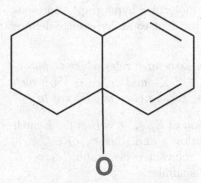

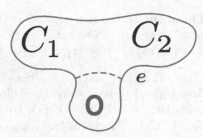

(a) Acyclic atom connected to two cycles by an unique edge

(b) Hyperedge e representing original edge. $e = (\{C_1, C_2\}, O)$.

Fig. 2. Special case where a graph can not encode the representation based on relevant cycle graph

of $V(C_{\mathcal{R}})$ and $E(C_{\mathcal{R}})$ in V and E. Therefore, in order to include all atom information into our relevant cycle hypergraph, V_{RC} is defined by the union of two subsets:

1. A first subset $C_{\mathcal{R}}$ corresponding to the set of relevant cycles,
2. and a second subset $V - V(C_{\mathcal{R}})$ corresponding to the set of atoms not included within a cycle.

Considering set of vertices V_{RC}, we define a function $p : V \to \mathcal{P}(V_{RC})$ defined as $p(u) = \{u\}$ if $u \notin V(C_{\mathcal{R}})$ and $\{c \in C_{\mathcal{R}} \mid u \in V(c)\}$ if not. This function p encodes the print of vertex $v \in V$ on V_{RC}. In the same way as for vertices, the set of hyperedges E_{RC} is composed of two subsets:

1. A set of edges E_{RC}^e composed of:
   - edges between relevant cycle vertices, corresponding to the set of edges $E_{C_{\mathcal{R}}}$,
   - edges $e = (p(u), p(v))$ such that $(u, v) \in E - E(C_{\mathcal{R}})$, $|p(u)| = 1$ and $|p(v)| = 1$. This set of edges corresponds to edges of molecular graph G connecting two acyclic atoms or connecting a single relevant cycle to another single relevant cycle (C_2 and C_4 in Figure 1) or an acyclic part of G (C_3 and N in Figure 1),
2. and a set of hyperedges $e = (p(u), p(v)) \in E_{RC}^h$ such that $(u, v) \in E - E(C_{\mathcal{R}})$, $|p(u)| > 1$ or $|p(v)| > 1$. This set of hyperedges corresponds to special cases where an edge connects at least two distinct relevant cycles to another part of the molecule (Figure 2). This edge is thus encoded by an hyperedge which connects the two sets of vertices $p(u)$ and $p(v)$.

This molecular hypergraph representation (Figure 3(c)) encodes all atoms $v \in V$ either by a vertex encoding a cycle or by v itself if $v \notin V(C_{\mathcal{R}})$. In the same way, each atomic bond $e \in E$ is encoded within our molecular hypergraph representation. In addition, we note that set of vertices incident to an hyperedge defines a clique:

Theorem 1. *Let be a graph* $G = (V, E)$ *and its associated relevant cycle hypergraph* $H_{RC}(G) = (V_{RC}, E_{RC})$. *If* $\exists e = (s_1, s_2) \in E_{RC}^h$ *and* $c_1, c_2 \in V_{RC}$ *such that* $\{c_1, c_2\} \subseteq s_1$ *or* $\{c_1, c_2\} \subseteq s_2$, *then* $(c_1, c_2) \in E_{RC}^e$, *i.e.* c_1 *is adjacent to* c_2.

Proof. If $c_1 \in s_1$ and $c_2 \in s_1$, then by construction of E_{RC}^h, $\exists e = (u, v) \in E$ such that $\{c_1, c_2\} \subseteq p(u) = s_1$. By definition of function p and since $c_1, c_2 \in C_R$, it holds that $u \in V(c_1) \cap V(c_2)$. By definition of relevant cycle graph, $(c_1, c_2) \in E_{C_R} \subset E_{RC}^e$. The proof for $c_1 \in s_2$ and $c_2 \in s_2$ is similar.

Algorithm 1 describes the different steps required to transform molecular graph G into its associated relevant cycle hypergraph H_{RC}. The first step consists in computing the relevant cycle graph of G, as described in [5], and initializing our hypergraph by this graph (Algo. 1, Lines 3 and 4). Then, the set of acyclic parts is included to the current graph representation (Algo. 1, Lines 6 and 7). Finally, hyperedges are included into our relevant cycle hypergraph (Algo. 1, Line 9).

4 Similarity between Relevant Cycle Hypergraphs

The previous section defines a molecular representation which provides a new way to encode adjacency relations between cyclic and acyclic parts of a molecule. In order to apply QSAR methods on this molecular representation, we have to define a similarity measure between relevant cycle hypergraphs. Graph kernels, such as treelet kernel [4], are only defined on molecular graphs and can not be applied directly on an hypergraph representation of a molecule. In this section, we propose to adapt treelet kernel to the comparison of relevant cycle hypergraphs.

Treelet kernel is a graph kernel defined as a kernel between two sets of patterns extracted from both graphs to be compared. The set of extracted patterns, denoted $\mathcal{T}$ and called treelets, is composed of all labeled trees with a number of

---

Algorithm 1. Computing relevant cycle hypergraph from molecular graph.

Require: $G = (V, E)$
Ensure: $H_{RC} = (V_{RC}, E_{RC}), E_{RC} = E_{RC}^e \cup E_{RC}^h$
  1: $G_C(C_R, E_{C_R}) = G_C(G)$ {Relevant cycle graph}
  2: {Adding all information included within cycles}
  3: $V_{RC} = C_R$
  4: $E_{RC}^e = E_{C_R}$
  5: {Adding information not included within a cycle}
  6: $V_{RC} = V_{RC} \cup \{v \notin V(C_R)\}$
  7: $E_{RC}^e = E_{RC}^e \cup \{(p(u), p(v)) \mid (u, v) \in E, |p(u)| = 1 \text{ AND } |p(v)| = 1\}$
  8: {Special case (Figure 2).}
  9: $E_{RC}^h = \{(p(u), p(v)) \mid (u, v) \in E, |p(u)| > 1 \text{ OR } |p(v)| > 1\}$
 10: **return** H_{RC}

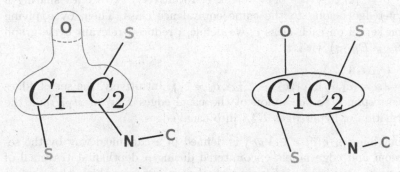

(a) Molecular graph G including cycles.   (b) Relevant cycle graph G_C.

(c) Relevant cycle hypergraph $H_{RC}(G)$. (d) Reduced   relevant   cycle   graph
$G_{RCR}(G)$.

Fig. 3. Different encodings of a same molecule

vertices less than or equal to 6. Based on the explicit enumeration of this set of
substructures, each graph G is associated to a vector $f(G)$ encoding the number
of occurrences of each treelet t in G:

$$f(G) = (f_t(G))_{t \in \mathcal{T}(G)} \text{ with } f_t(G) = |(t \trianglelefteq G)| \tag{1}$$

where $\mathcal{T}(G)$ denotes the set of treelets extracted from G and $\trianglelefteq$ the subgraph
isomorphism relationship. Using this vector representation, similarity between
treelet distributions is computed using a sum of subkernels between treelet's
number of occurrences:

$$K_{\mathcal{T}}(G, G') = \sum_{t \in \mathcal{T}(G) \cap \mathcal{T}(G')} k(f_t(G), f_t(G')) \tag{2}$$

where $k(.,.)$ defines any positive definite kernel between real numbers such as
linear kernel, Gaussian kernel or intersection kernel. Despite the fact that this
method may be applied on many kinds of graphs, it can not be directly applied
to hypergraphs.

An hypergraph encodes global relationships defined between sets of vertices. At the opposite, treelet kernel is defined on graphs where relationships are defined locally between elementary vertices. Therefore, in order to apply treelet kernel to our hypergraph representation, we have to transform global relationships defined within our hypergraph representation to local relationships between elementary vertices. This transformation is performed by merging all sets of vertices incident to an hyperedge. This merge operation relies to transform hyperedges to edges.

An equivalence relation $\sim$ between vertices $c \in V_{RC}$ is defined such that $c_1 \sim c_2$ if and only if $\exists e = (s_1, s_2) \in E_{RC}$ such that $\{c_1, c_2\} \subseteq s_1$ or $\{c_1, c_2\} \subseteq s_2$. Using equivalence relation $\sim$ previously defined, we can now define the equivalence class $\bar{c} = \{c'; c \sim c'\}$ of a vertex c. Intuitively, two cycles sharing a common hyperedge belong to the same equivalence class. Then, by applying a contraction kernel on each class $\bar{c}$, we define a reduced relevant cycle graph $G_{RCR} = (V_{RCR}, E_{RCR})$ with:

- $V_{RCR} = \{\bar{c}, c \in V_{RC}\}$,
- $E_{RCR} = \{e = (\bar{c}_1, \bar{c}_2), (c_1, c_2) \in E_{RC}, c_1 \not\sim c_2\}$. Intuitively, the set of edges E_{RCR} corresponds to the union of the usual edges E_{RC}^e of H_{RC} and the transformation of hyperedges E_{RC}^h into usual edges.

Labelling function $\mu_{RCR}(\bar{c}), c \in V_{RC}$, is defined in a canonical way by the sequence of atom and edge labels encountered during a depth first traversal of the spanning tree covering $\bar{c}$ and having the lowest lexicographic order. Such a spanning tree exists since any pair of vertices $\{c, c'\}$ sharing a same hyperedge is connected (Theorem 1).

Given this second representation of a molecule defined by the reduced relevant cycle graph, our new similarity measure based on treelet kernel is defined in two parts. A first step aims to extract the set of treelets $\mathcal{T}_1 = \mathcal{T}(V_{RC}, E_{RC}^e)$. (V_{RC}, E_{RC}^e) corresponds to a sub hypergraph of H_{RC} which does not include any hyperedge $e \in E_{RC}^h$. Therefore, the set of treelets $\mathcal{T}_1$ encodes information which does not include special cases depicted in Figure 2. Information corresponding to these special cases, encoded by hyperedges $e \in E_{RC}^h$, is included into our similarity measure by the set of treelets $\mathcal{T}_2$ extracted from the reduced relevant cycle Graph G_{RCR} built from the transformation of hyperedges into edges. In order to avoid redundancy, we reduce the set of treelets $\mathcal{T}_2$ to treelets containing at least one edge corresponding to an hyperedge $e_h \in E_{RC}^h$. Finally, we define the set of treelets $\mathcal{T}_{CR}(G)$ associated to a molecular graph G by $\mathcal{T}_1 \cup \mathcal{T}_2$. Similarity between molecules is then defined as a sum of subkernels comparing number of occurrences of each treelet $t \in \mathcal{T}_{CR}(G)$ (Equation 2). This approach allows us to use a set of patterns which encodes most of the adjacency relations between cyclic and acyclic parts.

5   Experiments

We have tested our new molecule representation on an experiment defined as a classification problem. This dataset is taken from the Predictive Toxicity

Table 1. Classification accuracy on PTC

Method	MM	FM	MR	FR
1 Treelet Kernel (TK)	208	205	209	212
2 TK on cycles (TC)	211	210	203	232
3 Treelet on relevant cycle hypergraph (TCH)	217	224	207	233
4 Cyclic Pattern Kernel [6]	209	207	202	228
5 Gaussian Edit Distance Kernel [10]	223	212	194	234
6 TK + MKL	218	224	224	250
7 TC + MKL	216	213	212	237
8 TCH + MKL	225	229	215	239
9 Combo TK - TC	219	226	**226**	251
10 Combo TK - TCH	**225**	**230**	224	**252**

Challenge [11] which aims to predict carcinogenicity of chemical compounds applied to female (F) and male (M) rats (R) and mice (M). This experiment is based on ten different datasets for each class of animal, each of them being composed of one train set and one test set. The amount of predicted molecules is equals to 336 for male mice, 349 for female mice, 344 for male rats and 351 for female rats. Table 1 shows the amount of correctly classified molecules over the ten test sets for each method and for each class of animal. The first three lines of Table 1 shows results obtained by a treelet kernel applied on differents molecular representations. Line 1 corresponds to treelet kernel applied on molecular graph, Line 2 to relevant cycle graph and Line 3 corresponds to kernel defined in Section 3. First, we can note that our new molecular representation obtains the best results among the three tested representations. This observation validates our hypothesis on the importance of relationships between cyclic and acyclic parts. This results can be compared with two other graph kernels. Line 4 shows results obtained by the kernel defined by Horváth based on the set of relevant cycles common to two molecules. As we can see, omitting relevant cycles relationships and adjacency relationships between cyclic and acyclic parts decreases the accuracy of this kernel. Line 5 corresponds to a graph kernel based on the notion of edit distance [10] between molecular graphs. This kernel obtains better results than treelet kernel applied on relevant cycle hypergraph for two classes over four. The second part of Table 1 shows results obtained by treelet kernels after a feature weighting step as defined in [5]. After this weighting step, treelet kernel applied on our new representation (Table 1, Line 8) obtains best results on two classes of animals and obtains second best results on the two other classes when only considering Lines 6 to 8. Note that our sparse feature weighting step reduces the number of treelets extracted from relevant cycle hypergraphs from 5700 to 25 relevant treelets. In comparison, treelet weighting step applied on molecular graph reduces the set of treelets from 3500 to 150 treelets. Note that this optimal weighting step selects both non linear treelets and treelets having 6 nodes which validates the relevance of using such substructures. Finally, treelet kernel has been combined with relevant cycle graph (Table 1, Line 9) and our new representation (Table 1, Line 10). This combination of two molecular representations obtains the best results on three classes over four of animals when compared to combination of relevant cycle graph and molecular graph representations, hence showing the relevance of our molecular representation.

6   Conclusion

In this article, we have defined a new molecular representation based on hypergraphs which is able to encode adjacency relationships between cyclic and acyclic parts of a molecule. In addition, we have proposed a method to apply treelet kernel on our hypergraph representation. Our experiments show that the adjacency information encoded by this molecular representation can lead to better results than methods applied on classic molecular graphs. One outlook of this work consists in including the relative positioning of bonds connecting acyclic parts of a molecule on a same cycle.

References

1. Berge, C.: Graphs and hypergraphs, vol. 6. Elsevier (1976)
2. Ducournau, A.: Hypergraphes: clustering, réduction et marches aléatoires orientées pour la segmentation d'images et de vidéo. PhD thesis, École Nationale d'Ingénieurs de Saint-Étienne (2012)
3. Fröhlich, H., Wegner, J.K., Sieker, F., Zell, A.: Optimal assignment kernels for attributed molecular graphs. In: Proceedings of the 22nd International Conference on Machine learning, ICML 2005, pp. 225–232. ACM Press (2005)
4. Gaüzère, B., Brun, L., Villemin, D.: Two New Graphs Kernels in Chemoinformatics. Pattern Recognition Letters 33(15), 2038–2047 (2012)
5. Gaüzère, B., Brun, L., Villemin, D., Brun, M.: Graph kernels based on relevant patterns and cycle information for chemoinformatics. In: Proceedings of ICPR 2012. IAPR, pp. 1775–1778. IEEE (November 2012)
6. Horváth, T.: Cyclic pattern kernels revisited. In: Ho, T.-B., Cheung, D., Liu, H. (eds.) PAKDD 2005. LNCS (LNAI), vol. 3518, pp. 791–801. Springer, Heidelberg (2005)
7. Horváth, T., Gartner, T., Wrobel, S.: Cyclic pattern kernels for predictive graph mining. In: Proceedings of the 2004 ACM SIGKDD International Conference on Knowledge Discovery and Data Mining, KDD 2004, pp. 158–167 (2004)
8. Kashima, H., Tsuda, K., Inokuchi, A.: Kernels for graphs, ch. 7, pp. 155–170. MIT Press (2004)
9. Mahé, P., Vert, J.-P.: Graph kernels based on tree patterns for molecules. Machine Learning 75(1), 3–35 (September 2008) (2009)
10. Neuhaus, M., Bunke, H.: Bridging the gap between graph edit distance and kernel machines. World Scientific Pub. Co. Inc. (2007)
11. Toivonen, H., Srinivasan, A., King, R., Kramer, S., Helma, C.: Statistical evaluation of the predictive toxicology challenge 2000-2001. Bioinformatics 19(10), 1183–1193 (2003)
12. Vert, J.-P.: The optimal assignment kernel is not positive definite, http://hal.archives-ouvertes.fr/hal-00218278
13. Vismara, P.: Union of all the minimum cycle bases of a graph. The Electronic Journal of Combinatorics 4(1), 73–87 (1997)

A Quantum Jensen-Shannon Graph Kernel Using the Continuous-Time Quantum Walk

Lu Bai[1], Edwin R. Hancock[1,*], Andrea Torsello[2], and Luca Rossi[2]

[1] Department of Computer Science, University of York, UK
{lu,erh}@cs.york.ac.uk
[2] Ca' Foscari University of Venice, Italy
{torsello,lurossi}@dsi.unive.it

Abstract. In this paper, we use the quantum Jensen-Shannon divergence as a means to establish the similarity between a pair of graphs and to develop a novel graph kernel. In quantum theory, the quantum Jensen-Shannon divergence is defined as a distance measure between quantum states. In order to compute the quantum Jensen-Shannon divergence between a pair of graphs, we first need to associate a density operator with each of them. Hence, we decide to simulate the evolution of a continuous-time quantum walk on each graph and we propose a way to associate a suitable quantum state with it. With the density operator of this quantum state to hand, the graph kernel is defined as a function of the quantum Jensen-Shannon divergence between the graph density operators. We evaluate the performance of our kernel on several standard graph datasets from bioinformatics. We use the Principle Component Analysis (PCA) on the kernel matrix to embed the graphs into a feature space for classification. The experimental results demonstrate the effectiveness of the proposed approach.

Keywords: Graph Kernels, Continuous-time Quantum Walk, Quantum Jensen-Shannon Divergence.

1   Introduction

There has been an increasing interest in learning graph structures using graph kernels. A graph kernel is usually defined in terms of a similarity measure between graphs [1]. To extend the large spectrum of kernel methods from the vectorial domain to the graph domain, Haussler [2] proposed a generic way, named R-convolution, to define a kernel between two graphs by decomposing them and measuring the pairwise similarities between the resulting substructures. For example, Kashima et al. [3] proposed a graph kernel where they compute the number of matchings random walks in a pair of graphs. Borgwardt et al. [4] proposed a shortest path kernel where they enumerate the shortest paths which possess the same length. Shervashidze et al. [5] developed a subtree kernel on limited size subtrees where they iteratively update the vertex labels in a pair

* Edwin R. Hancock is supported by a Royal Society Wolfson Research Merit Award.

W.G. Kropatsch et al. (Eds.): GbRPR 2013, LNCS 7877, pp. 121–131, 2013.

of graphs, and then count the numbers of matching vertex labels between pairs of subtrees in the two graphs. The main feature of these graph kernels is that they usually share the exploit the topological information on the arrangement of vertices and edges in a graph. An attractive alternative kernel measure between a pair of graphs is based on measuring the mutual information using the classical Jensen-Shannon divergence. In probability theory, the classical Jensen-Shannon divergence is a similarity measure between probability distributions, and it is symmetric, always well defined and bounded [6]. In [7], we have used the classical Jensen-Shannon divergence to define a Jensen-Shannon kernel for graphs. Here, the Jensen-Shannon kernel between a pair of graphs is defined using the classical Jensen-Shannon divergence between some suitably defined probability distributions over the vertices of the graphs. Since the entropy associated with a probability distribution of a graph can be directly computed without the need of decomposing the graph, the Jensen-Shannon kernel, unlike the aforementioned graph kernels, avoids the computational burdensome of comparing the similarities between all the pairs of substructures of the graphs.

To develop the work in [7] further, we aim to extend the classical Jensen-Shannon divergence measure of graphs into the context of quantum theory by using the quantum Jensen-Shannon divergence [6,8], and then use this as a means of defining a novel graph kernel. Unlike the classical divergence, which is defined as a similarity measure between probability distributions, the quantum Jensen-Shannon divergence is introduced in quantum theory as a distance measure between quantum states, where a quantum state is described by its density operator [8]. In order to compute the quantum Jensen-Shannon divergence between a pair of graphs, we first need to associate a quantum state with each graph. To this end, we propose to define a mixed quantum state which is based on the evolution of a continuous-time quantum walk on each graph.

Recently, the continuous-time quantum walk has been introduced as the natural quantum analogue of the classical random walk by Farhi and Gutmann in [9]. Similarly to the classical random walk on a graph, its state space is the set of vertices of the graph. However, unlike the classical random walk, whose state vector is real-valued and whose evolution is governed by a double stochastic matrix, the state vector of the continuous-time quantum walk is complex-valued and its evolution is governed by a time-varying unitary matrix. The continuous-time quantum walk possesses a number of interesting properties which are not exhibited by the classical random walk. For instance, the continuous-time quantum walk is reversible and non-ergodic, and does not have a limiting distribution. Hence, the continuous-time quantum walk can offer us an elegant way to design quantum algorithms on graphs which have some interesting properties. For further information on quantum walks, we refer the readers to the textbook [10].

In this paper we are interested in developing a quantum kernel for graphs using the quantum Jensen-Shannon divergence and the continuous-time quantum walk. Given a graph G, we start by evolving a continuous-time quantum walk on G. The quantum walk evolution can then be described by an ensemble of pure states each describing the state of the quantum walker at time t.

As a consequence, we can associate with G the resulting mixed state and its density operator. With the density operators of a pair of graphs to hand, the proposed graph kernel is finally defined as the quantum Jensen-Shannon divergence between their density operators. We evaluate the performance of our kernel on several standard graph datasets from bioinformatics. We use the kernel Principle Component Analysis (kPCA) on the kernel matrix to embed graphs into a feature space where we perform the classification. The experimental results demonstrate the effectiveness of the proposed framework.

2   Quantum Mechanical Background

In this section, we describe the quantum mechanical formalisms that will be used in this work. We commence by reviewing the fundamental concept of continuous-time quantum walk on a graph. We then describe how to associate a density operator with a given graph through a continuous-time quantum walk. With the density operator to hand, we finally show how to compute the von Neumann entropy of the graph.

2.1   The Continuous-Time Quantum Walk

The continuous-time quantum walk is the natural quantum analogue of the classical random walk [11,9]. Similarly to the classical random walk, the state space of the continuous-time quantum walk defined on a graph $G(V, E)$ is the set of the vertices V of $G(V, E)$. However, the evolution of the quantum walk is governed by an unitary matrix rather than a stochastic matrix.

In [9], the basis state corresponding to the walk being at vertex $u \in V$ in $G(V, E)$ is denoted, by Dirac notation, as $|u\rangle$. Here $|u\rangle$ are orthonormal vectors in a n-dimensional complex-valued Hilbert space $\mathcal{H}$. A general state of the walk is then a complex linear combination of the basis states, and the state of the walk at time t is defined as

$$|\psi_t\rangle = \sum_{u \in V} \alpha_u(t) |u\rangle \tag{1}$$

where the amplitude $\alpha_u(t) \in \mathbb{C}$ and $|\psi_t\rangle \in \mathbb{C}^{|V|}$ are both complex. The probability of the walk being at a particular vertex of the graph $G(V, E)$ is given by the square of the norm of the amplitude of the relative state. More formally, let X^t be a random variable giving the location of the walk at time t. Then the probability of the walk being at vertex $u \in V$ at time t is given by

$$\Pr(X^t = u) = \alpha_u(t)\alpha_u^*(t) \tag{2}$$

where $\alpha_u^*(t)$ is the complex conjugate of $\alpha_u(t)$. Moreover $\sum_{u \in V} \alpha_u(t)\alpha_u^*(t) = 1$ and $\alpha_u(t)\alpha_u^*(t) \in [0, 1]$, for all $u \in V$, $t \in \mathbb{R}^+$.

Let A is the adjacency matrix of $G(V, E)$, then the vertex degree matrix of $G(V, E)$ is a diagonal matrix D whose elements are given by $D(u, u) = d_u =$

$\sum_{v \in V} A(u, v)$. From the degree matrix D and the adjacency matrix A we can construct the Laplacian matrix $L = D - A$. Then the evolution of the walk at time t is given by Schrödinger equation, where we take the Hamiltonian of the system to be the graph Laplacian matrix L, as

$$\frac{\partial}{\partial t} |\psi_t\rangle = -iL |\psi_t\rangle \tag{3}$$

Given a initial state $|\psi_0\rangle$, Eq.(3) can be solved to calculate the state of the walk at a time, t, as

$$|\psi_t\rangle = e^{-iLt} |\psi_0\rangle \tag{4}$$

where $U_t = e^{-iLt}$ is the unitary matrix. To implement the simulation of the quantum walk evolution, we re-write Eq.(4) in terms of the spectral decomposition $L = \Phi^\top \Lambda \Phi$ of the Laplacian matrix L, where $\Lambda = diag(\lambda_1, \lambda_2, ..., \lambda_{|V|})$ is a diagonal matrix with the ordered eigenvalues as elements ($0 = \lambda_1 < \lambda_2 < ... < \lambda_{|V|}$) and $\Phi = (\phi_1|\phi_2|...|\phi_{|V|})$ is a matrix with the corresponding ordered orthonormal eigenvectors as columns. Hence, Eq.(4) can also be re-written as

$$|\psi_t\rangle = \Phi^\top e^{-i\Lambda t} \Phi |\psi_0\rangle \tag{5}$$

In this work, we define the initial state $|\psi_0\rangle$ of $G(V, E)$ as

$$|\psi_0\rangle = \sum_{u \in V} \frac{d_u}{\sqrt{\sum_{u \in V} d_u d_u^*}} |u\rangle \tag{6}$$

where $\frac{d_u}{\sqrt{\sum_{u \in V} d_u d_u^*}}$ is the initial amplitude on vertex u. In other words, the initial probability distribution induced by $|\psi_0\rangle$ is equal to the steady state of random walk on $G(V, E)$.

2.2   A Density Operator for Graphs

In quantum mechanics, the density operator ρ is a matrix that describes an ensemble of pure states, i.e. a mixed state. A pure state is a quantum state that can be described by a single state vector $|\psi\rangle$ and its density operator ρ can be written as $|\psi\rangle \langle\psi|$. On the other hand, we can think of a mixed quantum state as an ensemble of pure states described by a density operator ρ. More formally, consider a quantum system that can be found in a number of pure states $\{(|\psi_n\rangle, n)|(n = 1, 2, ..., N)\}$ each with a probability p_n. The density operator (i.e. density matrix) of the system is then defined as

$$\rho = \sum_n p_n |\psi_n\rangle \langle\psi_n| \tag{7}$$

Now we proceeed to show how to associate a density operator with a graph through a continuous-time quantum walk. Consider a graph $G(V, E)$ and continuous-time quantum walk $|\psi_t\rangle$ on G. We can see $|\psi_t\rangle$ as a pure state describing the state of the walk at time t. If we associate with each of these pure states a probability $p_t = 1/T$, we obtain a mixed state $\{(|\psi_t\rangle, t)|(t = 1, 2, \ldots, T)\}$ describing the quantum walk evolution on $G(V, E)$. Hence, the density operator ρ_G of $G(V, E)$ of this mixed state is defined as

$$\rho_G = \sum_{t=1}^{T} p_t |\psi_t\rangle \langle\psi_t| \tag{8}$$

2.3   The von Neumann Entropy of A Graph

In quantum mechanics, the von Neumann entropy is an extension of the classical Shannon entropy, and it is defined on a density matrix (i.e. density operator) ρ as $H_N(\rho) = -Tr(\rho \log \rho)$ [12]. Note that if ρ is the density matrix associated with a pure state, then the von Neumann entropy of ρ vanishes. Consider a graph $G(V, E)$ and its density operator ρ_G defined as in Eq.(8), its von Neumann entropy is

$$H_N(\rho_G) = -Tr(\rho_G \log \rho_G) \tag{9}$$

Note that computing $H_N(\rho_G)$ is a rather complex operation, since it involves taking the logarithm of the density operator matrix ρ_G. In practice, it is more convenient to firstly determine the spectral decomposition of $\rho_G = \Phi_{\rho;G}^{\top} \Lambda_{\rho;G} \Phi_{\rho;G}$, and then Eq.(9) can be re-written as

$$H_N(\rho_G) = -\sum_{j}^{|V|} \lambda_{j;\rho;G} \log \lambda_{j;\rho;G} \tag{10}$$

where $\lambda_{1;\rho:G}, \ldots, \lambda_{j;\rho;G}, \ldots, \lambda_{|V|;\rho;G}$ is the ordered eigenvalues of ρ_G.

3   A Quantum Jensen-Shannon Graph Kernel

In this section, we use the quantum Jensen-Shannon divergence to develop a novel kernel for graphs. We commence by reviewing the concept of quantum Jensen-Shannon divergence, which can be seen as an extension of the classical Jensen-Shannon divergence to the quantum realm. The quantum Jensen-Shannon divergence between a pair of density operators is defined as a function of the von Neumann entropy associated with the operators. With the density operators to hand, we show that the quantum kernel between the pair of graphs can be computed as a function of the quantum Jensen-Shannon divergence between their density operators.

3.1   Classical and Quantum Jensen-Shannon Divergence

The classical Jensen-Shannon divergence is a non-extensive mutual information measure defined between probability distributions over structured data, and it is related to the Shannon entropy [13]. Assume there are two (discrete) probability distributions $P = (p_1, p_2, \ldots, p_X)$ and $Q = (q_1, q_2, \ldots, q_Y)$, the classical Jensen-Shannon divergence between P and Q is defined as

$$D_{JS}(P, Q) = H_S\left(\frac{P+Q}{2}\right) - \frac{1}{2}H_S(P) - \frac{1}{2}H_S(Q) \tag{11}$$

where $H_S(P) = \sum_x p_x \log p_x$ is the Shannon entropy. The classical Jensen-Shannon divergence is always well defined, symmetric, negative definite and bounded, i.e., $0 \leq D_{JS} \leq 1$. By replacing the Shannon entropy H_S of a probability distribution with the von Neumann entropy of a density operator as defined in Eq.(10), in [6] the classical Jensen-Shannon divergence has been extended to the quantum realm to define the quantum Jensen-Shannon divergence between quantum states

$$D_{QJS}(\rho, \sigma) = H_N\left(\frac{\rho+\sigma}{2}\right) - \frac{1}{2}H_N(\rho) - \frac{1}{2}H_N(\sigma) \tag{12}$$

where ρ and σ are two density operators describing the corresponding quantum states, and $H_N(.)$ is the von Neumann entropy of a density operator. The quantum Jensen-Shannon divergence is always well defined, symmetric, positive definite and bounded, i.e., $0 \leq D_{QJS} \leq 1$ [6].

3.2 A Quantum Jensen-Shannon Kernel for Graphs

We propose a novel quantum kernel for graphs by using the quantum Jensen-Shannon divergence between the density operators associated with the graphs. To this end, we let a continuous-time quantum walk evolve on a pair of graphs $G_a(V_a, E_a)$ and $G_b(V_b, E_b)$ with time t $(t = 1, \ldots, T)$. Then the density operators $\rho_{G;a}$ and $\sigma_{G;b}$ of $G_a(V_a, E_a)$ and $G_b(V_b, E_b)$ can be computed from their mixed states using Eq.(8). With the density operators $\rho_{G;a}$ and $\sigma_{G;b}$, and the quantum Jensen-Shannon divergence $D_{QJS}(\rho_{G;a}, \sigma_{G;b})$ between $\rho_{G;a}$ and $\sigma_{G;b}$ computed using Eq.(12) to hand, the quantum Jensen-Shannon divergence $D_{QJS}(G_a, G_b)$ between the pair of graphs $G_a(V_a, E_a)$ and $G_b(V_b, E_b)$ is

$$D_{QJS}(G_a, G_b) = H_N\left(\frac{\rho_{G;a} + \sigma_{G;b}}{2}\right) - \frac{1}{2}H_N(\rho_{G;a}) - \frac{1}{2}H_N(\sigma_{G;a}) \tag{13}$$

Then, we define the quantum Jensen-Shannon kernel $k_{QJS}(G_a, G_b)$ between $G_a(V_a, E_a)$ and $G_b(V_b, E_b)$ as

$$k_{QJS}(G_a, G_b) = exp(\lambda D_{QJS}(G_a, G_b))$$
$$= exp(\lambda H_N\left(\frac{\rho_{G;a} + \sigma_{G;b}}{2}\right) - \frac{1}{2}\lambda H_N(\rho_{G;a}) - \frac{1}{2}\lambda H_N(\sigma_{G;a})) \tag{14}$$

where λ is a decay factor which satisfies $0 < \lambda < 1$, and $H_N(.)$ is the von Neumann entropy of the density operator associated to the graph. Here λ is used to ensure that the large values do not tend to dominant the kernel value. In particular, in this work we use $\lambda = 0.1$.

Lemma. *The quantum Jensen-Shannon graph kernel is positive definite* **pd**.
Proof. This follows the definitions in [6,8,14]. In [14], a diffusion kernel $k_s = \exp(\lambda s(G_p, G_q))$ associated with any symmetric similarity measure $s(G_p, G_q)$ has been proven to be **pd**. Since the quantum Jensen-Shannon divergence between a pair of density operators is symmetric [6,8], then the proposed quantum Jensen-Shannon graph kernel is **pd**.

Note that, a positive definite graph kernel is often called a *valid kernel*. Clearly, the positive definiteness constraint restricts the class of valid graph kernels which can be defined by different similarity measures on graphs. However, the property of positive definiteness is crucial for the definition of kernel machines and turns out to be sufficiently strong to implicate a considerable number of theoretical properties associated with graph kernels [15].

3.3 Algorithmic Complexity

The computational complexity of the proposed quantum Jensen-Shannon kernel of graphs depends on several factors. These factors include 1) the computation of the initial state of the continuous-time quantum walk, 2) the simulation of the continuous-time quantum walk evolution at each time step $t = 1, 2, \ldots, T$, 3) the construction of the density operator associated with the graph, and 4) the computation of the von Neumann entropy of the density operator. It is clear that these operations are all dependent on the graph size, and that the complexity is highly influenced by the termination time T of the continuous-time quantum walk evolution. For example, we can easily show that the cost of computing the von Neumann entropy of the density operator is $O(|V|^2)$, where $|V|$ is the number of vertices of the graph. However, it's easy to see that the overall complexity is dominated by that of computing the eigendecomposition of the graph Laplacian, which is cubic in the number of vertices of the graph, i.e. $O(|V|^3)$. This algorithmic complexity analysis reveals that our quantum Jensen-Shannon kernel between graphs can be computed in polynomial time.

4 Experimental Evaluations

In this section, we demonstrate the performance of our quantum Jensen-Shannon kernel, and then compare it to several state of the art graph based learning methods on three standard graph datasets abstracted from bioinformatics. These datasets include: MUTAG, ENZYMES and PPIs. The MUTAG dataset contains graphs representing 188 chemical compounds to predict mutagenicity. The ENZYMES dataset contains graphs representing protein tertiary structures consisting of 600 enzymes from the BRENDA enzyme database. The PPIs dataset consists of protein-protein interaction networks (PPIs). Here we select Proteobacteria40 PPIs and Acidobacteria46 PPIs as the testing graphs. Details about the datasets are shown in Table.1 and [5,16].

Table 1. Information of the Graph based Datasets

Datasets	MUTAG	ENZYMES	PPIs
Maximum # vertices	28	126	232
Minimum # vertices	10	2	3
Average # vertices	17.93	32.63	109.60
# testing graphs	188	600	86

4.1 Von Neumann Entropy Evaluation

We commence by exploring the relationship between the von Neumann entropy of a graph and its corresponding increasing time t. In our experiments, we utilize the testing graphs in the MUTAG, ENZYMES and PPIs datasets. For each graph, we let the continuous-time quantum walk evolve until a maximum time t, where we vary t from 1 to 50. For each time t we compute the density operator associated with the graph using Eq.(8). Then the von Neumann entropy of the graph at each time t can be computed from its corresponding density operator. The experimental results are shown in Fig.1. The left, middle and right subfigures of Fig.1 show the results of the evaluation on the MUTAG, Enzymes and PPIs datasets separately. The x-axis shows the time t which is from 1 to 50, and the y-axis shows the mean value of the von Neumann entropies of graphs belonging to the same class at each time t. Here the different lines represent the entropies of different classes of graphs separately. This evaluation suggests that the von Neumann entropies of different classes of graphs can be divided well, and tend to be stable with increasing time t.

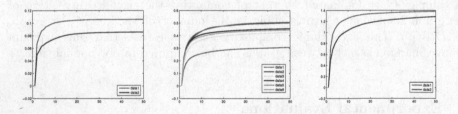

Fig. 1. Evaluations on the von Neumann Entropy for increasing time t

4.2 Experiments on Standard Graph Datasets from Bioinformatics

Experimental Setup. We evaluate the performance of our quantum Jensen-Shannon kernel on the graph datasets abstracted from bioinformatics databases, and then compare it with several alternative state of the art graph learning methods. These methods include 1) the Weisfeiler-Lehman subtree kernel [5], 2) the shortest path graph kernel [4], 3) the Shannon entropy associated with the information functionals FV and FP [17], and 4) the Ihara zeta function on graphs [18]. Based on the evaluation in Section 4.2, here we set time t from 1 to 50. For the kernel methods, we compute the kernel matrix of each graph kernel on each dataset, we then apply the PCA [19] on the kernel matrix to embed the graphs into principle component space as feature vectors. For other methods, we compute the characteristics values of graphs on each dataset. We perform 10-fold cross-validation using the Support Vector Machine (SVM) Classification associated with the Sequential Minimal Optimization (SMO) [20] on the graph feature vectors or characteristics values to evaluate the performance of our kernel and that of the alternative methods. We use nine samples for training and one for testing. All the SMO-SVMs and their parameters were performed and optimized on a Weka workbench [21]. We report the average classification accuracies of each method in Table 2.

Results. a) On the MUTAG dataset, the accuracies of all the methods are similar, our kernel overcomes or is competitive to the alternatives. b) On the ENZYMES dataset, the accuracy of our kernel is obviously lower than that of the Weisfeiler-Lehman subtree kernel, but is competitive to that of the shortest path graph kernel and outperforms that of others. c) On the PPIs dataset, the accuracy of our kernel is obviously higher than that of the alternatives. As a whole, our kernel outperforms or is competitive to the alternatives, only the Weisfeiler-Lehman subtree kernel and the shortest path graph kernel are competitive to our kernel on the ENZYMES dataset.

Table 2. Accuracies Comparisons on Graph Datasets abstracted from Bioninformatics

Datasets	MUTAG	ENZYMES	PPIs
quantum Jensen-Shannon kernel	84.04%	32.16%	76.20%
Weisfeiler-Lehman subtree kernel	84.57%	38.50%	76.16%
shortest path graph kernel	85.29%	31.16%	78.45%
Shannon entropy with FV	84.57%	24.17%	70.93%
Shannon entropy with FP	84.57%	24.17%	70.93%
Ihara zeta function on graphs	80.85%	32.00%	70.93%

5 Conclusion

In this paper, we developed a novel graph kernel by using the quantum Jensen-Shannon divergence and the continuous-time quantum walk on graphs. Given a graph, we evolved a continuous-time quantum walk on its structure and we showed how to associate a mixed quantum state to the graph and how to compute the von Neumann entropy of the corresponding density operator. With the von Neumann entropies to hand, the quantum Jensen-Shannon kernel between a pair of graphs was defined as a function of the quantum Jensen-Shannon divergence between the corresponding density operators. Finally, we used the Principle Component Analysis (PCA) on the kernel matrix to embed the graphs into a feature space where we performed the classification task. Experiments on several standard datasets demonstrate the effectiveness of the proposed graph kernel. Our future work is to extend the quantum graph kernel to a quantum hypergraph kernel. In [22], we have developed a hypergraph kernel by using the classical Jensen-Shannon divergence. In [23], we have explored the use of discrete-time quantum walks on a directed line graph which can be generated by transforming a hypergraph. It would thus be interesting to extend these works by using the quantum Jensen-Shannon divergence to compare the quantum walks of hypergraphs based on their directed line graphs.

References

1. Schölkopf, B., Smola, A.: Learning with Kernels. MIT Press (2002)
2. Haussler, D.: Convolution kernels on discrete structures. Technical Report UCS-CRL-99-10, Santa Cruz, CA, USA (1999)
3. Kashima, H., Tsuda, K., Inokuchi, A.: Marginalized kernels between labeled graphs. In: Proceedings of International Conference on Machine Learning, pp. 321–328 (2003)
4. Borgwardt, K.M., Kriegel, H.P.: Shortest-path kernels on graphs. In: Proceedings of the IEEE International Conference on Data Mining, pp. 74–81 (2005)
5. Shervashidze, N., Schweitzer, P., van Leeuwen, E.J., Mehlhorn, K., Borgwardt, K.M.: Weisfeiler-lehman graph kernels. Journal of Machine Learning Research 1, 1–48 (2010)
6. Lamberti, P., Majtey, A., Borras, A., Casas, M., Plastino, A.: Metric character of the quantum jensen-shannon divergence. Physical Review A 77, 052311 (2008)
7. Bai, L., Hancock, E.R.: Graph kernels from the jensen-shannon divergence. Journal of Mathematical Imaging and Vision (to appear)
8. Majtey, A., Lamberti, P., Prato, D.: Jensen-shannon divergence as a measure of distinguishability between mixed quantum states. Physical Review A 72, 052310 (2005)
9. Farhi, E., Gutmann, S.: Quantum computation and decision trees. Physical Review A 58, 915 (1998)
10. Dirac, P.: The Principles of Quantum Mechanics, 4th edn. Oxford Science Publications (1958)
11. Kempe, J.: Quantum random walks: an introductory overview. Contemporary Physics 44, 307–327 (2003)
12. Nielsen, M., Chuang, I.: Quantum computation and quantum information. Cambridge university press (2010)
13. Martins, A.F., Smith, N.A., Xing, E.P., Aguiar, P.M., Figueiredo, M.A.: Nonextensive information theoretic kernels on measures. Journal of Machine Learning Research 10, 935–975 (2009)
14. Konder, R., Lafferty, J.: Diffusion kernels on graphs and other discrete input spaces. In: Proceedings of International Conference on Machine Learning, pp. 315–322 (2002)
15. Neuhaus, M., Bunke, H.: Bridging the gap between graph edit distance and kernel machines. World Scientific (2007)
16. Escolano, F., Hancock, E.R., Lozano, M.A.: Heat diffusion: Thermodynamic depth complexity of networks. Physical Review E 85, 036206 (2012)
17. Dehmer, M.: Information processing in complex networks: Graph entropy and information functionals. Applied Mathematics and Computation 201, 82–94 (2008)

18. Ren, P., Wilson, R.C., Hancock, E.R.: Graph characterization via ihara coefficients. IEEE Transactions on Neural Networks 22, 233–245 (2011)
19. Schölkopf, B., Smola, A.J., Müller, K.R.: Kernel principal component analysis. In: Proceedings of International Conference on Artificial Neural Networks, pp. 583–588 (1997)
20. Platt, J.C.: Fast training of support vector machines using sequential minimal optimization. In: Schölkopf, B., Burges, C.J.C., Smola, A.J. (eds.) Advances in Kernel Methods, pp. 185–208 (1999)
21. Witten, I.H., Frank, E., Hall, M.A.: Data Mining: Practical Machine Learning Tools and Techniques. Morgan Kaufmann (2011)
22. Bai, L., Hancock, E.R., Ren, P.: A jensen-shannon kernel for hypergraphs. In: Gimel'farb, G., Hancock, E., Imiya, A., Kuijper, A., Kudo, M., Omachi, S., Windeatt, T., Yamada, K. (eds.) SSPR & SPR 2012. LNCS, vol. 7626, pp. 181–189. Springer, Heidelberg (2012)
23. Ren, P., Aleksic, T., Emms, D., Wilson, R., Hancock, E.: Quantum walks, ihara zeta functions and cospectrality in regular graphs. Quantum Information Processing 10, 405–417 (2011)

Treelet Kernel Incorporating Chiral Information

Pierre-Anthony Grenier[1], Luc Brun[1], and Didier Villemin[2]

[1] GREYC UMR CNRS 6072, Caen, France
[2] LCMT UMR CNRS 6507, Caen, France
{pierre-anthony.grenier,didier.villemin}@ensicaen.fr,
luc.brun@greyc.ensicaen.fr

Abstract. Molecules being often described using a graph representation, graph kernels provide an interesting framework which allows to combine machine learning and graph theory in order to predict molecule's properties. However, some of these properties are induced both by relationships between the atoms of a molecule and by constraints on the relative positioning of these atoms. Graph kernels based solely on the graph representation of a molecule do not encode this relative positioning of atoms and are consequently unable to predict accurately some molecule's properties. This paper presents a new method which incorporates spatial constraints into the graph kernel framework in order to overcome this limitation.

Keywords: Graph kernel, Chemoinformatics, Chirality.

1 Introduction

A molecular graph $G = (V, E, \mu, \nu)$ is a description of a molecule by a graph where the unlabeled graph (V, E) encodes the structure of the molecule, each vertex encoding an atom and each edge a bond between two atoms, μ associates to each vertex a label encoding the nature of the atom (carbon, oxygen, ...) and ν associates to each edge a type of bond (single, double, triple or aromatic). Several graph kernels [3,1] based on this representation have been proposed in order to predict molecule's properties. However, some molecules may have a same molecular formula, a same molecular graph but a different relative positioning of their atoms inducing different properties. Such molecules are said to be stereoisomers. However, usual graph kernels based on the molecular graph representation are not able to capture any dissimilarity between such molecules. From a more local point of view, an atom is called stereocenter if a permutation of the positions of two atoms belonging to its neighborhood produces a different stereoisomer. In the same way, two connected atoms form a stereocenter if a permutation of the positions of two atoms belonging to the union of their neighborhoods produces a different stereoisomer. According to chemical experts, stereoisomerism is represented to 98% by the geometrical isomerism of double connection and the asymmetry of carbons. We thus focus the remaining of this paper on those case.

W.G. Kropatsch et al. (Eds.): GbRPR 2013, LNCS 7877, pp. 132–141, 2013.
© Springer-Verlag Berlin Heidelberg 2013

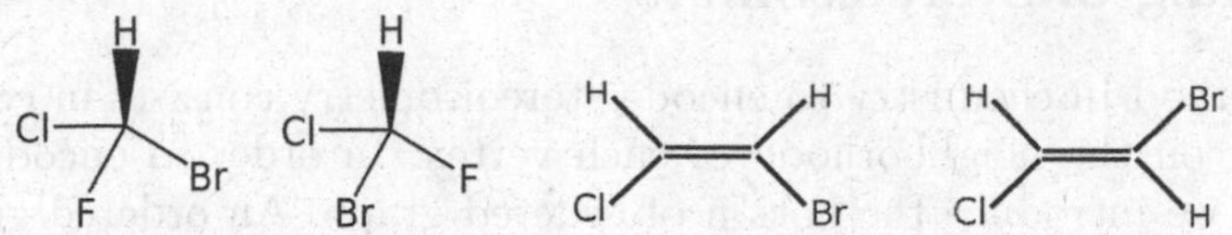

(a) Asymmetric carbon   (b) Double bond

Fig. 1. Two types of stereocenters

In order to get an intuition of stereoisomerism, let us consider an acyclic molecular graph rooted on an atom of carbon with four neighbors, each neighbor being associated to a different subtree. Such an atom, called an asymmetric carbon, is a stereocenter and has two different spatial configurations of its neighbors encoded by a same molecular graph (Fig. 1(a)). Using molecule represented in Fig. 1(a), one configuration corresponds to the case where the three atoms (Cl,Br,F) considered from the atom H are encountered in this order when turning clockwise around the central carbon atom. The alternative stereoisomer corresponds to the case where this sequence of atoms is encountered counter-clockwise when considered from the same position. Two carbons, connected by a double bond, can also define stereoisomers (Fig 1(b)). Indeed, on the left side of Fig.1(b) both hydrogen atoms are located on the same side of the double bond while they are located on opposite sides on the stereoisomer represented on the right. In this case both carbon atoms of the double bond correspond to a stereocenter.

Method described in [2] includes information about the spatial configuration of atoms within the tree-pattern kernel [3]. However, this method only considers the direct neighbors of a stereocenter while, as shown by Fig. 2, the difference between two subtrees of a stereocenter may not be located on the root of the subtrees. In this last case [2] considers as identical two different stereocenters.

In this paper we propose a method to incorporate the spatial configuration of atoms within a graph kernel based on a subtree enumeration [1]. This method remains valid even when the spatial configuration is not encoded in the direct neighborhood of a stereocenter. In Section 2, we define a graph encoding of stereoisomers and we introduce chiral vertices as vertices encoding stereocenters. Next, in Section 3, we restrict our attention to acyclic molecules. Such a restriction allows us to efficiently characterise a chiral vertex by a rooted tree. In Section 4, we define the smallest tree characterizing a chiral vertex and use this information to design a graph kernel between chiral molecules. Finally, we demonstrate the validity of our kernel through experiments in Section 5.

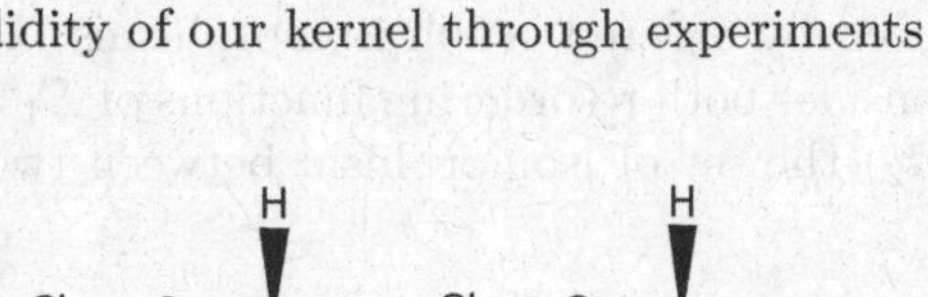

Fig. 2. Asymmetric carbons with identical neighborhood

2   Encoding of Stereoisomers

An usual method in chemistry to encode stereoisometry consists in considering a fixed order on the neighborhood of each vertex. In order to encode such an information, we introduce the notion of ordered graph. An ordered graph $G = (V, E, \mu, \nu, ord)$ is a molecular graph $\hat{G} = (V, E, \mu, \nu)$ together with a function ord which maps each vertex to an ordered list of its neighbors:

$$ord \begin{cases} V \to & V^* \\ v \to v_1 \dots v_n \end{cases} \tag{1}$$

where $V(v) = \{v_1, \dots, v_n\}$ denotes the neighborhood of v.

Two ordered graphs $G_1 = (V_1, E_1, \mu_1, \nu_1, ord_1)$ and $G_2 = (V_2, E_2, \mu_2, \nu_2, ord_2)$ are said to be isomorphic $G_1 \underset{o}{\simeq} G_2$ iff there is an isomorphism between both graphs which respects the order on the neighborhoods:

$$\exists f \in \text{Isom}(\hat{G}_1, \hat{G}_2) \; s.t. \; \forall v \in V_1 \; ord_1(v) = v_1 \dots v_n, \; ord_2(f(v)) = f(v_1) \dots f(v_n) \tag{2}$$

Note that, the ordered graph isomorphism induces an equivalence relationship as well as the usual graph isomorphism.

For exemple, in Fig. 1(a) the ordered list H,Cl,Br,F for the central carbon represents the molecule to the left (and H,Cl,F,Br represents its stereoisomer). But if we consider the molecule from the Cl atom, the list Cl,H,F,Br is a valid alternative encoding of the molecule. So, a spatial configuration of atoms within a neighborhood must be encoded by several equivalent orders. We thus introduce the notion of partially ordered graph which encodes all equivalent orderings of an ordered graph. A partially ordered graph (G, Σ) is an ordered graph G with a set of re-ordering functions Σ where $\sigma \in \Sigma$ associates to each vertex v a permutation on $\{1, \dots, |V(v)|\}$. Let $G = (V, E, \mu, \nu, ord)$ be an ordered graph, $\sigma(G) = (V, E, \mu, \nu, ord_\sigma)$ is defined as the application of σ on each ordered neighborhood of G:

$$\forall v \in V \; s.t. \; ord(v) = v_1, \dots, v_n, \; ord_\sigma(v) = v_{\varphi(1)}, \dots, v_{\varphi(n)} \text{ with } \varphi = \sigma(v). \tag{3}$$

Two partially ordered graphs (G_1, Σ_1) and (G_2, Σ_2) are said to be isomorphic iff:

$$G_1 \underset{po}{\simeq} G_2 \Leftrightarrow \begin{cases} \forall \sigma_1 \in \Sigma_1, \exists \sigma_2 \in \Sigma_2 \, | \, \sigma_1(G_1) \underset{o}{\simeq} \sigma_2(G_2) \\ \forall \sigma_2 \in \Sigma_2, \exists \sigma_1 \in \Sigma_1 \, | \, \sigma_1(G_1) \underset{o}{\simeq} \sigma_2(G_2) \end{cases} \tag{4}$$

The relationship induced by partially ordered isomorphisms is reflexive and transitive as the one induced by ordered graph isomorphisms. This relation is also symmetric since we consider both re-ordering functions of Σ_1 and Σ_2. We denote by IsomOrderP(G_1, G_2) the set of isomorphism between two partially ordered graph G_1 and G_2.

2.1   Partially Ordered Graph Encoding of a Molecule

The partial ordered graph of a molecule is defined by first defining its molecular graph $G_{unordered} = (V, E, \mu, \nu)$. Let us denote V_{C_1} the subset of V containing all atoms of carbon with four neighbors: $V_{C_1} = \{v \in V \mid \mu(v) = \text{'}C\text{'} \text{ and } |V(v)| = 4\}$. The subset of V containing all atoms of carbon which share a double bond with another carbon is noted V_{C_2}: $V_{C_2} = \{v \in V \mid \exists e(v,w) \in E, \nu(e) = 2, |V(v)| = |V(w)| = 3 \text{ and } \mu(v) = \mu(w) = \text{'}C\text{'}\}$. For each $v \in V_{C_2}$ we denote $w = n_=(v)$ the other carbon of its double bond. In order to encode spatial configurations, let us define an ordered graph $G_{Ordered}$ from $G_{unordered}$. Each vertex $v \in V - V_{C_1} - V_{C_2}$ does not require any encoding of the configuration of its neighborhood. The ordered list of its neighbors is thus set randomly. In order to set an order on the neighborhood of a vertex $v \in V_{C_1}$ we set randomly one of its neighbor v_1 at the first position. The three other neighbors of v are ordered in a way such that if we look at v from v_1, the three remaining neighbors are ordered clockwise (Section 1). One of the three orders (defined up to circular permutations) fulfilling this condition is chosen randomly (Fig. 3(a)). Finally, let us consider a vertex $v \in V_{C_2}$, with $n_=(v) = w$, $V(v) = \{w, a, b\}$ and $V(w) = \{v, c, d\}$. The order on neighborhoods of v and w are set as $ord(v) = w, a, b$ and $ord(w) = v, c, d$, whereby a, b, c, d are traversed clockwise when turning around the double bond for a given plane embedding. We add to this graph the set of re-ordering function Σ containing all the re-ordering functions σ such that: for each v in V_{C_1}, $\sigma(v)$ corresponds to an even number of transpositions on $\{1, \ldots, |V(v)|\}$ and for each v in V_{C_2}, with $n_=(v) = w$, $\sigma(v)$ and $\sigma(w)$ correspond to a same number of transpositions (Fig. 3(b)). Indeed, an additional transposition on one of the atoms of a double bond, would correspond to a permutation of the relative positioning of its neighbors hence encoding a different stereoisomer (Section 1).

Remark 1. Using the above construction scheme, the re-ordering functions of any partially order graph encoding a molecule satisfies the following properties:

- Given a sequence of neighbors of each vertex, we can always find a re-ordering such that the ordered list of each vertex starts by its selected neighbor.
- For any re-ordering functions, the permutations associated to two adjacent carbons belonging to V_{C_2} may be decomposed into a same number of transpositions.

A partially ordered graph encodes the spatial configuration of atoms within the neighborhood of each of its vertex. Let us now define a stereocenter (also called a chiral vertex).

(a) Element of V_{C_1} (b) Two elements of V_{C_2}

Fig. 3. Example of elements of V_{C_1} and of V_{C_2} with their ordered list (top) and the ordered lists obtained using two permutations $\sigma \in \Sigma$ and $\sigma' \in \Sigma$

Definition 1. Let $G = (V, E, \mu, \nu, ord, \Sigma)$ a partially ordered graph. A vertex $v \in V$ of degree n is a chiral vertex iff:

$$\forall (i,j) \in \{1, \ldots, n\}^2, \nexists f \in \text{IsomOrderP}(G, \tau_{i,j}(G)) \text{ with } f(v) = v$$

where $\tau_{i,j}$ is a re-ordering function equals to the identity on all vertices except v for which it permutes the vertices of index i and j in $ord(v)$.

In other words, a vertex is chiral if any permutation of its neighbors produces a different partially ordered graph (called a different stereoisomer within the chemistry framework).

3   Isomorphism between Labeled Partially Ordered Tree

Let us now restrict our attention to acyclic graphs in order to obtain a more efficient calculus of isomorphisms between partially ordered graphs. Given a rooted tree, the father of each node v is denoted p_v. We define an ordered rooted tree $T = (V, E, \mu, \nu, ord)$ as a rooted tree $\hat{T} = (V, E, \mu, \nu)$ with a function ord mapping each vertex to an ordered list of its children. Like the isomorphism between ordered graph presented in Sec. 2, there is an isomorphism between two labeled ordered tree $T_1 = (V_1, E_1, \mu_1, \nu_1, ord_1)$ and $T_2 = (V_2, E_2, \mu_2, \nu_2, ord_2)$ if there is an isomorphism between both trees which complies with their order :

$$\exists f \in \text{Isom}(\hat{T}_1, \hat{T}_2) \ s.t. \ \forall v \in V_1 \ ord_1(v) = v_1 \ldots v_n, \ ord_2(f(v)) = f(v_1) \ldots f(v_n) \tag{5}$$

where $\{v_1, \ldots, v_n\}$ denotes the children of v. Note that, an isomorphism between ordered tree maps roots of each tree one on the other and preserves father-child relationships.

Following [4], we associate to each ordered rooted tree, a unique depth-first string. This string is based on the sequence of node and edge labels obtained by traversing the tree in a depth-first order. As shown by [4](Lemma 2.2), two isomorphic ordered trees have the same depth-first string encoding and conversely.

Using the same approach than for partially ordered graphs, a partially ordered rooted tree (T, Σ) is an ordered rooted tree T associated to a set of re-ordering functions Σ on the children of each vertex. To define a partially ordered tree (T, Σ_T), from an acyclic partially ordered graph (G, Σ_G) encoding a molecule, we have to define a root and for each vertex an order and a set of permutations on its children encoding equivalent orders. Since the root has no parent, its children correspond to its set of neighbors and we set $ord_G(r) = ord_T(r)$. For any other vertex v, the list of its children is the list of its neighbors minus p_v. To define an order for each v in T, we apply one of the re-ordering function $\sigma \in \Sigma_G$ which puts p_v in the first position (Remark 1). The set of re-ordering functions Σ_T is defined by considering all re-ordering functions $\sigma \in \Sigma_G$ which, for each $v \in V$, keep p_v in the first position of the ordered list of v.

In order to define a unique code for each partially ordered tree we define, as in [4], the depth-first canonical form (DFCF*) of a partially ordered tree, as the ordered tree that gives the minimal depth-first string encoding among all possible ordered trees $\sigma(T)$ obtained by applying $\sigma \in \Sigma$ on T. The depth-first string encoding of the DFCF* is called the depth-first canonical string (DFCS*) of a partially ordered tree. Since, two isomorphic ordered trees have the same depth-first string encoding, two partially ordered trees are isomorphic if their DFCS* are identical.

Given a unique code associated to a partially ordered rooted tree, the chirality of a vertex may be efficiently tested if one can transpose definition 1 to partially ordered rooted trees:

Proposition 1. Let $T = (V, E, \mu, \nu, ord, \Sigma)$ be a partially ordered tree rooted in r. r is a chiral vertex if $\forall (i, j) \in \{1, \ldots, |V(r)|\}^2$, $T \underset{po}{\not\simeq} \tau_{i,j}(T)$,

where $\tau_{i,j}$ is a re-ordering function equals to the identity on any vertex but r where it permutes children of index i and j in the ordered list of r.

Proof. Using acyclic graphs, an isomorphism between partially ordered rooted trees corresponds to an isomorphism between partially ordered graphs with an additional constraint on the mapping of both roots. If we can find an isomorphism between T and $\tau_{i,j}(T)$ such an isomorphism f satisfies $f(r) = r$ and also corresponds to an isomorphism between partially ordered graphs. Conditions of Def. 1 are thus violated and r is not chiral. The reverse implication may be demonstrated using the same type of reasoning. □

A partially ordered tree (T, Σ) can have two isomorphic subtrees whose roots have the same parent. In that case a permutation exchanging those subtrees on the DFCF* leads to an isomorphic ordered tree. In such a case, the root of these subtrees are said to be equivalent:

$$v_i \sim v_j \Leftrightarrow \begin{cases} \exists v \in V \ s.t. \ p_{v_i} = p_{v_j} = v \text{ and} \\ \exists \sigma \in \Sigma \mid \varphi(i) = j, \ \text{DFCF*}(\sigma(T)) \underset{o}{\simeq} \text{DFCF*}(T) \text{ with } \varphi = \sigma(v) \end{cases}$$

(6)

Since all equivalent nodes are the children of a same parent, the representative of each class is defined as the vertex with the minimal index within the ordered list of children of its parent:

$$\forall i \in \{1, \ldots, n\} \ rep(v_i) = min\{j \mid v_j \sim v_i\}. \tag{7}$$

4   From a Global to a Local Characterization of Chirality

Proposition 1 provides a global characterization of chirality. However, such a proposition does not allow to characterize the minimal subgraph of a molecule which induces the chiral property of a vertex. Using acyclic graphs, such a minimal subgraph corresponds to the smallest partially rooted tree, rooted on a chiral vertex v which allows to characterize the chirality of v using Proposition 1.

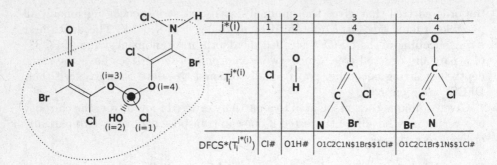

Fig. 4. Left: An asymmetric carbon ⊙ with its minimal chiral subtree (*surrounded by a dotted line*). Right: minimal subtrees rooted on its children.

4.1 Minimal Chiral Subtree of an Asymmetric Carbon

Let v be a chiral vertex representing an asymmetric carbon. We denote its neighbors $v_1, \ldots, v_4$. We consider the partially ordered tree (T, Σ) rooted in v and described in Sec. 3. We note $T_1, \ldots, T_4$ the subtrees of T rooted on the children of v. For any $i \in \{1, 2, 3, 4\}$ we denote T_i^j the subtree of T_i composed of all nodes with a depth lower than j. According to Proposition 1, the chirality of v may be characterized from its subtrees T_i^j iff all pairs of subtrees are non isomorphic. Indeed, in such a case no transposition of two subtrees T_i^j and $T_k^{j'}$ can induce an isomorphic partially ordered rooted tree. Therefore for each $i \in \{1, 2, 3, 4\}$, we define the minimal subtree associated to v_i as $T_i^{j^*(i)}$ with $j^*(i) = \min\{j \mid \forall k \in \{1, \ldots, 4\} - \{i\}, \ T_i^j \not\simeq T_k^j\}$. For exemple in Fig. 4, the root of T_1 is a Cl atom and the root of each other T_i is an oxygen atom, thus $j^*(1) = 1$. The minimal chiral subtree of v is the subtree of T rooted on v, where v has for children $T_1^{j^*(1)}, \ldots, T_4^{j^*(4)}$. The asymmetric carbon is then represented by the DFCS* of this tree.

To find $j^*(i)$, we increase j for each T_i^j until $T_i^j \not\simeq T_k^j$ for each $k \in \{1, \ldots, 4\}$, $k \neq i$. At each iteration we compute the DFCS* of each tree. Therefore the calculus of the minimal chiral subtree of v is performed in $\mathcal{O}((\max_i |T_i^{j^*(i)}|)^2)$ which is bounded by $\mathcal{O}(|V|^2)$.

4.2 Minimal Chiral Subtree of Double Bond

Let us consider a double bond $e = (v_a, v_b)$ and let us denote by v_a^1 and v_a^2 the two remaining neighbors of v_a. Considering the partially ordered tree T rooted on v_a, v_a is chiral only if the subtrees rooted on the children of v_a are not isomorphic (Proposition 1). This implies that the two subtrees rooted on v_a^1 and v_a^2 are not isomorphic. This necessary condition is however not sufficient. Indeed if the subtrees rooted on the remaining neighbors v_b^1 and v_b^2 of v_b are isomorphic, then one can apply a re-ordering function $\sigma \in \Sigma$ on T which simultaneously

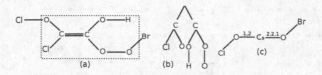

Fig. 5. A double bond (*a*), its minimal chiral subtree (*b*) and its contraction (*c*)

permutes the subtrees rooted on v_a^1 and v_a^2 and the subtrees rooted on v_b^1 and v_b^2 (Remark 1). The resulting rooted tree $\sigma(T)$ is isomorphic to T by definition but also to $\tau(T)$, where τ permutes only vertices v_a^1 and v_a^2 in the ordered list of children of v_a. In such a case, v_a is non chiral (Proposition 1). Therefore, the non chirality of v_b induces the non chirality of v_a and conversely.

Hence v_a and v_b are chirals, only if the two following conditions are satisfied: subtrees rooted on v_a^1 and v_a^2 are non isomorphic and subtrees rooted on v_b^1 and v_b^2 are also non isomorphic.

In order to encode this constraints, we define as in Section 4.1 the minimal non isomorphic subtrees rooted on v_a^1 (T_a^1) and v_a^2 (T_a^2) together with the minimal non isomorphic subtrees rooted on v_b^1 (T_b^1) and v_b^2(T_b^2). We denote by T_a and T_b the two partially ordered rooted trees rooted on v_a and v_b. The subtrees of these two roots being respectively (T_a^1, T_a^2) and (T_b^1, T_b^2).

The tree encoding the chirality of the double bond is then defined as a partially ordered rooted tree, whose root corresponds to a virtual vertex (not corresponding to any atom) connected to the two subtrees T_a and T_b. As for Sec. 4.1, the computation of the minimal chiral subtree is bounded by $\mathcal{O}(|V|^2)$. Fig. 5a represents a double bond between two carbon atoms with its minimal chiral subtree (Fig. 5b).

4.3   Graph Contraction

Using results in Section 4.1 and 4.2, each stereocenter may be associated to a minimal chiral subtree and a DFCS* code representing it (Section 3). However, properties of a molecule are both determined by its set of minimal chiral subtrees and by relationships between these trees and the remaining part of the molecule. In order to obtain a local characterization of such relationships, we propose to contract the minimal chiral subtree of each stereocenters.

Let us consider a stereocenter s and its minimal chiral subtree $(T = (V_T, E_T), \Sigma)$ associated to a DFCS* code c_s. We define for this tree a set of connection vertices $V_{\text{con}} = \{v \in \text{Leaf}(T) \mid d(v) > 1\}$ and a set of edges to contract $K_T = E_T - E_{\text{con}}$ where $E_{\text{con}} = \{(v, p_v) \in V_{\text{con}} \times V_T\}$. The contraction of K_T creates a new graph $G_s = (V_s, E_s)$, with a contracted node n_s labeled by c_s and $V_s = V - (V_T - V_{\text{con}}) \cup \{n_s\}$; $E_s = E - K_T$ (Fig. 5c).

Each edge of E_{con} connects an element l of V_{con} to n_s in G_s. The label of $e = (n_s, l)$ has to encode the position of l in the minimal chiral subtree. We thus consider the path connecting r to l in the minimal chiral subtree: $CP(l) = v_1, .., v_n$ where $v_1 = r$ and $v_n = l$. Let us denote i_j the index of v_j in the ordered list of children of p_{v_j}. The sequence $i_2 \ldots i_n$ defines a unique path

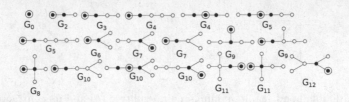

Fig. 6. The set of stereotreelet with n_s (⊙), elements of V_{con} (•), elements of $V - V_{con}$ (○)

in the chiral subtree associated to n_s. Such a sequence may thus be considered as a proper label of edge e. However as mentioned in Section 3, some paths may pass through equivalent subtrees and should thus be considered as equivalent. In order to encode such equivalence relationship we define the label of e as

$$\nu(e) = \overset{n}{\underset{i=2}{\bigodot}} rep(v_i) \text{ with } rep \text{ defined in Eq. 7.}$$

4.4   StereoTreelet

For each stereocenter s we have a graph G_s. The stereotreelets of G_s are defined as all subtrees of G_s whose size is lower than 6 and which include n_s. Since each neighbors v of n_s corresponds to a leaf of the minimal chiral tree of s, the edge (v, n_s) is already encoded within the code c_s of n_s. Consequently, we impose that each neighbor v of n_s in a stereotreelet must have at least another neighbor (different of n_s). This constraint induces the set of stereotreelets represented in Fig. 6. The set of stereotreelet $\mathcal{T}(G)$ of G is defined as the union of stereotreelets of each G_s.

When all stereotreelets of G have been enumerated, we compute its spectrum $s(G)$ which corresponds to a vector representing the treelet distribution. Each component of this vector is equal to the frequency of a given stereotreelet t: $s(G) = (f_t(G))_{t \in \mathcal{T}(G)}$ with $f_t(G) = |(t \subseteq G)|$. The kernel between two graphs G and G' is defined as a sum of kernels between the different number of treelets common to both graphs: $k(G, G') = \sum\limits_{t \in \mathcal{T}(G) \cap \mathcal{T}(G')} K(f_t(G), f_t(G'))$.

5   Experiments

We have tested our method on a dataset of acyclic chiral molecules [5] related to a regression problem. This dataset is composed of 90 molecules together with their optical rotations. In practice, we only select 35 molecules, since almost all molecules have only one stereocenter, and for 55 molecules this stereocenter is unique in the dataset. Such molecules correspond to a property represented only once in the dataset which can thus not be accurately predicted. The property to predict, the optical rotation, is connected with chirality and has a standard deviation of 38.25 for the 35 selected molecules.

For our experiment we use a leave-one-out cross-validation on the dataset to predict the optical rotation of each molecule. The predicted rotations are

Table 1. Optical rotation prediction for the acyclic chiral dataset

Method	Kernel Ridge Average Error	RMSE	Weighted Average Average Error	RMSE	Gram's matrix computations (s)
Random Kernel	31.7	39.5	32.0	39.3	0.03
KMean[6]	31.0	38.7	32.3	39.6	153.84
Treelet Kernel[1]	26.0	33.9	28.9	37.4	0.49
Stereotreelet Kernel	21.0	25.6	11.6	16.3	0.13

computed by using both kernel ridge regression and the weighted mean of known values using the similarity measure provided by the kernel $\hat{y} = \dfrac{\sum\limits_i k(G_i,G) \times y_i}{\sum\limits_i k(G_i,G)}$. We present in Table 1 the average errors, Root Mean Squared Errors (RMSE) and computation times of the Gram matrix for our method and the ones of [6,1] which do not take into account chirality. Results obtained by using a random Gram matrix are also shown.

Weighted mean provides much better results for our kernel since on this dataset each molecule has a non null similarity with a reduced number of molecules (less than 10). Such a reduced number of data do not allow kernel ridge regression to perform reliable prediction. Other methods provide similar results than those obtained using a random Gram matrix. These results are also comparable with the variance of the dataset. Such a result may be explained by the fact that optical rotation is connected to chirality which is not encoded by these kernels.

6   Conclusion

In this paper we proposed a graph kernel for chemoinformatics that considers the spatial constraints of atoms within molecules. Our experiments show promising results and our future work will consist to create larger datasets and to extend our method to graphs including cycles.

References

1. Gaüzère, B., Brun, L., Villemin, D.: Two new graphs kernels in chemoinformatics. Pattern Recognition Letters (April 2012)
2. Brown, J., Urata, T., Tamura, T., Arai, M., Kawabata, T., Akutsu, T.: Compound analysis via graph kernels incorporating chirality. J. Bioinform. Comp. Bio., S1:63–81 (2010)
3. Mahé, P., Vert, J.-P.: Graph kernels based on tree patterns for molecules. Machine Learning 75(1), 3–35 (2008)
4. Chi, Y., Yang, Y., Muntz, R.R.: Canonical forms for labeled trees and their applications in frequent subtree mining. KAIS 8(2), 203–234 (2005)
5. Zhu, H.J., Ren, J., Pittman Jr., C.U.: Matrix model to predict specific optical rotations of acyclic chiral molecules. Tetrahedron 2007 63, 2292–2314 (2007)
6. Suard, F., Rakotomamonjy, A., Bensrhair, A.: Kernel on bag of paths for measuring similarity of shapes. In: ESANN (2002)

A Novel Software Toolkit for Graph Edit Distance Computation

Kaspar Riesen[1], Sandro Emmenegger[1], and Horst Bunke[2]

[1] Institute for Information Systems, University of Applied Sciences and
Arts Northwestern Switzerland,
Riggenbachstrasse 16, CH-4600 Olten, Switzerland
{kaspar.riesen,sandro.emmenegger}@fhnw.ch
[2] Institute of Computer Science and Applied Mathematics, University of Bern,
Neubrückstrasse 10, CH-3012 Bern, Switzerland
bunke@iam.ch

Abstract. Graph edit distance is one of the most flexible mechanisms
for error-tolerant graph matching. Its key advantage is that edit distance
is applicable to unconstrained attributed graphs and can be tailored to
a wide variety of applications by means of specific edit cost functions.
The computational complexity of graph edit distance, however, is exponential in the number of nodes, which makes it feasible for small graphs
only. In recent years the authors of the present paper introduced several powerful approximations for fast suboptimal graph edit distance
computation. The contribution of the present work is a self standing
software tool integrating these suboptimal graph matching algorithms.
It is about being made publicly available. The idea of this software tool
is that the powerful and flexible algorithmic framework for graph edit
distance computation can easily be adapted to specific problem domains
via a versatile graphical user interface. The aim of the present paper is
twofold. First, it reviews the implemented approximation methods and
second, it thoroughly describes the features and application of the novel
graph matching software.

1 Introduction to Graph Edit Distance

Graph matching refers to the process of evaluating the structural similarity of
graphs. A large number of methods for graph matching have been proposed
in recent years [1–5]. Compared to other graph matching methods, graph edit
distance is very flexible. Due to its ability to cope with arbitrary structured
graphs with unconstrained label alphabets for both nodes and edges. Therefore,
graph edit distance has been used in the context of classification and clustering
tasks in diverse applications [6–8].

Given two graphs, the source graph g_1 and the target graph g_2. The basic idea
of graph edit distance is to transform g_1 into g_2 using some distortion operations.
A standard set of distortion operations is given by *insertions*, *deletions*, and
substitutions of both nodes and edges. We denote the substitution of two nodes

W.G. Kropatsch et al. (Eds.): GbRPR 2013, LNCS 7877, pp. 142–151, 2013.

u and v by $(u \to v)$, the deletion of node u by $(u \to \varepsilon)$, and the insertion of node v by $(\varepsilon \to v)$. For edges we use a similar notation. A sequence of edit operations $e_1, \ldots, e_k$ that transform g_1 completely into g_2 is called an *edit path* between g_1 and g_2.

Obviously, for every pair of graphs (g_1, g_2), there exist an infinite number of different edit paths transforming g_1 into g_2. Let $\Upsilon(g_1, g_2)$ denote the set of all possible edit paths between two graphs g_1 and g_2. To find the most suitable edit path out of $\Upsilon(g_1, g_2)$, one introduces a cost for each edit operation, measuring the strength of the corresponding operation. The idea of such a cost function is to define whether or not an edit operation represents a strong modification of the graph. Usually, the cost is defined with respect to the underlying node or edge labels, i.e. the cost $c(e)$ is a function depending on the edit operation e.

Clearly, between two similar graphs, there should exist an inexpensive edit path, representing low cost operations, while for dissimilar graphs an edit path with high cost is needed. Consequently, the *edit distance* of two graphs is defined by the minimum cost edit path between two graphs. Formally, the graph edit distance between g_1 and g_2 is defined by

$$d(g_1, g_2) = \min_{(e_1, \ldots, e_k) \in \Upsilon(g_1, g_2)} \sum_{i=1}^{k} c(e_i)$$

The possibility to parametrize graph edit distance by means of a cost function crucially amounts for the versatility of this particular dissimilarity model. That is, by means of graph edit distance it is possible to integrate domain specific knowledge about object similarity when defining the cost of the elementary edit operations. Thus, the concept of edit distance can be tailored to specific applications.

Traditionally, the computation of edit distance is carried out by means of a tree search algorithm which explores the space of all possible mappings of the nodes and edges of g_1 to the nodes and edges of g_2. Yet, a spate of other graph edit distance computation algorithms have been developed during the last years. The present paper introduces a flexible software package for various graph edit distance computation variants (including the traditional tree search algorithm). The graph edit distance software will be made publicly available soon under

http://www.fhnw.ch/wirtschaft/iwi/gmt

In Fig. 1 the main window of our novel graph matching software tool is shown. In ① in Fig. 1 the user of the framework is asked to define the *source graph set* $S = \{g_1, \ldots, g_n\}$ the *target graph set* $T = \{g'_1, \ldots, g'_m\}^1$, the folder where the individual graphs are locally stored (*graph folder*) and a *results folder* where the computed distance matrix $\mathbf{D} = (d(g_i, g_j))_{n \times m}$ is saved ($g_i \in S$ and $g_j \in T$). For more detailed and technical descriptions of the input formats of both *graph sets* and *graphs* as well as the output format of the distance matrix we refer to the above mentioned website. In ② in Fig. 1 the user defines whether or not to log meta information about the graphs being processed and the corresponding edit paths.

---

[1] Clearly, the source and target set might be the same sets in some applications.

The remainder of the present paper reviews different versions of graph edit distance available in our software tool and describes the user-defined parameters and options for graph edit distance computation in detail.

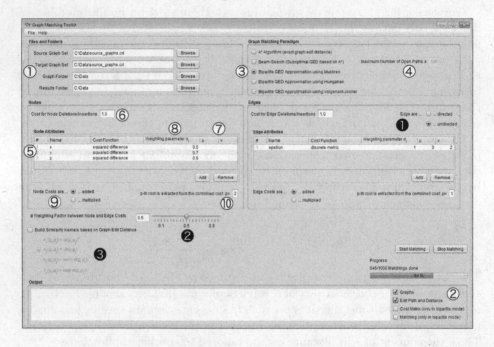

Fig. 1. The main window of our novel graph matching software tool

2   Graph Edit Distance Algorithms

In ③ in Fig. 1 the graph edit distance algorithm actually to be applied can be chosen by the user. The five available algorithms are briefly described in the next paragraphs.

A*-Algorithm with Bipartite Heuristic (exact graph edit distance). A widely used method for exact graph edit distance is based on the A* algorithm [9] which is a best-first search algorithm. The basic idea is to organize the underlying search space as an ordered tree. The root node of the search tree represents the starting point of our search procedure, inner nodes of the search tree correspond to partial solutions, and leaf nodes represent complete – not necessarily optimal – solutions. Such a search tree is constructed dynamically at runtime by iteratively creating successor nodes linked by edges to the currently considered node in the search tree. In order to determine the most promising node in the current search tree often a heuristic function is used. Formally, for a node p in the search tree, we use $g(p)$ to denote the cost of the optimal path from the root node to the current node p, and we use $h(p)$ for denoting the estimated cost from p to a leaf node. The sum $g(p) + h(p)$ gives the total cost assigned to an open node in the search tree. Given that the estimation of the future cost $h(p)$ is lower than, or equal to, the real cost, it is guaranteed that the algorithm finds an optimal edit path [9]. To solve the problem of estimating a lower bound $h(p)$ for the future costs, one can map the unprocessed nodes and edges of graph g_1 to the unprocessed nodes and edges of graph g_2 such that the resulting costs are minimal. In [10] it is proposed to use a fast bipartite assignment algorithm of the unprocessed nodes and edges of the two graphs as heuristic function $h(p)$. This specific heuristic function $h(p)$ described in [10] is actually implemented in our software framework.

Beam Search. The method described in the previous paragraph finds an optimal edit path between two graphs g_1 and g_2 and thus returns the exact graph edit distance $d(g_1, g_2)$. Unfortunately, the computational complexity of any exact graph edit distance algorithm is exponential in the number of nodes of the involved graphs (whether or not a heuristic function $h(p)$ is deployed to govern the tree traversal process). This means that the running time and space complexity may be huge even for reasonably small graphs[2]. In [11] the issue of efficient graph edit distance computation is addressed by simple variants of a standard A* algorithm. One method presented in [11] is based on the idea of *beam search*. Instead of expanding all successor nodes in the search tree, only a fixed number s of nodes to be processed are kept in the set of open nodes at all times. Whenever a new partial edit path is added, only the s partial edit paths p with the lowest costs $g(p) + h(p)$ are kept, and the remaining partial edit paths are removed. This means that not the full search space is explored, but only those nodes are expanded that belong to the most promising partial matches.

For similar graphs, it is clear that edit operations of an optimal path have low costs. Therefore if only the partial edit paths with lowest costs are considered, we will obtain an edit path that is nearly optimal, which will result in a suboptimal distance close to the exact distance. For dissimilar graphs, the suboptimal distance will remain large. Note that this method requires the user to define the maximum number of open paths s (cp. ④ in Fig. 1). The parameter s controls both the degree of suboptimality and the computation time of the procedure. That is, increasing the parameter s simultaneously augments the probability of finding the true graph edit distance and the running time.

Bipartite Graph Edit Distance using Assignment Algorithms. In [12–14] the authors of the present paper introduced a novel algorithmic framework which allows us to approximately compute edit distance in a substantially faster way than traditional methods. The proposed algorithms consider only local, rather than global, edge structure during the optimization process. The method is based on an (optimal) fast bipartite assignment procedure mapping nodes and their local structure of one graph to nodes and their local structure of another graph.

---

[2] In practice we are able to compute the edit distance of graphs typically containing 12 nodes at most.

In [12] the algorithmic framework first substitutes all nodes of the smaller graph and the nodes remaining in the larger graph are either deleted (if they belong to g_1) or inserted (if they belong to g_2). In [13] this idea is extended by allowing insertions or deletions to occur not only in the larger, but also in the smaller of the two graphs under consideration. To this end, for two graphs g_1 and g_2 to be matched with nodes $V_1 = \{u_1, \ldots, u_n\}$ and $V_2 = \{v_1, \ldots, v_m\}$, respectively, a cost matrix $\mathbf{C}$ is defined as follows:

$$\mathbf{C} = \begin{bmatrix} c_{11} & c_{12} & \cdots & c_{1m} & c_{1\varepsilon} & \infty & \cdots & \infty \\ c_{21} & c_{22} & \cdots & c_{2m} & \infty & c_{2\varepsilon} & \ddots & \vdots \\ \vdots & \vdots & \ddots & \vdots & \vdots & \ddots & \ddots & \infty \\ c_{n1} & c_{n2} & \cdots & c_{nm} & \infty & \cdots & \infty & c_{n\varepsilon} \\ c_{\varepsilon 1} & \infty & \cdots & \infty & 0 & 0 & \cdots & 0 \\ \infty & c_{\varepsilon 2} & \ddots & \vdots & 0 & 0 & \ddots & \vdots \\ \vdots & \ddots & \ddots & \infty & \vdots & \ddots & \ddots & 0 \\ \infty & \cdots & \infty & c_{\varepsilon m} & 0 & \cdots & 0 & 0 \end{bmatrix}$$

where c_{ij} denotes the cost of a node substitution $c(u_i \to v_j)$, $c_{i\varepsilon}$ denotes the cost of a node deletion $c(u_i \to \varepsilon)$, and $c_{\varepsilon j}$ denotes the cost of a node insertion $c(\varepsilon \to v_j)$.

Obviously, the left upper corner of the cost matrix represents the costs of all possible node substitutions, the diagonal of the right upper corner the costs of all possible node deletions, and the diagonal of the bottom left corner the costs of all possible node insertions. Note that each node can be deleted or inserted at most once. Therefore any non-diagonal element of the right-upper and left-lower part is set to ∞. The bottom right corner of the cost matrix is set to zero since substitutions of the form $(\varepsilon \to \varepsilon)$ should not cause any costs.

In the definition of cost matrix $\mathbf{C}$, to each entry c_{ij}, i.e. to each cost of a node substitution $c(u_i \to v_j)$, the minimum sum of edge edit operation costs, implied by node substitution $u_i \to v_j$, is added. That is, using a bipartite optimization procedure the cost of an optimal assignment of the adjacent edges of u_i and v_j is computed and added to entry c_{ij}. Clearly, to entry $c_{i\varepsilon}$ the cost of the deletion of all adjacent edges of u_i is added, and to the entry $c_{\varepsilon j}$ the cost of all insertions of the adjacent edges of v_j is added. Note that in ② in Fig. 1 one can define whether or not to log the cost matrix $\mathbf{C}$ in the output window.

On the basis of the quadratic cost matrix $\mathbf{C}$ any bipartite assignment algorithm can be executed. The result returned by bipartite optimization procedures applied to $\mathbf{C}$ corresponds to the minimum cost mapping of the nodes and their local edge structure of g_1 to the nodes and their local edge structure of g_2. In ② in Fig. 1 one can choose to log the optimal mapping of local structures found on matrix $\mathbf{C}$ in the output window. Given the optimal mapping between local structures, the edit operations on nodes and the implied edit operations of the edges can be inferred, and the accumulated costs of the individual edit operations on both nodes and edges can be computed. Note that assignment algorithms are not able to consider the global edge structure during the matching process. Hence, optimal matchings of nodes (considering the local edge structure) do not necessarily lead to an optimal (i.e. minimum cost) edit path. That is, this procedure leads to a suboptimal graph edit distance, which is equal to or greater than the exact edit distance.

In [12, 13] we make use of Munkres' algorithm [15] as basic bipartite optimization procedure. In [14] not only *Munkres' Algorithm* but also a modern version of the *Hungarian Algorithm* as well as the algorithm of *Volgenant and Jonker* [16] are incorporated to solve the assignment problem. In our software package all three assignment algorithms are implemented.

3   Defining the Cost Function

The definition of adequate and application-specific cost functions is a key task in edit distance based graph matching. The definition of the cost is usually depending on the underlying label alphabets for nodes and edges. In our algorithmic framework, the labels for both nodes and edges can be given by the set of integers $L = \{1, 2, 3, \ldots\}$, real numbers $L = \mathbb{R}$, a set of symbolic labels $L = \{\alpha, \beta, \gamma, \ldots\}$, strings defined over an alphabet V $L = \{V^*\}$, or an arbitrary combination of different labels. Unlabeled graphs are obtained as a special case by assigning the same symbolic label λ to all nodes and edges. In our software tool a single node can be labeled with up to five different node attributes which can be arbitrarily named by the user (cp. ⑤ in Fig. 1). In Fig. 2 (a), for instance, the nodes are labeled with three attributes x, y and z (the nodes represent points in a three-dimensional space $\mathbb{R}^3$). In Fig. 2 (b) the nodes are labeled with two attributes, viz. a symbolic attribute named *type* and a string named *sequence*. Note that the label alphabets are implicitly defined by the distance function to be applied on them (see next paragraph).

The first step in cost definition is to define a non-negative parameter representing the cost of a deletion or insertion $c(u \to \varepsilon)$ or $c(\varepsilon \to u)$, respectively, of an arbitrary node u (cp. ⑥ in Fig. 1). For the sake of symmetry, an identical cost for deletions and insertions has to be defined. Second, for each attribute a distance function for node substitutions has to be chosen by the user. Typically, the cost of a node substitution $(u \to v)$ is measured by means of some distance function $d : L \times L \to \mathbb{R}$ defined on the node label alphabet L. For now we assume that the nodes are labeled with a single attribute from alphabet L. The attribute values of u and v are $u.A \in L$ and $v.A \in L$, respectively. In our software tool four different distance functions can be defined on each node attribute. Note that the first two distance functions are applicable to numerical attributes only. The last distance function is applicable to string attributes:

1. *absolute value of difference*: $d(u.A, v.A) = |u.A - v.A|$
2. *squared difference*: $d(u.A, v.A) = (u.A - v.A)^2$
3. *discrete metric*: $d(u.A, v.A) = \begin{cases} \mu, & \text{if } u.A = v.A \\ \nu, & \text{else} \end{cases}$

    where μ, ν are non-negative real values ($\mu, \nu \in \mathbb{R}^+$) to be defined by the user (cp. ⑦ in Fig. 1).

4. *Levenshtein distance*: $d(u.A, v.A)$ = minimal number of single-character edit operations (deletions, insertions, substitutions) required to change string $u.A$ into string $v.A$, also known as *string edit distance* (*sed*).

Assuming that the nodes are labeled with $k > 1$ attributes, the i-th attribute values of u and v are $u.A_i \in L_i$ and $v.A_i \in L_i$, respectively. For each attribute an individual distance function $d_i : L_i \times L_i \to \mathbb{R}$ has been defined by the user. The weighting parameter $\sigma_i \in]0, 1]$ (cp. ⑧ in Fig. 1) can be defined in order to scale the relative importance of an attribute distance value by means of

$$\sigma_i \cdot d_i(u.A_i, v.A_i)$$

In our framework there are two ways of combining the k individual weighted distance values $(\sigma_i \cdot d_i(u.A_i, v.A_i))_{1 \le i \le k}$, viz. by building the sum or the product (cp. ⑨ in Fig. 1). Finally, in ⑩ in Fig. 1 the user is asked to define a parameter p indicating that the p-th root is extracted from the combined node cost. Depending on whether the individual node costs are added or multiplied, we thus get the cost $c(u \to v)$ for node substitution $(u \to v)$ as

$$\left(\sum_{i=1}^{k} \sigma_i \cdot d_i(u.A_i, v.A_i) \right)^{1/p} \quad \text{or} \quad \left(\prod_{i=1}^{k} \sigma_i \cdot d_i(u.A_i, v.A_i) \right)^{1/p}$$

In Fig. 2 two examples of node cost functions are shown. In Fig. 2 (a) the cost for a node substitution is given as a weighted Euclidean distance between the nodes:

$$c(u \to v) = \sqrt{0.5 \cdot (u.x - v.x)^2 + 0.7 \cdot (u.y - v.y)^2 + 0.9 \cdot (u.z - v.z)^2}$$

In Fig. 2 (b) the cost for a node substitution is defined by:

$$c(u \to v) = \begin{cases} 0.5 \cdot sed(u.sequence, v.sequence), & \text{if } u.type = v.type \\ sed(u.sequence, v.sequence), & \text{if } u.type \neq v.type \end{cases}$$

The edge attributes and their distance function can be defined analogously. Additionally, for edges the user has to define whether the edges are directed or undirected (cp. ❶ in Fig. 1). The weighting parameter $\alpha \in [0, 1]$ (cp. ❷ in Fig. 1) controls whether the edit operation cost on the nodes or on the edges is more important. That is, each cost of every node operation (deletion, insertion, substitution) is multiplied by α. In the case of edge operations the costs are multiplied by $(1 - \alpha)$. The default setting is $\alpha = 0.5$ leading to balanced importance between node and edge operation cost.

4   Similarity Kernel from Edit Distance

Kernel machines constitute a very powerful class of algorithms [17, 18]. As a matter of fact, kernel methods have become a rapidly emerging sub-field in intelligent information processing. As any kernel function can be regarded as a

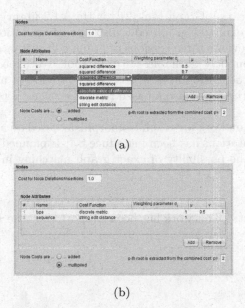

(a)

(b)

Fig. 2. Two different parameter settings for defining the cost functions on node attributes

object similarity measure, the edit distance of graphs can also be interpreted as a pattern similarity measure in the context of kernel machines, which makes a large number of powerful methods applicable to graphs [19], including support vector machines for classification and kernel principal component analysis for feature space transformation and dimensionality reduction. In our algorithmic framework we provide four different transformations of graph edit distance $d(g_1, g_2)$ to a similarity measure $\kappa_i(g_1, g_2)$ (cp. ❸ in Fig. 1):

- $\kappa_1(g_1, g_2) = -d(g_1, g_2)^2$
- $\kappa_2(g_1, g_2) = -d(g_1, g_2)$
- $\kappa_3(g_1, g_2) = \tanh(-d(g_1, g_2))$
- $\kappa_4(g_1, g_2) = \exp(-d(g_1, g_2))$

Note that these similarity kernels are not positive definite and are therefore not valid kernels in the strict sense. Yet, there is theoretical and practical evidence that using kernel machines in conjunction with indefinite kernels may be both reasonable and beneficial [19, 20].

5   Conclusion and Future Work

In comparison with the great variety of software tools for statistical pattern recognition, the number of tools for structural pattern recognition is rather limited. There are some software tools available for manipulating graphs or exact graph matching (e.g. the iGraph tool [21] or the VF2 library [22]), yet, a software tool for (approximate) graph edit distance computation is still missing. The present paper reviews three versions of graph edit distance which have been integrated in one publicly available software tool. We expect that the graph matching software tool introduced in this paper provides a major contribution towards promoting the use of graph based representations in pattern recognition and related fields.

In [23] a novel framework for graph isomorphism based on approximate graph edit distance computations has been introduced. It is planned to integrate these methods and thus the possibility of exact graph matching in our software tool in future work.

Acknowledgements. This work has been supported by the *Hasler Foundation* Switzerland.

References

1. Conte, D., Foggia, P., Sansone, C., Vento, M.: Thirty years of graph matching in pattern recognition. Int. Journal of Pattern Recognition and Artificial Intelligence 18(3), 265–298 (2004)
2. Luo, B., Wilson, R., Hancock, E.R.: Spectral embedding of graphs. Pattern Recognition 36(10), 2213–2223 (2003)
3. Wilson, R., Hancock, E.R.: Levenshtein distance for graph spectral features. In: Kittler, J., Petrou, M., Nixon, M. (eds.) Proc. 17th Int. Conf. on Pattern Recognition, vol. 2, pp. 489–492 (2004)
4. Boeres, M.C., Ribeiro, C.C., Bloch, I.: A randomized heuristic for scene recognition by graph matching. In: Ribeiro, C.C., Martins, S.L. (eds.) WEA 2004. LNCS, vol. 3059, pp. 100–113. Springer, Heidelberg (2004)
5. Sorlin, S., Solnon, C.: Reactive tabu search for measuring graph similarity. In: Brun, L., Vento, M. (eds.) GbRPR 2005. LNCS, vol. 3434, pp. 172–182. Springer, Heidelberg (2005)
6. Neuhaus, M., Bunke, H.: An error-tolerant approximate matching algorithm for attributed planar graphs and its application to fingerprint classification. In: Fred, A., Caelli, T.M., Duin, R.P.W., Campilho, A.C., de Ridder, D. (eds.) SSPR&SPR 2004. LNCS, vol. 3138, pp. 180–189. Springer, Heidelberg (2004)
7. Ambauen, R., Fischer, S., Bunke, H.: Graph edit distance with node splitting and merging and its application to diatom identification. In: Hancock, E., Vento, M. (eds.) GbRPR 2003. LNCS, vol. 2726, pp. 95–106. Springer, Heidelberg (2003)
8. Robles-Kelly, A., Hancock, E.R.: Graph edit distance from spectral seriation. IEEE Transactions on Pattern Analysis and Machine Intelligence 27(3), 365–378 (2005)
9. Hart, P.E., Nilsson, N.J., Raphael, B.: A formal basis for the heuristic determination of minimum cost paths. IEEE Transactions of Systems, Science, and Cybernetics 4(2), 100–107 (1968)
10. Riesen, K., Fankhauser, S., Bunke, H.: Speeding up graph edit distance computation with a bipartite heuristic. In: Frasconi, P., Kersting, K., Tsuda, K. (eds.) Proc. 5th. Int. Workshop on Mining and Learning with Graphs, pp. 21–24 (2007)

11. Neuhaus, M., Riesen, K., Bunke, H.: Fast suboptimal algorithms for the computation of graph edit distance. In: Yeung, D.-Y., Kwok, J.T., Fred, A., Roli, F., de Ridder, D. (eds.) SSPR & SPR 2006. LNCS, vol. 4109, pp. 163–172. Springer, Heidelberg (2006)

12. Riesen, K., Neuhaus, M., Bunke, H.: Bipartite graph matching for computing the edit distance of graphs. In: Escolano, F., Vento, M. (eds.) GbRPR. LNCS, vol. 4538, pp. 1–12. Springer, Heidelberg (2007)

13. Riesen, K., Bunke, H.: Approximate graph edit distance computation by means of bipartite graph matching. Image and Vision Computing 27(4), 950–959 (2009)

14. Fankhauser, S., Riesen, K., Bunke, H.: Speeding up graph edit distance computation through fast bipartite matching. In: Jiang, X., Ferrer, M., Torsello, A. (eds.) GbRPR 2011. LNCS, vol. 6658, pp. 102–111. Springer, Heidelberg (2011)

15. Munkres, J.: Algorithms for the assignment and transportation problems. Journal of the Society for Industrial and Applied Mathematics 5, 32–38 (1957)

16. Jonker, R., Volgenant, T.: A shortest augmenting path algorithm for dense and sparse linear assignment problems. Computing 38, 325–340 (1987)

17. Schölkopf, B., Smola, A.: Learning with Kernels. MIT Press (2002)

18. Shawe-Taylor, J., Cristianini, N.: Kernel Methods for Pattern Analysis. Cambridge University Press (2004)

19. Neuhaus, M., Bunke, H.: Bridging the Gap Between Graph Edit Distance and Kernel Machines. World Scientific (2007)

20. Haasdonk, B.: Feature space interpretation of SVMs with indefinite kernels. IEEE Transactions on Pattern Analysis and Machine Intelligence 27(4), 482–492 (2005)

21. Csárdi, G., Nepusz, T.: The igraph software package for complex network research. Inter. Journal Complex Systems 1695 (2006)

22. Cordella, L.P., Foggia, P., Sansone, C., Vento, M.: An improved algorithm for matching large graphs. In: Proc. of the 3rd IAPR TC-15 Workshop on Graphbased Representations in Pattern Recognition, pp. 149–159 (2001)

23. Riesen, K., Fankhauser, S., Bunke, H., Dickinson, P.: Efficient suboptimal graph isomorphism. In: Torsello, A., Escolano, F., Brun, L. (eds.) GbRPR 2009. LNCS, vol. 5534, pp. 124–133. Springer, Heidelberg (2009)

Map Edit Distance vs. Graph Edit Distance for Matching Images

Camille Combier[1,2], Guillaume Damiand[2,3], and Christine Solnon[2,3]

[1] Université Lyon 1, LIRIS, UMR 5205 CNRS, 69622 Villeurbanne, France
[2] Université de Lyon, France
[3] INSA de Lyon, LIRIS, UMR 5205 CNRS, 69621 Villeurbanne, France
{guillaume.damiand,christine.solnon}@liris.cnrs.fr

Abstract. Generalized maps are widely used to model the topology of nD objects (such as 2D or 3D images) by means of incidence and adjacency relationships between cells (0D vertices, 1D edges, 2D faces, 3D volumes, ...). Recently, we have introduced a map edit distance. This distance compares maps by means of a minimum cost sequence of edit operations that should be performed to transform a map into another map. In this paper, we introduce labelled maps and we show how the map edit distance may be extended to compare labeled maps. We experimentally compare our map edit distance to the graph edit distance for matching regions of different segmentations of a same image.

1 Motivations

In many computer vision applications we have to match interest points or regions extracted from different images in order to, *e.g.*, recognize objects or reconstitute 3D models from 2D images. When looking for such matchings, graph-based approaches offer a good compromise between local approaches, which match each point with the most similar point of the other image independently from its relationships with other points, and global approaches such as RANSAC, which consider rigid transformations. Indeed, graph-based approaches are able to exploit local relationships while being more tolerant to deformations than global approaches such as RANSAC based ones [1].

There exist different kinds of graph matchings [2], ranging from subgraph isomorphism to more error-tolerant matchings such as the graph edit distance. The graph edit distance is a generic measure, which is parametrized by edition costs, and it is widely used to match graphs. It defines the distance between two graphs G_1 and G_2 as the minimum cost sequence of edit operations for transforming G_1 into G_2. Edit operations are vertex and edge deletion, insertion and substitution. A vertex matching may be derived from the sequence of edit operations in a straightforward way: Any vertex v_1 of G_1 which is substituted to a vertex v_2 of G_2 is actually matched with v_2 [3].

Graphs are well suited to model binary relationships such as point proximity or region adjacency. However, graphs are less well suited to model the topology of the subdivision of a plane in faces, edges, and vertices. Combinatorial maps

W.G. Kropatsch et al. (Eds.): GbRPR 2013, LNCS 7877, pp. 152–161, 2013.

are very nice data structures to model this kind of topological information: they model the topology of nD objects subdivided into cells (*e.g.*, vertices, edges, faces, volumes, ...) by means of incidence and adjacency relationships between these cells. Combinatorial maps have been extended to generalized maps in [4], which are fully homogeneous in any dimension, thus simplifying algorithms and the development of computer libraries. In 2D, combinatorial and generalized maps may be used to model the topology of an embedding of a planar graph in a plane. In particular, these models are very well suited for scene modeling [5], for 2D and 3D image segmentation [6], and there exist efficient algorithms to extract maps from images [7].

We have defined a map edit distance in [8]. This map edit distance is a straight-forward extension of the graph edit distance: it defines the distance between two maps as a minimum cost sequence of edit operations, and a matching may be derived from this edit operation sequence. However, this map edit distance has been defined for non labelled maps. In this paper, we introduce labelled maps, such that cells may be associated with labels which describe their properties, and we extend our map edit distance to labelled maps. Another goal of this paper is to compare our map edit distance with the graph edit distance for matching regions of different segmentations of a same image, and therefore answer the following question: Does the topology of the subdivision of the image in regions (besides region adjacency relationships) help to match image regions?

Outline of the Paper. In Section 2, we recall definitions related to generalized maps and to the map edit distance. In Section 3, we introduce labelled maps and show how the map edit distance may be extended to handle labels. In Section 4, we experimentally compare our map edit distance to the graph edit distance for matching regions of segmented images. In Section 5, we discuss further work.

2 Recalls on Generalized Maps and the Map Edit Distance

In this work we consider generalized maps, and we refer the reader to [4] for more details.

Definition 1 (nG-map). *Let $n \geq 0$. An n-dimensional generalized map (or nG-map) is defined by a tuple $G = (D, \alpha_0, \ldots, \alpha_n)$ such that*

1. *D is a finite set of darts;*
2. *$\forall i \in \{0, \ldots, n\}$, α_i is an involution on D (i.e., it is a bijection such that $\forall d \in D, \alpha_i(\alpha_i(d)) = d$);*
3. *$\forall i, j \in \{0, \ldots, n\}$ such that $i + 2 \leq j$, $\alpha_i \circ \alpha_j$ is an involution.*

We say that a dart d is i-sewn with a dart d' whenever $d' = \alpha_i(d)$ and $d \neq d'$, whereas it is i-free whenever $d = \alpha_i(d)$. We say that a dart d is free if it is i-free for every dimension.

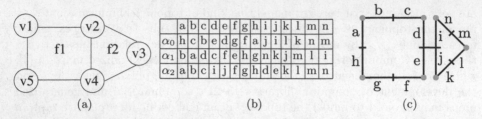

Fig. 1. (a) A plane graph. (b) The corresponding 2G-map. (c) Its graphical representation: darts are represented by segments labeled with letters, consecutive darts separated with a little segment are 0-sewn (*e.g.*, $\alpha_0(b) = c$ and $\alpha_0(c) = b$), consecutive darts separated with a dot are 1-sewn (*e.g.*, $\alpha_1(a) = b$ and $\alpha_1(b) = a$), parallel darts are 2-sewn (*e.g.*, $\alpha_2(d) = i$ and $\alpha_2(i) = d$).

2G-maps may be used to model the embedding of a planar graph into a plane. For example, Fig. 1 displays a plane graph, composed of 5 vertices, 6 edges and 2 faces, and the corresponding 2G-map, composed of 14 darts.

Cells are implicitly defined by sets of darts corresponding to orbits: the i-cell incident to a dart d is defined by $cell_i(d) =< \{\alpha_0, \ldots, \alpha_n\} \setminus \{\alpha_i\} > (d)$. Let us consider, for example, the 2G-map of Fig. 1. The cells incident to dart e are:

- $cell_0(e) =< \{\alpha_1, \alpha_2\} > (e) = \{e, f, j, k\}$, corresponding to vertex v_4;
- $cell_1(e) =< \{\alpha_0, \alpha_2\} > (e) = \{d, e, i, j\}$, corresponding to edge (v_2, v_4);
- $cell_2(e) < \{\alpha_0, \alpha_1\} > (e) = \{a, b, c, d, e, f, g, h\}$, corresponding to face f_1.

The map edit distance is based on edit operations which are used to transform maps. These edit operations allow one to add/delete free darts, and to sew/unsew darts. More precisely, let $G = (D, \alpha_0, \ldots, \alpha_n)$ be an nG-map.

- Let $d \in D$ be a free dart of G (*i.e.*, $\forall i \in \{0, \ldots, n\}, \alpha_i(d) = d$). The $del_d(G)$ operation removes d from D.
- Let $d \notin D$ be a dart. The $add_d(G)$ operation adds d to D so that d becomes a free dart of G, *i.e.*, $\forall i \in \{0, \ldots, n\}, \alpha_i(d) = d$.
- Let S be a set of triples (d, i, d') such that $d \in D$ and $d' \in D$ are i-free (*i.e.*, $d \neq d', \alpha_i(d) = d$, and $\alpha_i(d') = d'$). The $sew_S(G)$ operation i-sews d to d' for every triple $(d, i, d') \in S$, *i.e.*, it sets $\alpha_i(d)$ to d' and $\alpha_i(d')$ to d.
- Let S be a set of triples (d, i, d') such that $d \in D$ and $d' \in D$ are i-sewn darts (*i.e.*, $d \neq d', \alpha_i(d) = d'$, and $\alpha_i(d') = d$). The $unsew_S(G)$ operation i-unsews d to d' for every triple $(d, i, d') \in S$, *i.e.*, it sets $\alpha_i(d)$ to d and $\alpha_i(d')$ to d'.

When comparing these edit operations to classical graph edit operations, the *del* and *add* operations are related to vertex deletion and addition operations, whereas the *sew* and *unsew* operations are related to edge deletion and addition operations. A main difference is that sew/unsew operations operate on sets of darts instead of sewing/unsewing darts one by one. Indeed, sewing/unsewing a single dart may lead to a non valid nG-map. Let us consider for example the nG-map of Fig. 1. We cannot 2-unsew darts d and i without also 2-unsewing darts e and j (otherwise $\alpha_0 \circ \alpha_2$ no longer is an involution so that Property 3 of definition 1 no longer is satisfied). The map edit distance is then defined as the cost of the minimal cost edit path.

Definition 2 (edit path). *Let G be an nG-map and $\Delta =< \delta_1, \ldots, \delta_k >$ be a sequence of k edit operations. Δ is an edit path for G if $\delta_k(\delta_{k-1}(\ldots(\delta_1(G))))$, denoted $\Delta(G)$, is an nG-map (according to definition 1).*

Definition 3 (map edit distance). *Let c be a function which associates a cost $c(\delta) \in \mathbb{R}^+$ with every operation δ. The edit distance between the two nG-maps G and G' is $d_c(G, G') = \sum_{\delta_i \in \Delta}(c(\delta_i))$ where Δ is an edit path such that $\Delta(G) = G'$ and $\sum_{\delta_i \in \Delta}(c(\delta_i))$ is minimal.*

3   Extension of the Map Edit Distance to Labelled Maps

Generalized maps describe the topology of the subdivision of a space into cells. However, they do not express other information such as, for example, geometry, texture or colour information. This kind of information may be added by means of labels associated with cells. In generalized maps, cells are implicitly defined by sets of darts and correspond to orbits. Therefore, to associate a label with an i-cell c, we propose to label every dart of c, *i.e.*, every dart d such that $cell_i(d) = c$. Note that, a dart belongs to exactly one cell for every dimension $i \in \{0, \ldots, n\}$. For consistency reasons, we impose that all darts of a same i-cell have the same label for dimension i.

Definition 4 (Labelled nG-maps). *Let $n \geq 1$. A labelled nG-map is a tuple $G = (D, \alpha_0, \ldots, \alpha_n, L, l)$ such that $(D, \alpha_0, \ldots, \alpha_n)$ is an nG-map, L is a set of labels, and $l : D \times \{0, \ldots, n\} \to L$ is a labelling function such that $\forall d, d' \in D, \forall i \in \{0, \ldots, n\}, cell_i(d) = cell_i(d') \Rightarrow l(d, i) = l(d', i)$.*

In other words, $l(d, i)$ is the label associated with the i-cell incident to d. Let us consider, for example, the 2G-map of Fig. 1. To associate the label x with vertex v_4, we define $l(e, 0) = l(f, 0) = l(j, 0) = l(k, 0) = x$; to associate the label y with edge (v_2, v_4), we define $l(d, 1) = l(e, 1) = l(i, 1) = l(j, 1) = y$; to associate the label z with face f_1, we define $l(a, 2) = l(b, 2) = l(c, 2) = l(d, 2) = l(e, 2 = l(f, 2) = l(g, 2) = l(h, 2) = z$.

The map edit distance of [8] has been defined for non labelled maps. To extend it to labelled maps, we introduce a new edit operation that substitutes dart labels. Let $G = (D, \alpha_0, \ldots, \alpha_n, L, l)$ be a labelled nG-map, $d \in D$ be a dart, $i \in \{0, ..., n\}$, and $l' \in L$ be a label. The $subs_{(d,l',i)}(G)$ operation substitutes the label $l(d, i)$ of dart d with the new label l'. The cost function c must also be extended so that the cost of an edit operation ($subs, del, add, sew$, or $unsew$) depends on dart labels.

4   Experimental Comparison

In this section, we compare the map edit distance with the graph edit distance for matching regions of different segmentations of a same image. Our goal is to

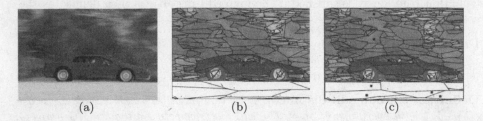

Fig. 2. (*a*) An image. (*b*) A segmentation of (a) in 389 regions. (*c*) A finer segmentation of (a) in 415 regions.

evaluate the interest of using maps, which model the topology of the subdivision of the image in regions, instead of using graphs, which only model region adjacency relationships.

4.1 Test Suite

For this very first experimental comparison, we compare different segmentations of a same image. This allows us to have a ground truth for evaluating matchings: we consider that two regions coming from two different segmentations of the same image are correctly matched if their intersection is not empty. We have considered 6 different images (2 cars, 2 cows and 2 motorbikes) extracted from the ETHZ benchmark[1]. For each image, we have generated different segmentations, using the algorithm of [9] with different threshold values, so that the number of regions varies from 240, for the coarser segmentations, to 460, for the finer ones. Fig. 2 gives an example of two segmentations of a same image. Note that all segmentations are recomputed from the same initial image so that regions of finer segmentations are not necessarily subdivisions of regions of coarser segmentations.

For each segmentation, we have built a graph and a 2G-map which represent it. The graph is a classical region adjacency graph (RAG), which associates a vertex with each region, and an edge with every pair of adjacent regions. The 2G-map associates a face with every region, the edges of the map describe the adjacency relations between the regions and the vertices of the map describe the adjacency relations between the edges.

The sizes of the resulting RAGs and 2G-maps are given in Table 1. Note that if the number of faces of the 2G-maps corresponds to the number of vertices of the RAGs, the 2G-maps have slightly more edges than RAGs as two regions having multiple adjacency relationships are linked by a single edge in RAGs (multiple adjacency occurs when two regions are adjacent several times).

Each region r of the segmented images is described by two basic descriptors: A color descriptor, $color(r)$, which is the average color of the pixels of r

---

[1] available at http://pascallin.ecs.soton.ac.uk/challenges/VOC/databases.html

Table 1. Comparison of the sizes of RAGs and 2G-maps, for the coarser segmentations (Min) and the finer ones (Max)

RAGs				2G-maps							
Nb of vertices		Nb of edges		Nb of darts		Nb of vertices		Nb of edges		Nb of faces	
Min	Max	Min	Max	Min	Max	Min	Max	Min	Max	Min	Max
240	460	688	1217	2792	4948	463	816	698	1237	240	460

(a value ranging between 0 and 255 as we consider gray-level colours), and an area descriptor, $area(r)$, which is the number of pixels of r.

First experiments showed us that the graph edit distance can hardly correctly match regions when RAGs are not labelled. Actually, RAGs usually have many automorphisms (*i.e.*, symetries), so that there exist many different matchings between two isomorphic RAGs which preserve adjacency relationships (but of course, only one of these matchings correctly matches regions). In order to improve the matching process, we have added structural labels to RAGs. Therefore, for RAGs:

- each edge (u, v) is labelled with $adj(u, v)$, the number of adjacency relationships between the two regions associated with u and v;
- each vertex u corresponding to a region r_u is labelled with a triple $(totAdj(u), color(r_u), area(r_u))$, where $totAdj(u) = \sum_v(adj(u, v))$ is the total number of adjacency relationships of all edges (u, v) incident to u.

The structural labels $adj(u, v)$ and $totAdj(u)$ greatly improve results for RAGs. We do not add these structural labels to 2G-maps as this information is already available in 2G-maps. Therefore, for every dart d of 2G-maps, we define $l(d, 2) = (color(r_d), area(r_d))$, where r_d is the region associated with the dart d. As no information is associated with vertices and edges of the 2G-maps, we define $l(d, 0) = l(d, 1) = \epsilon$.

4.2 Cost Functions

For RAGs, we define the cost of substituting a vertex u whose label is $(totAdj(u), color(r_u), area(r_u))$ with a vertex v whose label is $(totAdj(v), color(r_v), area(r_v))$ by

$$c(subs(u, v)) = \omega_{struct} \cdot |totAdj(u) - totAdj(v)|$$
$$+ \omega_{color} \cdot \frac{|color(r_u) - color(r_v)|}{255}$$
$$+ \omega_{area} \cdot (1 - \frac{min(area(r_u), area(r_v))}{max(area(r_u), area(r_v))})$$

where ω_{struct}, ω_{color}, and ω_{area} are 3 parameters which determine the relative weights of structural, color and area information. The cost of adding or deleting a vertex or an edge is set to 1.

For 2G-maps, we define the cost of substituting a dart u whose label is $l(u, 2) = (color(r_u), area(r_u))$ with a dart v whose label is $l(v, 2) = (color(r_v), area(r_v))$ by

$$c(subs(u,v)) = \omega_{color} \cdot \frac{|color(r_u) - color(r_v)|}{255}$$
$$+ \omega_{area} \cdot (1 - \frac{min(area(r_u), area(r_v))}{max(area(r_u), area(r_v))})$$

where ω_{color} and ω_{area} are 2 parameters which determine the relative weights of color and area information. The cost of adding or deleting a dart is set to 1. The cost of sewing/unsewing operations is equal to the number of triples (d, i, d') added/removed (as sew/unsew operations add/remove sets of seams for consistency reasons).

4.3   Matching Algorithms

Computing edit distances is a NP-hard problem, both for graphs and maps. Exact algorithms do not scale well and cannot compute edit distances within a reasonable amount of time for the graphs and maps considered here. Therefore, we use heuristic algorithms, which compute approximate solutions.

For the graph edit distance, we use the algorithm proposed in [3]: The graph matching problem is approximated by an assignment problem which is solved by the Munkres algorithm [10].

For the map edit distance, we use an extension to labelled maps of the greedy algorithm described in [11]: Starting from an empty matching, this algorithm iteratively matches darts until no more darts can be matched; at each iteration the pair of darts to be matched is chosen in order to minimize the corresponding edit costs.

Both algorithms have polynomial time complexities: $\mathcal{O}(v^3)$ for the graph edit distance, where v is the number of vertices of the largest graph, and $\mathcal{O}(d^2 \cdot log(d))$ for the map edit distance, where d is the number of darts of the largest map. However, as d is more than ten times larger than v on our benchmark, the matching process is faster for graphs than for maps: for the coarsest segmentations, having 240 regions, graphs (having 240 vertices) are matched in 2 seconds or so whereas maps (having 2792 darts) are matched in 8 seconds or so; for the finest segmentations, having 460 regions, graphs (having 460 vertices) are matched in 5 seconds or so whereas maps (having 4948 darts) are matched in 25 seconds or so.

4.4   Experimental Results

Let us now compare graph and map edit distances for matching regions of two different segmentations of a same image. Comparing different segmentations of a same image allows us to have a ground truth: we consider that two regions coming from two different segmentations of the same image are correctly matched if their intersection is not empty. When images are modelled with RAGs, we measure the percentage of vertices which are correctly matched, *i.e.*, whose associated regions are correctly matched. When images are modelled with 2G-maps, we measure the percentage of darts which are correctly matched, *i.e.*, whose associated 2-cells are correctly matched.

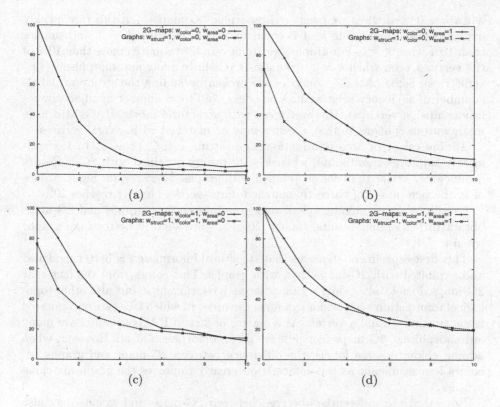

Fig. 3. Average percentage of correctly matched darts/vertices (on the y-axis) with respect to the difference k of segmentation levels (on the x-axis). (a) Using structural information only, *i.e.*, $\omega_{colour} = 0$ and $\omega_{area} = 0$. (b) Using structural and area information, *i.e.*, $\omega_{colour} = 0$ and $\omega_{area} = 1$. (c) Using structural and colour information, *i.e.*, $\omega_{colour} = 1$ and $\omega_{area} = 0$. (d) Using structural, colour and area information, *i.e.*, $\omega_{colour} = 1$ and $\omega_{area} = 1$.

For each image, we sort its different segmentations from the coarsest one (which has 240 regions) to the finest one (which has 460 regions). For each segmentation i, we measure the percentage of correctly matched vertices/darts when comparing it with segmentation $i + k$ of the same image, where the difference k of segmentation levels ranges from 0 up to 10 (it may be less than 10 if segmentation $i + 10$ does not exist): when $k = 0$, we actually compare a segmentation with itself; the larger k, the more different the two segmentations.

Fig. 3 displays the evolution of the percentage of correctly matched vertices/darts with respect to the difference k of segmentation levels (on average for the 6 images and each segmentation level i). For the graph edit distance, we always set ω_{struct} to 1 as this always improves results. We consider different combinations of the two other weight parameters ω_{colour} and ω_{weight}. Let us first compare graphs and 2G-maps when colour and area information is ignored, *i.e.*, $\omega_{colour} = 0$ and $\omega_{area} = 0$. For 2G-maps, we are able to correctly match all darts when $k = 0$ (*i.e.*, when we compare isomorphic 2G-maps), but this rate quickly decreases when increasing k: it is equal to 20% or so when $k = 3$, and smaller than 10% when $k \geq 5$. For graphs, we are never able to match more than 10% of the vertices, even when $k = 0$. Actually, RAGs have many automorphisms (*i.e.*, symetries). Some of these symetries are broken by adding the structural labels (number of adjacency relationships on edges, and total number of adjacency relationships on vertices). However, even with structural labels, RAGs still have many automorphisms so that a vertex may be matched with several vertices.

Adding colour or area information, *i.e.*, setting ω_{colour} or ω_{area} to 1, significantly improves results and, when $k \leq 5$, results obtained with 2G-maps are significantly better than those obtained with graphs. However, for higher values of k, the percentage of correctly matched darts/vertices hardly reaches 20%.

Finally, when combining colour and area information, graphs and 2G-maps obtain rather similar results, though 2G-maps are slightly better than graphs when $k \leq 4$.

This first experiment shows us that structural information is better modelled and exploited with 2G-maps than with graphs. This comes from the fact that 2G-maps do not only model region adjacency relationships, but also other topological information such as, for example, the order in which faces are encountered when turning around a vertex. As a matter of fact, if RAGs usually have many automorphisms, 2G-maps usually have no automorphism at all. However, when adding colour or area labels, the difference between 2G-maps and graphs becomes less significant as this information greatly improves the graph matching process.

Note that the differences observed between 2G-maps and graphs may also come from the matching algorithms we have considered: these matching algorithms are heuristic algorithms which compute approximate solutions. The graph matching algorithm of [3] considers only local, rather than global, edge structure during the optimization process. Also, the greedy map matching algorithm of [11] considers local seams to choose the next pair of darts to match. It is not possible to compute exact solutions within a reasonable amount of time, considering the fact that graphs (resp. maps) have hundreds (resp. thousands) of vertices (resp. darts). Therefore, we cannot assess the quality of the approximations computed by the heuristic algorithms.

5   Conclusion

In this paper, we have extended generalized maps to labelled nG-maps, thus allowing us to add information on cells in every dimension, and we have extended the map edit distance to handle these labels by adding a dart substitution operation. We have compared the map edit distance with the graph edit distance for matching regions of different segmentations of a same image. We have shown that regions are better matched when we use 2G-maps for modelling the topology of the subdivision of the image in regions rather than when we use RAGs for modelling region adjacency relationships.

We have performed similar experiments on 2D meshes modelling 3D objects. We have generated different degradations of a same mesh (obtained by merging nearly co-planar adjacent faces), and compare the percentage of correctly matched faces when using 2G-maps and when using graphs. We observed similar results, *i.e.*, 2G-maps allow us to better match faces than graphs.

As future works, we plan to study different domains of application which use graphs and similarity measures in order to see if we can improve existing solutions by using nG-maps and the map edit distance. We also would like to improve our algorithm in order to speed-up the computation times and propose more heuristics to guide the choice of the best pair of darts to be matched. Lastly we plan to study other types of similarity measures. We think for example to extend graph kernels to generalized maps.

References

1. Fischler, M.A., Bolles, R.C.: Random sample consensus: A paradigm for model fitting with applications to image analysis and automated cartography. Comm. of the ACM 6(24), 381–395 (1981)
2. Conte, D., Foggia, P., Sansone, C., Vento, M.: Thirty Years Of Graph Matching In Pattern Recognition. International Journal of Pattern Recognition and Artificial Intelligence (2004)
3. Riesen, K., Bunke, H.: Approximate graph edit distance computation by means of bipartite graph matching. Image Vision Comput. 27, 950–959 (2009)
4. Lienhardt, P.: N-dimensional generalized combinatorial maps and cellular quasi-manifolds. Computational Geometry and Applications 4(3), 275–324 (1994)
5. Fradin, D., Meneveaux, D., Lienhardt, P.: A hierarchical topology-based model for handling complex indoor scenes. Computer Graphics Forum 25(2), 149–162 (2006)
6. Braquelaire, J.P., Brun, L.: Image segmentation with topological maps and inter-pixel representation. Visual Communication and Image Representation 9(1), 62–79 (1998)
7. Damiand, G.: Topological model for 3d image representation: Definition and incremental extraction algorithm. Computer Vision and Image Understanding 109(3), 260–289 (2008)
8. Combier, C., Damiand, G., Solnon, C.: From maximum common submaps to edit distances of generalized maps. Pattern Recognition Letters 33(15), 2020–2028 (2012)
9. Dupas, A., Damiand, G.: First results for 3D image segmentation with topological map. In: Coeurjolly, D., Sivignon, I., Tougne, L., Dupont, F. (eds.) DGCI 2008. LNCS, vol. 4992, pp. 507–518. Springer, Heidelberg (2008)
10. Munkres, J.: Algorithms for the assignment and transportation problems. Journal of the Society of Industrial and Applied Mathematics 5(1), 32–38 (1957)
11. Combier, C., Damiand, G., Solnon, C.: Measuring the distance of generalized maps. In: Jiang, X., Ferrer, M., Torsello, A. (eds.) GbRPR 2011. LNCS, vol. 6658, pp. 82–91. Springer, Heidelberg (2011)

An Algorithm for Maximum Common Subgraph of Planar Triangulation Graphs

Yao Lu[1], Horst Bunke[2], and Cheng-Lin Liu[1]

[1] National Lab of Pattern Recognition
Institute of Automation, Chinese Academy of Sciences
yaolubrain@gmail.com, liucl@nlpr.ia.ac.cn
[2] Institute of Computer Science and Applied Mathematics (IAM),
University of Bern
bunke@iam.unibe.ch

Abstract. We propose a new fast algorithm for solving the Maximum Common Subgraph (MCS) problem. MCS is an NP-complete problem. In this paper, we focus on a special class of graphs, i.e. Planar Triangulation Graphs, which are commonly used in computer vision, pattern recognition and graphics. By exploiting the properties of Planar Triangulation Graphs and restricting the problem to connected MCS, for two such graphs of size n and m and their maximum common subgraph of size k, our algorithm solves the MCS problem approximately with time complexity $O(nmk)$.

Keywords: Planar Triangulation Graphs, Delauney Triangulation, Maximum Common Subgraph.

1 Introduction

Images and many other objects can be represented as graphs. The graph representation of an object characterizes its local features and their spatial relationship. Its theoretical properties, applications and efficient algorithms have been studied for decades [1]. Maximum Common Subgraph (MCS) is an important problem in pattern recognition [2]. It incorporates graph isomorphism and subgraph isomorphism as special cases. It has applications in computer vision and pattern recognition such as video indexing [3] and document classification [4]. However, MCS is known to be NP-complete in general. In order to obtain an efficient algorithm of practical value, we need to specialize the problem and/or look for approximate solutions.

In this paper, we focus on a special class of graphs, i.e. Planar Triangulation Graphs. Planar Triangulation Graphs are commonly used in pattern recognition, computer vision, and graphics. Perhaps the best known procedure to obtain such a graph is Delaunay triangulation. It has important properties such as sparseness, locality, and avoiding skinny angles in the triangulation. It has been shown that not much spatial information is lost after Delaunay triangulation of a set of points [5]. Moreover, Delaunay triangulation can be efficiently computed

W.G. Kropatsch et al. (Eds.): GbRPR 2013, LNCS 7877, pp. 162–171, 2013.

with time complexity $O(n \log n)$ by various methods [6]. And there are graph matching algorithms specialized for Delaunay triangulation graphs [7, 8].

The main contribution of in this paper is a new algorithm for the MCS problem of Planar Triangulation Graphs that has a practical high execution speed. By exploiting the properties of Planar Triangulation Graphs and restricting the problem to connected MCS, we are able to derive an algorithm with time complexity $O(nmk)$, for approximately solving the MCS problem of two Planar Triangulation Graphs of size n and m and their maximum common subgraph of size k. Given two graphs, our algorithm will return a common subgraph, but it is not guaranteed that it is a maximum common subgraph. However, our experimental verification showed that most of the common subgraphs returned in our experiments are in fact maximum common subgraphs, and those that are not are missing only a small fraction of nodes and edges.

2   Connected Maximum Common Subgraphs

2.1   Basic Definitions

Definition 1. *A graph is an ordered pair $G = (V, E)$ comprising a set V of nodes together with a set $E \subseteq V \times V$ of edges.*

Remark. The graphs we assume in this paper are unweighted and undirected graphs without node or edge attributes.

Definition 2. *A subgraph of a graph G is a graph whose node set is a subset of that of G, and whose adjacency relation is a subset of that of G restricted to this subset.*

Definition 3. *An isomorphism of graphs G and H is a bijective function between the node sets of G and H, $f : V(G) \rightarrow V(H)$ such that any two nodes u and v of G are adjacent in G if and only if $f(u)$ and $f(v)$ are adjacent in H.*

Definition 4. *A subgraph isomorphism from G to H is an injective function $f : V(G) \rightarrow V(H)$ such that if there exists a subgraph S of H and f is a graph isomorphism from G to S.*

Definition 5. *Let G, G_1, and G_2 be graphs. G is a common subgraph of G_1 and G_2 if there exists a subgraph isomorphisms from G to G_1 and from G to G_2.*

Definition 6. *A common subgraph G of G_1 and G_2 is maximal if there exists no other common subgraph G' of G_1 and G_2 that has more edges than G.*

Remark. This definition is given in [15] as Maximum Common Edge Subgraph. The work described in this paper will be based on this kind of MCS. It is preferred over Maximum Common Induced Subgraph for the reasons explained in [10, 15].

Definition 7. *A graph is connected if there is a path from any node to any other node in the graph.*

Finding connected MCS is in general NP-complete since subgraph isomorphism remains NP-complete even for connected graphs of bounded treewidth [9], which is a special case of connected MCS.

2.2   Related Work

Many exact algorithms for finding MCS have been proposed in the literature.
Two early examples are [10] and [11], which are based on backtrack search and
maximum clique detection in an association graph, respectively. Later algorithms
are described in [12–14]. All these algorithms have exponential time complexity.
Moreover, there are algorithms specialized for chemical structures [15], which
can achieve faster solutions in this domain than general MCS algorithms.

In this paper, we specialize on Planar Triangulation Graphs, a class of graphs
which are widely used in computer vision, pattern recognition and graphics.

3   Planar Triangulation Graphs

A Planar Triangulation Graph is obtained by triangulation of points in a plane.
See [6, 16] for references on triangulation. In Fig.1, we show an example obtained
by Delaunay triangulating a set of points in the two-dimensional plane.

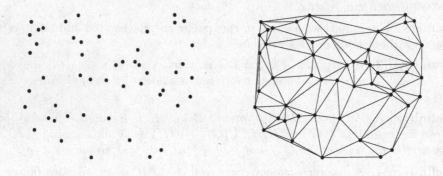

Fig. 1. Delaunay Triangulation of a set of points in the two-dimensional plane

3.1   Properties

Planar triangulation graphs have two important properties [6].

Property 1. *Each triangle of a Planar Triangulation Graph has at most three
adjacent triangles.*

Property 2. *A Planar Triangulation Graph of size n has $O(n)$ triangles and
$O(n)$ edges.*

These properties allow us to design a fast algorithm for connected MCS of Planar
Triangulation Graphs, as will be shown in the following sections.

3.2   Breadth-First Traversal of Triangles

In this subsection, we describe a method for traversing the triangles of a Planar Triangulation Graph. Note that it is a traversal of *triangles* rather than of *nodes*.

Starting from a given ordered triangle of a Planar Triangulation Graph with a triangle visiting order (e.g. clock-wise), there is a unique breadth-first traversal of triangles of the graph. The breadth-first traversal process works as follows. For example, in Fig. 2(a), let us take triangle **e** (with order **6-3-4**) as the root of the traversal. Then the adjacent triangles of the root triangle are visited in clock-wise order. From edge **6-3**, triangle **d** is visited, from edge **3-4**, triangle **b** is visited, and from edge **4-6**, triangle **f** is visited. Then we continue the process with triangle **d**, and so on. Except for the root triangle, for each visit of a triangle, at most one new node is encountered. After the traversal is finished, the visiting order of nodes **1** to **10** is: 4,5,2,3,8,1,7,9,10. The breadth-first traversal of triangles can be represented as a tree, as shown in Fig. 2(b). By property 1, this tree has a special structure: it has at most 3 children at its root node and at most 2 children at the other nodes.

Given a triangle, its adjacent triangles can be found by using a hash table with $O(1)$ operations. And the hash table can be built with time complexity $O(n)$ since there are only $O(n)$ triangles and each triangle has at most three adjacent triangles. Therefore, for a Planar Triangulation Graph of size n, the breadth-first traversal of triangles has time complexity $O(n)$.

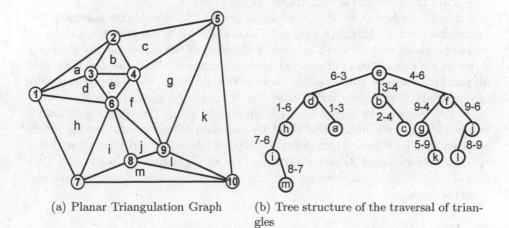

(a) Planar Triangulation Graph   (b) Tree structure of the traversal of triangles

Fig. 2. Breadth-First Traversal of Triangles

With these special properties and structures, we can derive an efficient heuristic algorithm for finding connected MCS of a given pair of Planar Triangulation Graphs, as shown in the next section.

4   Algorithm

The algorithm works as follows: starting with an arbitrarily chosen pair of ordered triangles as root in graphs G_1 and G_2, do a pair of breadth-first traversals of triangles on both graphs simultaneously. The visiting orders of the nodes of the initial pair triangles are 1, 2, 3. Two triangles of two graphs are matched and added to their traversal if their corresponding nodes have the same node visiting order. If a node has not been visited in the traversal yet, its node visiting order is set to be 0. The process continues until no triangles can be added to the pair of traversals. Finally, the graphs composed of the matched triangle pairs are the MCS. A pseudo-code description of our algorithm is given in Algorithm 1.

As an example, consider finding the MCS of the graphs in Fig. 3(a). In Fig. 3(b), we start with ordered triangles **2-3-4** and **b-c-e** as the roots of the pair of traversals. The visiting orders of nodes **2**, **3**, **4** and nodes **b**, **c**, **e** are 1, 2, 3, respectively. Triangles **2-3-4** and **b-c-e** are matched and added to the traversal since their corresponding nodes have the same node visiting order (all 0 now and 1,2,3 after). From ordered triangle **2-3-4**, triangles **2-3-1**, **3-4-6** and **4-2-5** are visited, in that order. And from ordered triangle **b-c-e**, triangles **b-c-a**, **c-e-f** and **e-b-d** are visited, in that order. Next, triangles **2-3-1** and **b-c-a** are matched since the the condition is satisfied, then triangles **3-4-6** and **c-e-f**, and finally triangles **4-2-5** and **e-b-d** (Fig. 3(c)). However, triangle **4-5-6** cannot be matched to either **d-e-g** or **e-f-g** because node **6**'s node visiting order is 6 and node **g**'s node visiting order is 0. Finally, as shown in Fig. 3(d), the graphs composed of **2-3-4**, **2-3-1**, **3-4-6** and **4-2-5** and of **b-c-e**, **b-c-a**, **c-e-f** and **e-b-d** are the MCS of the two graphs in Fig. 3(a).

For each such a pair of traversals, it takes $O(k)$ operations assuming the maximum common subgraph of size k. To find the MCS, we have to consider all pairs of ordered triangles as the roots of the pairs of the traversals. By Property 2, there are $O(nm)$ pairs of ordered triangles in total. Consequently, if we take all pairs of ordered triangles as the roots of the traversals, there are $O(nm)$ pairs of traversals. Hence, the overall time complexity of the algorithm is $O(nmk)$.

At first glance, our algorithm looks similar to String Growing algorithm for subgraph isomorphism [17]. But there are two main differences: (1) our algorithm is specialized for Planar Triangulation Graphs while String Growing algorithm is specialized for Region Adjacent Graphs. (2) Our algorithm has worst case time complexity $O(nmk)$ while String Growing algorithm has worst case exponential time complexity.

5   Experiments

In the experiments, n random points in the two-dimensional plane were generated to obtain point set S_1 and triangulated by Delaunay triangulation to obtain graph G_1. m points around the center of S_1 were selected to obtain point set S_2 and then triangulated by Delaunay triangulation to obtain graph G_2. Due to the boundary effect of Delaunay triangulation, G_2 is not necessarily a subgraph of G_1 in general.

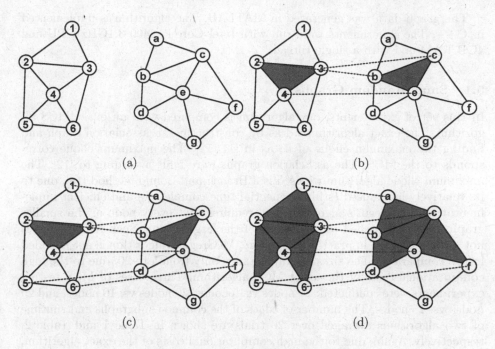

Fig. 3. Illustration of the algorithm

Algorithm 1. MCS of Planar Triangulation Graphs

---

1 **foreach** *Pair of ordered triangles (T_i,T_j)* **do**
2   Initialize empty set M
3   Initialize empty queues Q_1 and Q_2
4   Enqueue(Q_1,T_i)
5   Enqueue(Q_2,T_j)
6   **while** *Q_1 and Q_2 are not empty* **do**
7          $T_1 = $ Dequeue(Q_1)
8          $T_2 = $ Dequeue(Q_2)
9          $M = M \cup \{(T_1,T_2)\}$
10   **foreach** *Pair of triangles (T_1^{adj},T_2^{adj}) adjacent to (T_1,T_2) and not in M* **do**
11   **if** *their corresponding nodes have the same node visiting order* **then**
12   Enqueue(Q_1,T_1^{adj})
13   Enqueue(Q_2,T_2^{adj})
14   **end**
15   **end**
16   **end**
17   Record M with its cardinality
18 **end**
19 **return** M with the maximum cardinality

The graph data was generated in MATLAB. The algorithm is implemented in C++. The experiments were run with Intel Core i5-2400 3.1GHz CPU and 4GB RAM and with a single thread.

5.1 Small Random Graphs

In this set of experiments, our algorithm is compared with an exact MCS algorithm. The exact algorithm works by constructing an association graph and finding the maximum clique of it, as in [11, 12]. The maximum clique corresponds to the MCS. The association graphs were built according to [12]. The maximum clique algorithm we used is a Branch-and-Bound method [18], due to its relatively high speed (still exponential time complexity) and efficient implementation. McGregor's algorithm [10] requires that every node of the smaller graph must be matched to some node of the larger graph. Such requirement is not always satisfied in practice. Therefore, McGregor's algorithm is not included in the comparison. The sizes of graphs range only from 5 to 20, due to the high computational costs of the exact MCS algorithm. Three cases of MCS testing experiments were conducted: 20 nodes vs. 5 nodes, 20 nodes vs. 10 nodes, and 20 nodes vs. 15 nodes. The number of edges of the common subgraphs and runtime of two algorithms averaged over 20 trials are shown in Table 1 and Table 2, respectively. Again, due to the high computational costs of the exact algorithm, only 20 trials in each case were conducted.

Table 1. Average edges of MCS dependent on graph size (nodes)

Algorithm	20 vs. 5	20 vs. 10	20 vs. 15
Exact	7.5	20.25	32.95
Ours	7.4	19.5	32.7

Table 2. Runtime (sec) dependent on graph size (nodes)

Algorithm	20 vs. 5	20 vs. 10	20 vs. 15
Exact	0.002	6.35	2202.123
Ours	0.005	0.032	0.087

5.2 Large Random Graphs

In this set of experiments, the performance of our algorithm in finding MCS of relatively large graphs is shown. We vary the size of G_1 and G_2 to record the runtime of our algorithm in Fig. 4. The size of graphs range from 50 nodes to 2000 nodes. See Fig. 5 for visualization.

In Fig. 4(a), the size of G_2 is kept constant at 50 nodes, while the size of G_1 varies from 50 to 2000 nodes. In contrast, in Fig. 4(b), the size of G_2 varies from 50 to 300 nodes at constant size of G_1 at 500 nodes. The computation time

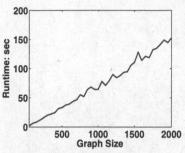 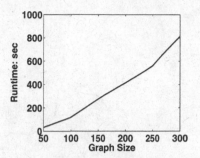

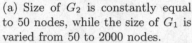

(a) Size of G_2 is constantly equal to 50 nodes, while the size of G_1 is varied from 50 to 2000 nodes.

(b) Size of G_1 is constantly equal to 500 nodes, while the size of G_2 is varied from 50 to 300 nodes.

Fig. 4. Runtime (sec) dependent on graph size (nodes)

measured in these experiments (averaged over 10 trials) confirms the theoretical complexity mentioned in Section 4. In Fig. 4(a), a linear increase of the computation time in terms of the size of G_1 of is observed, while the behavior in Fig. 4(b) is superlinear in the size of G_2 since the increase of the size of the smaller graph would also increase the size of the maximum common subgraph.

To the knowledge of the authors, there exists no other specialized algorithm for computing connected MCS of Planar Triangulation Graphs. Therefore there exists no direct competitor against which the proposed algorithm could be evaluated. Of course one could benchmark the new algorithm against other algorithms that were developed for general graphs. Here we note, however, that the algorithms in [10, 12–14] were tested on much smaller graphs ($\leq$ 100 nodes) than the ones considered in this set of experiments. Therefore, it is computationally prohibitive to run comparison experiments.

6 Conclusion and Future Work

We present a fast algorithm for approximately solving the MCS problem of Planar Triangulation Graphs. In its present version, the algorithm can only cope with unweighted graphs without node attributes. However, it is straightforward to include weights on edges and attributes on nodes. Theoretical analysis on the quality of the approximation and more systematic experimental comparison with other MCS algorithms, such as one using approximate maximum clique detection methods [19], will be explored in the future. Also the application of the proposed algorithm to Planar Triangulation Graphs obtained from real images will be an interesting topic to be explored in future research.

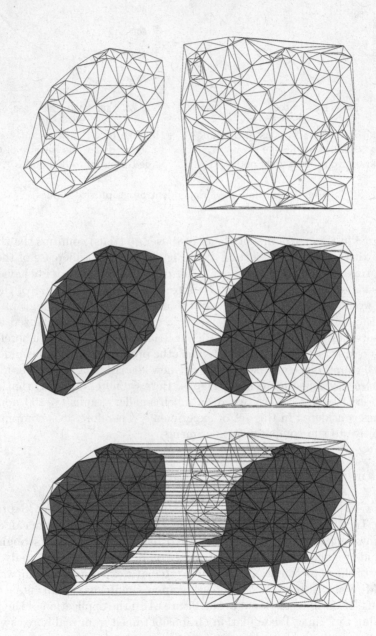

Fig. 5. 100 nodes vs. 200 nodes: 39 sec

References

1. Conte, D., Foggia, P., Sansone, C., Vento, M.: Thirty Years of Graph Matching in Pattern Recognition. International Journal of Pattern Recognition and Aritificial Intelligence 18(3), 265–298 (2004)
2. Bunke, H., Shearer, K.: A Graph Distance Metric Based on the Maximal Common Subgraph. Pattern Recognition Letters 19(3), 255–259 (1998)
3. Shearer, K., Bunke, H., Venkatesh, S.: Video Indexing and Similarity Retrieval by Largest Common Subgraph Detection Using Decision Trees. Pattern Recognition 34(5), 1075–1091 (2001)
4. Schenker, A., Last, M., Bunke, H., Kandel, A.: Classification of Web Documents Using Graph Matching. International Journal of Pattern Recognition and Artificial Intelligence 18(03), 475–496 (2004)
5. Dobkin, D., Friedman, S., Supowit, K.: Delaunay Graphs Are Almost as Good as Complete Graphs. Discrete and Computational Geometry 5(4), 399–407 (1990)
6. Berg, M., Cheong, O., Kreveld, M., Overmars, M.: Computational geometry: algorithms and applications. Springer (2008)
7. Finch, A., Wilson, R., Hancock, E.: Matching Delaunay Graphs. Pattern Recognition 30(1), 123–140 (1997)
8. Shin, D., Tjahjadi, T.: Similarity Invariant Delaunay Graph Matching. In: da Vitoria Lobo, N., Kasparis, T., Roli, F., Kwok, J.T., Georgiopoulos, M., Anagnostopoulos, G.C., Loog, M. (eds.) SSPR & SPR 2008. LNCS, vol. 5342, pp. 25–34. Springer, Heidelberg (2008)
9. Matouěk, J., Thomas, R.: On the Complexity of Finding Iso-and other morphisms for Partial k-trees. Discrete Mathematics 108(1), 343–364 (1992)
10. McGregor, J.: Backtrack Search Algorithms and the Maximal Common Subgraph Problem. Software Practice and Experience 12(1), 23–34 (1982)
11. Levi, G.: A Note on the Derivation of Maximal Common Subgraphs of Two Directed or Undirected Graphs. Calcolo 9(4), 341–352 (1972)
12. Koch, I.: Enumerating All Connected Maximal Common Subgraphs in Two Graphs. Theoretical Computer Science 250(1), 1–30 (2001)
13. Bunke, H., Foggia, P., Guidobaldi, C., Sansone, C., Vento, M.: A Comparison of Algorithms for Maximum Common Subgraph on Randomly Connected Graphs. In: Caelli, T.M., Amin, A., Duin, R.P.W., Kamel, M.S., de Ridder, D. (eds.) SSPR & SPR 2002. LNCS, vol. 2396, pp. 123–132. Springer, Heidelberg (2002)
14. Conte, D., Foggia, P., Vento, M.: Challenging Complexity of Maximum Common Subgraph Detection Algorithms: A Performance Analysis of Three Algorithms on a Wide Database of Graphs. Journal of Graph Algorithms and Applications 11(1), 99–143 (2007)
15. Raymond, J., Willett, P.: Maximum Common Subgraph Isomorphism Algorithms for the Matching of Chemical Structures. Journal of Computer-Aided Molecular Design 16(7), 521–533 (2002)
16. Bern, M., Eppstein, D.: Mesh Generation And Optimal Triangulation. Computing in Euclidean Geometry 1, 23–90 (1992)
17. Lladós, J., Martí, E., Villanueva, J.: Symbol Recognition by Error-Tolerant Subgraph Matching between Region Adjacency Graphs. IEEE Trans. Pattern Analysis and Machine Intelligence 23(10), 1137–1143 (2001)
18. Konc, J., Janezic, D.: An Improved Branch and Bound Algorithm for the Maximum Clique Problem. Communications in Mathematical and in Computer Chemistry/MATCH 58(3), 569–590 (2007)
19. Pelillo, M., Torsello, A.: Payoff-Monotonic Game Dynamics and the Maximum Clique Problem. Neural Computation 18(5), 1215–1258 (2006)

Graph Characteristics
from the Schrödinger Operator

Pablo Suau[1], Edwin R. Hancock[2], and Francisco Escolano[1]

[1] Mobile Vision Research Lab, University of Alicante, Spain
{pablo,sco}@dccia.ua.es
[2] Department of Computer Science, University of York, UK
edwin.hancock@york.ac.uk

Abstract. In this paper, we show how the Schrödinger operator may
be applied to the problem of graph characterization. The motivation is
the similarity of the Schrödinger equation to the heat difussion equation, and the fact that the heat kernel has been used in the past for
graph characterization. Our hypothesis is that due to the quantum nature of the Schrödinger operator, it may be capable of providing richer
sources of information than the heat kernel. Specifically the possibility of complex amplitudes with both negative and positive components,
allows quantum interferences which strongly reflect symmetry patterns
in graph structure. We propose a graph characterization based on the
Fourier analysis of the quantum equivalent of the heat flow trace. Our
experiments demonstrate that this new method can be succesfully applied to characterize different types of graph structures.

Keywords: graph characterization, heat flow, Schrödinger equation,
quantum walks.

1 Introduction

The analysis of graph and network topology is widely used in fields including
computer vision, biology, data mining and linguistics. In all these areas, effective methods for characterizing or distinguishing different graph structures are
essential, and as a result, many approaches to the graph characterization problem have been proposed, including algorithms based on random walks [1], the
Ihara zera function [2] or the spectral radius [3]. Another family of graph characterization technique, introduced by Escolano *et al.* [4], can be derived from the
analysis of the heat flow. Heat flow accounts for information transfer between
nodes of a graph and it is determined by the heat kernel [5], and this in turn is
the solution of the heat difussion equation.

Mathematically, the heat diffussion equation is similar to Schrödinger equation, which characterizes the dynamics of a particle in a quantum system [6].
However, similarities are superficial since both underlying physics and the dynamics induced by the Schrödinger equation are different to those induced by the
heat difussion equation. In this paper we demonstrate that the solution of the

W.G. Kropatsch et al. (Eds.): GbRPR 2013, LNCS 7877, pp. 172–181, 2013.

Schrödinger equation, i.e. the Schrödinger operator, may provide a useful tool for characterizing graph structure. The quantum nature of the Schrödinger operator gives rise to several interesting non-classical effects including quantum interferences produced by the negative components of the complex amplitudes arising from the solution of Schrödinger equation. Moreover, interferences proved to be useful in several applications, e.g. detection of symmetric motifs in graphs via continuous-time quantum walks [7] or graph embedding by means of quantum commute times [8]. Furthermore, the quantum nature of the amplitude must also be taken into account when designing new methods and algorithms based on the Schrödinger operator. For instance, dynamic systems based on the Schrödinger operator, e.g. continuous-time quantum walks, are non-ergodic. In this paper, we propose a new graph characterization method that exploits features due to non-ergodicity.

The remainder of this paper is structured as follows. In Section 2 we summarize the concept of heat flow for graph characterization. In Section 3 the Schrödinger operator is introduced. The main contributions of this paper are presented in Section 4, in which we formally analyze the Schrödinger operator and propose a new graph characterization technique based on an equivalent of the heat flow. Then, in Section 5, we show some experimental results. Finally we draw some conclusions and point out ways in which this work can be further extended.

2   Heat Flow

Let $G = (V, E)$ be an undirected graph where V is its set of nodes and $E \subseteq V \times V$ is its set of edges. The Laplacian matrix $L = D - A$ is constructed from the $|V| \times |V|$ adjacency matrix A, in which the element $A(u,v) = 1$ if $(u,v) \in E$ and 0 otherwise, where the elements of the diagonal $|V| \times |V|$ degree matrix are $D(u,u) = \sum_{v \in V} A(u,v)$. The $|V| \times |V|$ heat kernel matrix K_t is the fundamental solution of the heat equation

$$\frac{\partial K_t}{\partial t} = -LK_t, \tag{1}$$

and depends on the Laplacian matrix L and time t. It describes how information flows across the edges of a graph with time, and its solution is $K_t = e^{-Lt}$.

The heat kernel K_t is a doubly stochastic matrix. Double stochasticity implies that diffusion conserves heat. In [4], a graph is characterized from the constraints it imposes to heat diffusion due to its structure. This characterization is based on the normalized instantaneous flow $F_t(G)$ of graph G, that accounts the edge-normalized heat flowing through the graph at a given instant t, and it is defined as:

$$F_t(G) = \frac{2|E|}{n} \sum_{i=1}^{n} \sum_{j \neq i} A(i,j) \left(\sum_{k=1}^{n} \phi_k(i)\phi_k(j)e^{-\lambda_k t} \right). \tag{2}$$

A more compact definition of the edge-normalized instantaneous flow is $F_t(G) = (2|E|/n)A : K_t$, where $X : Z = trace(XZ^T)$ is the Frobenius inner product. The heat flow trace describing the graph is constructed by computing Eq. 2 on the interval $[0, t_{max}]$.

3   Heat Kernel vs. Schrödinger Operator

The Schrödinger equation describes how the complex state vector $|\psi_t\rangle \in \mathbb{C}^{|V|}$ of a continuous-time quantum walk varies with time [9]:

$$\frac{\partial |\psi_t\rangle}{\partial t} = -iL|\psi_t\rangle. \tag{3}$$

Given an initial state $|\psi_0\rangle$ the latter equation can be solved to give $|\psi_t\rangle = \Psi_t |\psi_0\rangle$, where $\Psi_t = e^{-iLt}$ is a complex $|V| \times |V|$ unitary matrix. In this paper we refer to Ψ_t as the *Schrödinger operator*. Our attention in this paper will be focused on the operator itself and not on the quantum walk process. As can be seen, Eq. 3 is similar to Eq. 1. However, the physical dynamics induced by the Schrödinger equation are totally different, due to the existence of oscillations and interferences.

In this section we address the question of whether the Schrödinger operator may be used to characterize the structure of a graph. Empirical analysis on different graph structures shows that both the heat kernel and the Schrödinger operator evolve with time in a manner which strongly depends on graph structure.[1] However, the underlying physics and the dynamics are different (see Fig. 1). In the case of heat flow heat diffuses between nodes through the edges, eventually creating transitive links (energy exchanges between nodes that are not directly connected by an edge), until reaching a stationary energy equilibrium state. The Schrödinger operator yields a faster energy distribution through the system (e.g. for a 100 nodes line graph, it takes $t = 50$ time steps for the Schrödinger operator to reach every possible position on the graph, taking more than twice this time in the case of the heat kernel). Moreover, due to negativecomponents of the complex amplitudes, interferences are created, producing energy waves. The main difference is that the Schrödinger operator never reaches an equilibrium state. In other words, it is non-ergodic. Graph connectivity imposes constraints on the distribution of energy. In the case of the heat kernel, a higher number of energy distribution constraints implies the creation of more transitive links with time [4]. This is also true in the case of the Schrödinger operator, for which lower frequency and more symmetrical energy distribution patterns are also observed.

3.1   Analysis of the Schrödinger Operator

Further formal analysis of the Schrödinger operator supports the empirical evidence stated above. We first consider the Schrödinger operator when t tends to zero. Its Taylor expansion is given by:

$$\Psi t = e^{-iLt} = \cos Lt - i \sin Lt = I_{|V|} - iLt - \frac{t^2}{2!}L^2 + i\frac{t^3}{3!}L^3 + \frac{t^4}{4!}L^4 \cdots, \tag{4}$$

---

[1] Videos showing the evolution of both heat kernel and Schrödinger operator are available at http://www.dccia.ua.es/~pablo/downloads/schrodinger_operator.zip

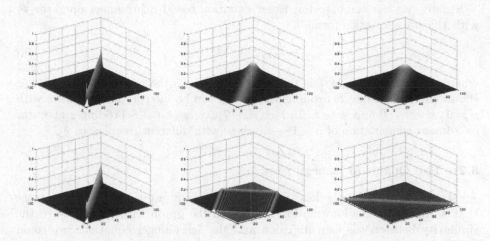

Fig. 1. Evolution of the keat kernel (top) and the Schrödinger operator (bottom) with time ($t = 1, 25$ and 50) for a 100 node line graph

where $I_{|V|}$ is the $|V| \times |V|$ identity matrix. Hence

$$\lim_{t \to 0} \Psi_t \approx I_{|V|} - iLt, \tag{5}$$

where $\Psi_t = K_t$ when $t = 0$. At this time instant every node conserves its energy (as in the case of the Heat Kernel). The role of the identity matrix is to make the Schrödinger operator unitary. Due to the $-iLt$ term, it can be seen that energy spreads as a wave even for t values close to zero. Thus, the Schrödinger operator causes energy to distribute in a waveform from the initial time instant.

In order to explore the ergodicity of the Schrödinger operator we consider both its spectral decomposition and that of the heat kernel:

$$K_t = \sum_{p=1}^{n} e^{-t\lambda_p} \phi_p \phi_p^T \text{ and} \tag{6}$$

$$\Psi_t = \sum_{p=1}^{n} e^{-it\lambda_p} \phi_p \phi_p^T, \tag{7}$$

where λ_p is the p-th eigenvalue of the Laplacian L and ϕ_p its corresponding eigenvector.

The spectral decomposition of the heat kernel demonstrate that it is dominated by the lowest eigenvalues, due to the fact that $e^{-t\lambda_p}$ tends to zero as t tends to infinity. However, $e^{-it\lambda_p}$ is indefinite when t tends to infinity. Thus, there are two importante differences with the heat kernel. Firstly, the Schrödinger operator never converges (it is non-ergodic), and secondly, it is not dominated by any particular eigenvalue (i.e. there is more dependence on global graph structure as t tends to infinity).

Finally, we can compare the Euler equation based Schrödinger operator Ψ_t with the wave equation formula

$$\psi = ve^{i(kx - wt + \epsilon)}, \tag{8}$$

where v is the amplitude, ϵ is the initial phase, k is the wavenumber, and w is the angular frequency. Schrödinger operator can be interpreted as a wave with $v = 1$, $k = \epsilon = 0$ and $w = L$. In fact, Eq. 7 expresses the Schrödinger operator as a linear combination of $p = 1 \cdots n$ waves with different frequencies λ_p.

3.2   The Quantum Energy Flow

As stated in Section 2 the heat flow characterizes a graph by means of a trace that accounts for the information flowing on the graph with time. Due to the similarity between the heat diffusion and the Schrödinger equations, we could define quantum energy flow (QEF) as

$$Q_t(G) = A : \Psi_t, \tag{9}$$

and the quantum energy trace (the equivalent of heat flow) as the evolution of Q_t with time. It must be noted that the hamiltonian of the quantum system defined by Ψ_t is given by the graph laplacian L. The adjacency matrix A in Eq. 9 causes the QEF to only account for the energy distributing through edges. In Fig. 2 we compare the heat flow and the QEF traces for two different types of graph. In [4], graph structure is characterized by the heat flow's phase transition point (PTP). The overall information transmited in the system increases until reaching a PTP, and then decreases until convergence. This is ilustrated inf Fig. 2 (left). A PTP based caracterization can not be applied in the case of the Schrödinger operator, due to its non-ergodicity and the existence of several PTPs. However, we observe again a difference in phase transition frequency depending on the structure of the graph.

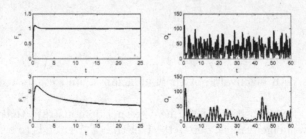

Fig. 2. Heat flow (left) and QEF (right) for two different 10 node graphs: a random graph (top) and a line graph (bottom). In both cases, the x axis represents time.

3.3   Frequency Domain Analysis of the Schrödinger Operator

The results and analysis above suggest a correlation between graph structure and both the Schrödinger operator and the QEF frequency patterns. We therefore propose a graph characterization based on the QEF in the frequency domain. In order to obtain this characterization, we consider the QEF as a non-periodic signal: we select a time interval $[0, T]$ and we apply the Fast Fourier Transform to the QEF. We refer to this representation as the *frequency domain trace*. The frequency domain trace for the graphs in Fig. 2 can be seen in Fig. 3. The first conclusion from these plots is that the more complex graphs are characterized by the presence of higher frequencies.

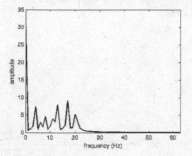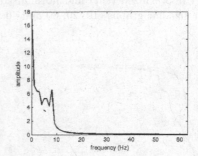

Fig. 3. Frequency domain trace obtained from the quantum energy flow of the random graph (left) and the line graph (right) in Fig. 2

However, this representation depends on graph size. Fig. 4 (left) shows the frequency domain trace for four differently sized line graphs. This plot demonstrates that the maximum spectral amplitude is proportional to the graph size. In order to compare arbitrarely sized graphs we apply a simple frequency domain trace normalization based on its maximum amplitude. The result of this normalization can be seen in Fig. 4 (right).

During our experiments we will represent graphs by means of a *cumulative frequency domain trace*, obtained by accumulating the normalized amplitudes from lower to higher frequencies of their corresponding frequency domain traces. In Fig. 5 we compare the cumulative frequency domain trace obtained from five graphs and their corresponding heat flows. In the case of the cumulative frequency domain trace, the area under the curve provides a good estimate of graph complexity. Simpler graphs yield larger areas. The PTPs of the corresponding heat flow traces also provide a good complexity estimate. In this case, the PTP for simple graphs is reached later in time. However, in this particular example, the heat flow trace estimates the complexity of the line graph to be lower than that of the circle graph. That is not the case of the cumulative frequency domain trace, for which the complexity of the line graph is higher.

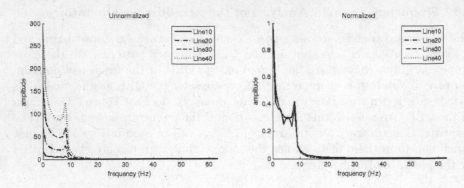

Fig. 4. Unnormalized (left) and normalized (right) frequency domain traces for four different size line graphs (10, 20, 30 and 40 nodes)

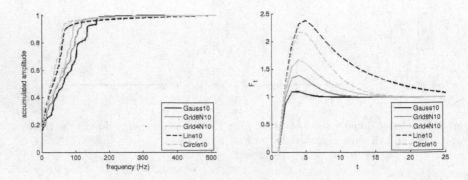

Fig. 5. Cumulative frequency domain traces (left) and heat flow (right) for five simple 10 node graphs: a random graph (Gauss10), a 8-connected 2x5 grid (Grid8N10), a 4-connected 2x5 grid (Grid4N10), a line graph (Line10) and a circular graph (Circle10)

4   Experimental Results

4.1   Noise Sensitivity

The aim of this first experiment is to show the sensitivity of frequency domain traces to graph noise. We first constructed a base 400 nodes random graph by means of the Erdös-Rényi model [10]. We then compared the frequency domain trace of the base graph to those obtained after applying random edit operations on it. In this experiment we only applied edge removal operations, and thus, in each iteration, we remove a random edge from the base graph and we compute the Euclidean trace between the unnormalized traces. The results are shown in Fig. 6. Four experiments were performed, using four different time intervals $[0..T]$ to construct the frequency domain traces.

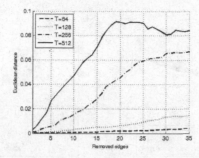

Fig. 6. Results of the noise sensitivity experiment. Number of edit operations (edge removals) versus distance between edited graph's frequency domain trace and base graph's one, for four different T values.

From Fig. 6, it is clear that the final trace is not strongly affected by small disturbances. For larger time intervals there appears to be a significant sensitivity to noise. However, difference between traces is still low. The remainder of the experiments in this paper are conducted after setting $T = 1024$.

4.2   Graph Characterization

In order to test the discriminative power of our characterization we constructed a dataset of synthetic graphs. The dataset consists of three groups of 32 graphs, each group characterized by a different graph structure. All of the graphs in the dataset have 90 nodes. The graphs in the first group are random graphs constructed using the Erdös-Rényi [10] model, in which each pair of nodes is linked by an edge with probability given by p. In our experiments we set $p = 0.1$. The graphs in the second group belong to the category of scale free graphs (i.e. graphs for which its degree distribution follows a power law), and were constructed using the Barabási and Albert's model [11].In this model we have set $m_0 = 5$ for the initial size of the graphs and $m = 2$ for the number of links to add during each iteration, following the addition of a node. Finally, the graphs in the third group correspond to small world graphs (i.e. graphs in which most nodes are not neighbours of each other, but in which average path length between a graph pair of nodes is small). These small world graphs are generated by means of the Watts and Strogatz algorithm [12]. In this case we set the mean degree value to $K = 10$ and the rewiring probability to $p = 0.2$.

A cumulative frequency domain trace was computed for all graphs in the set, and the results are shown in Fig. 7. The first conclusion of our experiment is that these traces clearly discriminate between different graph structures. This conclusion is supported by a Multidimensional Scaling analysis (MDS) of the traces (also shown in Fig. 7). The aim of MDS is to apply dimensionality reduction on data while preserving relative distances between patterns. If we project the traces onto a 2D space, the graphs in the three groups are clearly split into three different clusters with high intra-cluster homogeneity and high-inter cluster separability.

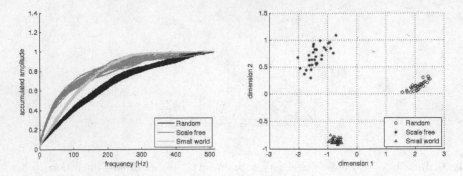

Fig. 7. Characterization of synthetic graphs. Left: cumulative frequency domain traces. Right: MDS results.

In Fig. 7 we explore the relationship between frequency and structure. The frequency spectrum of random graphs is characterized by higher amplitudes at high frequencies. In the case of small world graphs, the predominant frequencies are in the middle part of the spectrum. Scale free graphs are characterized by higher amplitudes at lower frequencies. These results suggest that the structure of random graphs is more complex in the sense that it imposes more constraints to the distribution of energy on the graph. As a consequence, energy waves exhibit higher frequency as they propagate. Scale free and small world graphs impose less restrictions on the distribution of energy through the graph, and are associated with lower frequency patterns.

5   Conclusions and Future Work

Heat flow, based on the heat kernel, has been succesfully used to characterize graph structure. The aim of the present paper was to answer the question of whether the Schrödinger operator (the solution to the Schrödinger equation) can be used also to characterize graph structure. After analyzing energy distribution through the graph based on the Schrödinger operator, we introduced a new characterization method based on the analysis in the frequency domain. Our experiments show that the *cumulative frequency domain trace* is a useful tool for graph analysis, that is not sensitive to small changes in graph structure.

However, based on these promising preliminary results, further in depth analysis is required. Firstly, and similarly to heat flow, the cumulative frequency domain trace does not provide us with a quantitative measure to directly compare graph structures. A first step in this direction could be to apply this trace as part of the thermodynamic depth complexity measurement framework, in order to obtain a numerical representation of graph structure [4][13]. Secondly, during our analysis of the Schrödinger operator we detected the presence of symmetric energy distribtuion patterns on the graph. We could analyze how this symmetry depends on graph structure and whether the results of this analysis are related to previous work on symmetry detection based on quantum walks [7]. Finally, it would be interesting to relate the Schrödinger operator and the cumulative frequency domain trace representation to the structure of complex network systems such as social or biological networks.

References

1. Aziz, F., Wilson, R.C., Hancock, E.R.: Graph Characterization via Backtrackless Paths. In: Pelillo, M., Hancock, E.R. (eds.) SIMBAD 2011. LNCS, vol. 7005, pp. 149–162. Springer, Heidelberg (2011)
2. Peng, R., Wilson, R., Hancock, E.: Graph Characterization vi Ihara Coefficients. IEEE Transactions on Neural Networks 22(2), 233–245 (2011)
3. Das, K.C.: Extremal Graph Characterization from the Bounds of the Spectral Radius of Weighted Graphs. Applied Mathematics and Computation 217(18), 7420–7426 (2011)
4. Escolano, F., Hancock, E., Lozano, M.A.: Heat Diffusion: Thermodynamic Depth Complexity of Networks. Physical Review E 85(3), 036206(15) (2012)
5. Xiao, B., Hancock, E., Wilson, R.: Graph Characteristics from the Heat Kernel Trace. Pattern Reognition 42(11), 2589–2606 (2009)
6. Aubry, M., Schlickewei, U., Cremers, D.: The Wave Kernel Signature: A Quantum Mechanical Approach To Shape Analysis. In: IEEE International Conference on Computer Vision (ICCV), Workshop on Dynamic Shape Capture and Analysis (4DMOD) (2011)
7. Rossi, L., Torsello, A., Hancock, E.R.: Approximate Axial Symmetries from Continuous Time Quantum Walks. In: Gimel'farb, G., Hancock, E., Imiya, A., Kuijper, A., Kudo, M., Omachi, S., Windeatt, T., Yamada, K. (eds.) SSPR & SPR 2012. LNCS, vol. 7626, pp. 144–152. Springer, Heidelberg (2012)
8. Emms, D., Wilson, R.C., Hancock, E.R.: Graph Embedding Using Quantum Commute Times. In: Escolano, F., Vento, M. (eds.) GbRPR. LNCS, vol. 4538, pp. 371–382. Springer, Heidelberg (2007)
9. Farhi, E., Gutmann, S.: Quantum Computation and Decision Trees. Physical Review A 58, 915–928 (1998)
10. Erdös, P., Rényi, A.: On Random Graphs. I. Publicationes Mathematicae 6, 290–297 (1959)
11. Barabási, A.L., Albert, R.: Emergence of Scaling in Random Networks. Science 286(5439), 509–512 (1999)
12. Watts, D.J., Strogatz, S.H.: Collective Dynamics of 'Small-World' Networks. Nature 393(6684), 440–442 (1998)
13. Han, L., Escolano, F., Hancock, E., Wilson, R.: Graph Characterizations From Von Neumann Entropy. Pattern Recognition Letters 33(15), 1958–1967 (2012)

Persistent Homology in Image Processing*

Herbert Edelsbrunner

Institute of Science and Technology Austria
Am Campus 1, 3400 Klosterneuburg
edels@ist.ac.at

Taking *images* is an efficient way to collect data about the physical world. It can be done fast and in exquisite detail. By definition, *image processing* is the field that concerns itself with the computation aimed at harnessing the information contained in images [10]. This talk is concerned with topological information. Our main thesis is that persistent homology [5] is a useful method to quantify and summarize topological information, building a bridge that connects algebraic topology with applications. We provide supporting evidence for this thesis by touching upon four technical developments in the overlap between persistent homology and image processing.

1. **Hierarchically Represented Images.** Algorithms for persistent homology are fast but can be challenged by the size of high-resolution images. Hierarchical data structures, such as quad-trees for 2- and oct-trees for 3-dimensional images, reduce the size and access time [9]. They can also be used to speed up the computation of persistence diagrams by orders of magnitudes [1].
2. **Adaptive Topology.** The connectivity on a microscopic level (between pixels or voxels) has an influence on the global topological information as computed for example by persistent homology. The adaptive topology is defined on the microscopic level to guarantee global results that are consistent with basic symmetries in topology, including Lefschetz and Alexander duality [7].
3. **Stable Measurements.** A fundamental result in persistence is its stability under perturbations of the input data [3,4]. We leverage this property to obtain stable length estimates of tube-like shapes, such as root systems of plants [6]. Such estimates are desirable in phenotype-genotype studies as they limit the influence of noise in the data acquisition on the conclusions we draw.
4. **Connection to Scale Space.** We use persistent homology to quantify the effect of diffusion on critical points in an image [8]. The main result is an upper bound on the persistence moment, showing that it goes to zero as the image diffuses [2]. It sheds light on the experimentally observed phenomenon that the creation of critical points during diffusion is a rare event.

The four technical developments differ from each other, ranging from foundational to applied and from algorithmic to mathematical. This is evidence for the

---

* This research is partially supported by the European Science Foundation (ESF) under the Research Network Programme, the European Union under the Toposys Project FP7-ICT-318493-STREP, the Russian Government under the Mega Project 11.G34.31.0053.

versatility of persistent homology, which is a consequence of its role as a bridge that connects fundamental concepts in mathematics and general phenomena in our physical reality.

References

1. Bendich, P., Edelsbrunner, H., Kerber, M.: Computing robustness and persistence for images. IEEE Trans. Visual. Comput. Graphics 16, 1251–1260 (2010)
2. Chen, C., Edelsbrunner, H.: Diffusion runs low on persistence fast. In: Proc. 13th Internat. Conf. Comput. Vision, pp. 423–430 (2011)
3. Cohen-Steiner, D., Edelsbrunner, H., Harer, J.: Stability of persistence diagrams. Discrete Comput. Geom. 37, 103–120 (2007)
4. Cohen-Steiner, D., Edelsbrunner, H., Harer, J., Mileyko, Y.: Lipschitz functions have L_p-stable persistence. Found. Comput. Math. 10, 127–139 (2010)
5. Edelsbrunner, H., Letscher, D., Zomorodian, A.J.: Topological persistence and simplification. Discrete Comput. Geom. 28, 511–533 (2002)
6. Edelsbrunner, H., Pausinger, F.: Stable length estimates of tube-like shapes. Manuscript, IST Austria, Klosterneuburg, Austria (2013)
7. Edelsbrunner, H., Symonova, O.: The adaptive topology of a digital image. In: 9th Internat. Sympos. Voronoi Diagrams Sci. Engin., pp. 41–48 (2012)
8. Iijima, T.: Basic theory on normalization of a pattern (in case of typical one-dimensional pattern) – in Japanese. Bull. Electrotechn. Lab. 26, 368–388 (1962)
9. Samet, H.: The Design and Analysis of Spatial Data Structures. Addison-Wesley, Reading (1990)
10. Sonka, M., Hlavak, V., Boyle, R.: Image Processing, Analysis and Machine Learning, 2nd edn. PWS Publishing, Pacific Grove (1999)

Towards Minimal Barcodes[*]

Rocío González-Díaz[1], María-José Jiménez[1], and Hamid Krim[2]

[1] Applied Math Dept., School of Computer Engineering, U. Seville, Spain
[2] ECE Dept., NCSU, Raleigh, NC USA
{rogodi,majiro}@us.es, ahk@ncsu.edu

Abstract. In the setting of persistent homology computation, a useful tool is the persistence barcode representation in which pairs of birth and death times of homology classes are encoded in the form of intervals. Starting from a polyhedral complex K (an object subdivided into cells which are polytopes) and an initial order of the set of vertices, we are concerned with the general problem of searching for filters (an order of the rest of the cells) that provide a minimal barcode representation in the sense of having minimal number of "k-significant" intervals, which correspond to homology classes with life-times longer than a fixed number k. As a first step, in this paper we provide an algorithm for computing such a filter for $k = 1$ on the Hasse diagram of the poset of faces of K.

Keywords: Persistent homology, persistence barcodes, graphs, polyhedral complexes.

1   Introduction

The persistence barcode representation, which encodes pairs of cells meaning birth and death of homology classes in persistent homology computation, depends on the filter considered for such computation. Although, as we will see later, the total number of intervals remains invariant, the lengths of these intervals depend on the selected filter. Since, non-significant intervals (i.e. intervals with short length) do not imply relevant homological information, we are interested in providing good properties to be satisfied by the selected filter, so that the number of non-significant intervals in the corresponding persistence barcode is maximized (i.e. the number of significant intervals is minimized). Motivated by practical applications of persistent homology computation, our starting point is a given polyhedral complex and an initial order of the set of vertices. From an information-theoretic viewpoint, and if we interpret the number of "significant intervals" as the coding length of a complex, our goal is to then select the most "parsimonious" representation (also by Occam's razor principle). As is also well known, the coding length is also intimately related to the notion of entropy (i.e. a topological entropy of a complex in our case). While ideally, one would want to balance this minimization with a penalty term of the number of "insignificant"

---

[*] Partially supported under grant MTM2012-32706. Authors listed alphabetically.

W.G. Kropatsch et al. (Eds.): GbRPR 2013, LNCS 7877, pp. 184–193, 2013.

intervals. This development is under way, and requires the statistical distribution of the long intervals. Our aim is to look for a way to insert the rest of the cells along the filter in order to minimize the number of significant intervals.

The remainder of this paper is organized as follows. Section 2 covers the relevant background material. In Section 3, we prove that the number of intervals in a persistence barcode does not depend on the selected filter, we give the definition of minimal barcode and show two practical examples in which computation of minimal barcodes can be useful. In Section 4, an algorithm for computing on the Hasse diagram of the poset of faces of the polyhedral complex, a filter that produces a minimal barcode is given. Section 5 is devoted to relations between minimal barcodes and the optimal discrete Morse function. Conclusions and future work are presented in Section 6.

2   Background

This paper is developed in a combinatorial algebraic topology setting, where the objects of interest are (geometric) polyhedral complexes [14]. A *polyhedral complex* K is a collection of convex polytopes such that: (1) every face of a polytope in K is itself a polytope in K; (2) the intersection of any two polytopes in K is a face of each of them. The set of cells of dim. i (or i-cells), σ^i (superscript specifies its dimension), will be denoted by K^i, and the number of cells in the set of cells S by $|S|$. An n-dim. polyhedral complex satisfies that $K^n \neq \emptyset$ and $K^{n+1} = \emptyset$, where $n > 0$. Particular cases arise when the polytopes belong to a specified set of polyhedra, such as simplicial complexes (vertices, edges, triangles, tetrahedra, up to dim. 3) or cubical complexes (vertices, edges, squares, cubes, up to dim. 3). In general, we will always refer to a finite polyhedral complex K (i.e. with m cells, m being a finite number). Several polytopes associated with combinatorial optimization problems have surprisingly small extended formulations (see [3,20,10]). It may not be very surprising that no polynomial size extended formulations of polytopes associated with NP-hard optimization problems like the traveling salesman polytope are known.

Homology theory uses algebraic groups to encode the topological structure of K. Finite formal sums of elements of K^i (called i-*chains*) define an additive abelian group structure on K^i. A *proper face* of $\sigma \in K^i$ is a face of σ of dim. $i - 1$. The *boundary* of σ, denoted by $\partial(\sigma)$ is the formal sum (with coefficients in $\mathbb{Z}/\mathbb{Z}2$) of the proper faces of σ. The boundary operator is extended to all chains of K by linearity. An i-chain a is an i-*cycle* if $\partial(a) = 0$; it is an i-*boundary* if there is an $(i + 1)$-chain b such that $\partial(b) = a$. Two i-cycles a and a' are *homologous* if $a + a'$ is an i-boundary. The quotient of i-cycles over i-boundaries is the ith *homology group of K*. The i-*Betti number* that is the rank of the ith homology group of K will be denoted by β^i. Then, the basic topological structure of K is quantified by the number of independent cycles in each homology group. See [17,12].

Persistent homology [4,21] studies homology classes and their lifetimes (persistence). While homology characterizes an object, persistent homology characterizes a sequence of growing object-instances, i.e. an object together with an

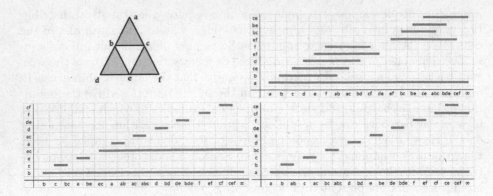

Fig. 1. An example of a 2-dim. simplicial complex and three persistence barcodes corresponding to different filters (fixed by the order in the set of cells given along the horizontal axis of the representations). Bottom: two minimal barcodes. Right: two barcodes with same order of the 0-cells.

order of the cells $\{\sigma_1, \ldots, \sigma_m\}$ (where $\sigma_s < \sigma_t$ iff $s < t$) called a *filter*, such that if σ_i is a proper faces of σ_j then $i < j$.

Given a filter of K, the algorithm for computing persistent homology that appears in [4], marks an i-cell σ_t as positive (birth) if it belongs to an i-cycle in K_t (σ_t creates a new homology class at time t) and negative (death) otherwise (σ_t destroys the homology class created at some time s for $0 \leq s < t$).

Given a filter $\{\sigma_1, \ldots, \sigma_m\}$, a *persistence barcode* [2] is a graphical representation of pairs of birth and death times as a collection of horizontal line segments (*intervals*) in a plane. If a cell σ_s creates a homology class at time s, and it is destroyed at time t, $0 \leq s < t \leq m$ then the interval $[s, t)$ is added to the corresponding persistence barcode (see [2]); If a cell σ_s, $0 \leq s \leq m$ creates a homology class at time s and it survives along the process, then the interval $[s, \infty)$ is added to the persistence barcode.

3 Minimal Barcodes

In this section, a formal definition of minimal barcode is presented along with two practical examples for which a computation of minimal barcodes can be useful.

For a fixed i, we refer to *i-barcode* the set of intervals of a given persistence barcode corresponding to the pairs of positive i-cells and negative $(i+1)$-cells of K. The following result holds.

Lemma 1. *The number of intervals in an i-barcode, $0 \leq i \leq n$, is constant, independently of the selected filter.*

Proof. First, the number of intervals of infinite length in the i-barcode is independent on the filter since it coincides with β^i. Second, each i-cell σ_t^i in the given filter, $0 \leq t \leq m$, is marked as positive or negative. No cell can remain unmarked after the whole process. This is easy to prove using AT-models [6]: σ_t^i is marked as positive if $f^{i-1}\partial(\sigma_t^i) = 0$ and negative otherwise (see [7]). Third, let B^i (resp. D^i) be the number of positive (resp. negative) i-cells. Then, we have that $|K^0| = B^0$, $|K^n| = B^n + D^n$, $\beta^n = B^n$, and, for $0 < i < n$, $|K^i| = B^i + D^i$ and $\beta^i = B^i - D^{i+1}$. Therefore, we obtain that $D^0 = 0$, $B^0 = |K^0|$, and for $0 < i \leq n$, $D^i = |K^{i-1}| - \beta^{i-1} - D^{i-1}$ and $B^i = |K^i| - D^i$. We conclude that, for $0 \leq i \leq n$, nor B^i neither D^i (which coincides with the number of finite intervals in the $(i-1)$-barcode) depend on the selected filter. $\square$

A general idea in the study of topological persistence is that significant topological attributes must have long life-times, and topological features with short life-times are considered to be "noise". Following this idea, in the definition below, k-significant intervals correspond to homology classes whose life-times are longer than a fixed number k.

Definition 1. *Fixed $k > 0$, an interval $[s, t)$ is k-significant if $k < t - s$.*

Our general aim is to find, under some constraints, depending on the nature of the application, filters that minimize the number of long-life homological classes which are associated with significant intervals.

Nevertheless, filters are, in many cases, totally determined. Examples of this is when objects are presented as point cloud data and Rip or Cech complexes are constructed to fill in the higher-dimensional simplices of the proximity graph whose edges are determined by proximity, i.e. vertices within some specified distance ϵ (see [8]). But, in other cases, only order of 0-cells are given. For example, when a continuous function (e.g. a height function or barycentric distance) is provided and the 0-cells of K are ordered by the function values at them.

We briefly present here two particular examples of this last case:

1. Application of persistent homology to the evaluation of a 3D reconstruction process (carving voxel) of human models from images captured from a set of cameras placed around the subject. In fact, we refer to the visual hull that is constructed from images of cameras from different viewpoints. This problem can be seen as a view planning problem (see [19], a survey of computer vision sensor planning, [18], a more recent survey of view planning for 3-D vision). In our case, [9], starting from a compact block of voxels, each time a camera is added, a set of voxels are deleted (carved) from the 3D reconstruction, so the sequence of 3D reconstructions along decreasing number of cameras gives place to a filter of the corresponding cubical complexes. This allows to analyze the topological evolution of the reconstruction process. Only k-significant intervals are considered, where k is the distance (in number of cells) from one reconstruction to the next one. An initial partial order is hence considered in the set of vertices that have to be added in the computation along the process. See Fig. 2.
2. In [11] an image/video application using topological invariants for human gait recognition is shown. Using a background subtraction approach, a stack of silhouettes is extracted and glued through their gravity centers, forming

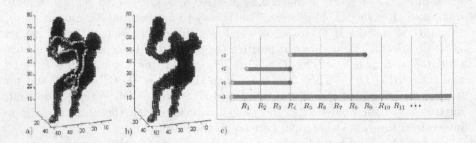

Fig. 2. Examples of 3D reconstructions using a) 4 cameras and b) 10 cameras. Representative cycles of homology are highlighted in both cases. c) Persistence barcode associated to the whole sequence of 3D reconstructions with increasing number of cameras (from 1 to 50) is shown.

a 3D digital image I. From this 3D representation, the boundary simplicial complex $\partial K(I)$ is obtained. Four filters are computed preserving four different given orders of the vertices of $\partial K(I)$ depending on four directions of view. The persistence barcodes associated with the previous filters are then computed (see Fig. 3). These filters capture relations among the parts of the human body when walking. Only intervals with long life-times are considered. Finally, a topological gait signature is extracted from the persistence barcodes according to the filters.

From now on, suppose a bijective function $h : K^0 \to \{1, 2, ..., |K^0|\}$ (i.e. an order of the 0-cells of K) is given. Let us denote by $\mathcal{F}$ the set of filters F of K such that for any two 0-cells $\sigma_s^0, \sigma_t^0 \in F$, $s < t$, it is satisfied that $h(\sigma_s^0) < h(\sigma_t^0)$.

Definition 2. *A persistence barcode associated with a filter $F \in \mathcal{F}$ is minimal if the persistence barcode associated with any other filter in $\mathcal{F}$ contains greater or equal number of significant intervals.*

Observe that a filter $F \in \mathcal{F}$ with a minimal barcode always exists and might not be unique (see Fig. 1 as examples of minimal barcodes).

4 Hasse Diagrams for the Poset of Faces and Minimal Barcodes

Our aim in this section is to construct a filter $F \in \mathcal{F}$ with a minimal barcode.

Consider the poset given by the set of cells of K together with the partial order induced by the coface relation, that is, $\tau < \sigma$ if τ is a face of σ. The *Hasse diagram H* of this poset (poset of faces) is the directed graph whose vertex set is the set of cells and whose arcs are the covering pairs (τ, σ) in the poset, that is, $\tau < \sigma$ and there is no ρ such that $\tau < \rho < \sigma$ (it is said that σ covers τ). We draw the Hasse diagram in the plane in such a way that, if τ is a face of

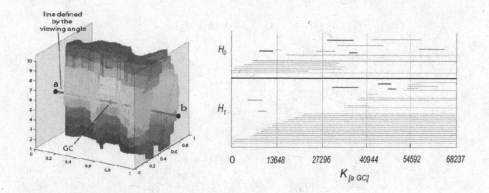

Fig. 3. Left: A simplicial complex $\partial K(I)$ corresponding to a gait, vertical direction of view (defined by the segment $[a, b]$) and the gravity center GC. Right: persistence barcode according to the vertical direction of view, corresponding to a filter of the subcomplex $K_{[a, GC]}$ (from a to GC) of $\partial K(I)$.

σ (σ covers τ), then the point representing σ is in a lower level than the point representing τ, corresponding the level with the dimension of the cells. Then no arrows are required in the drawing, since the directions of the arrows are implicit. V^i denotes the set of points v^i at level i, for $0 \leq i \leq n$.

In Alg. 1, a matching (or independent set of edges) M in H is provided, together with a vertex-labeling of H (see Fig. 4). The resulting labeling and matching will produce a filter of K.

A weight $w^i(v^i)$ for each $v^i \in V^i$, $1 \leq i \leq n$, will also be assigned along the process as follows. First, $\ell^0 : V^0 \to \{1, \ldots, |K^0|\}$ is defined for each vertex $v^0 \in V^0$ by $\ell^0(v^0) = h(\sigma^0)$, where σ^0 is the 0-cell represented by the point v^0. Second, for $i = 1$ to $i = n$, the weight of each point $v^i \in V^i$ will be

$$w^i(v^i) = \max\{\ell^{i-1}(v^{i-1}) \text{ such that } v^{i-1} \text{ is adjacent to } v^i\}.$$

Observe that more than one point in V^i can have the same weight. Then, a matching between vertices of V^{i-1} and vertices of V^i is given satisfying that if v^{i-1} is matched with v^i then $\ell(v^{i-1}) = w(v^i)$. Observe that fixing a weight w, only one point of the set $W = \{v^i \in V^i \text{ such that } w^i(v^i) = w\}$ is matched with some point in V^{i-1}. At the end of the process, a bijective function $\ell^i : V^i \to \{1, \ldots, |K^i|\}$, $1 \leq i \leq n$, is obtained, satisfying:

P1 if $w^i(u^i) < w^i(v^i)$, then $\ell^i(u^i) < \ell^i(v^i)$ for any $u^i, v^i \in V^i$;

P2 if $w^i(v^i) = w^i(u^i)$ and $(v^{i-1}, v^i) \in M$ for some $v^{i-1} \in V^{i-1}$, $v^i, u^i \in V^i$ then $\ell^i(v^i) < \ell^i(u^i)$ (any point in V^i matched with a point in V^{i-1} always precedes the other points in V^i with same weight);

P3 if $w^i(v^i) = w^i(u^i)$, $v^i, u^i \in V^i$ and $(v^i, v^{i+1}) \in M$ for some $v^{i+1} \in V^{i+1}$ then $\ell^i(u^i) < \ell^i(v^i)$ (points in V^i matched with points in V^{i+1} go after other points in V^i not matched with any point in V^{i+1}, with same weight);

P4 Points in V^i with same weight have consecutive labels.

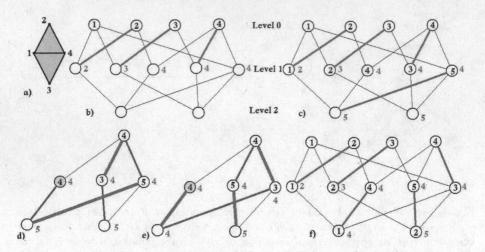

Fig. 4. a) An example of a 2-dim. simplicial complex; b) 1st step ($i = 1$) of Alg. 1 (red edges represent matchings); c) 2nd step ($i = 1$) of Alg. 1; d) subgraph H_{v_3} and an augmenting path (blue and red edges); e) a maximal matching for H_{v_3} (red edges); f) final matching M on H.

P1, P2 and P3 will guarantee that the corresponding filter is correct in the sense that a cell is not added to the filter until all its faces are added.

Algorithm 1. *Computing a vertex-labeling and a matching M of H.*

a. *Labeling $\ell^0 : V^0 \to \{1, 2, \ldots, |K^0|\}$ is given by the corresponding initial order of the 0-cells of K.*

b. *Assign the weight $w^1(v^1)$ to each point $v^1 \in V^1$.*

c. *$M = \{(v^0, v^1)$ satisfying that $v^0 \in V^0$, $v^1 \in V^1$, $\ell^0(v^0) = w^1(v^1)\}$. If there are more than one point in V^1 with the same weight, select one of them to match.*

d. *Construct $\ell^1 : V^1 \to \{1, 2, \ldots, |K^1|\}$ satisfying P1, P2, P4.*

Repeat the following process for $i = 1$ to $i = n - 1$:

1. *Assign the weight $w^{i+1}(v^{i+1})$ to each point $v^{i+1} \in V^{i+1}$.*

2. *Update matching:*

$$M := M \cup \{(v^i, v^{i+1}) \text{ such that } v^i \in \bar{M}^i, v^{i+1} \in V^{i+1}, \ell^i(v^i) = w^{i+1}(v^{i+1})\},$$

where $\bar{M}^i$ is the set of points of V^i not-matched with any point of V^{i-1}. If there are more than one point in V^{i+1} with the same weight, select one of them to match.

3. *For each $v^{i-1} \in V^{i-1}$, consider the subgraph $H_{v^{i-1}}$ whose set of vertices is: $A \cup B \cup C$ where $A = \{v^{i-1}\}$, $B = \{v^i \in V^i \text{ such that } w^i(v^i) = \ell^{i-1}(v^{i-1})\}$ and $C = \{v^{i+1} \in V^{i+1} \text{ such that } w^{i+1}(v^{i+1}) = \ell^i(v^i) \text{ for some point } v^i \in V^i \text{ satisfying that } w^i(v^i) = \ell^{i-1}(v^{i-1})\}$.*

3.1. If the matching $M|_{H_{v^{i-1}}}$ (i.e., M restricted to $H_{v^{i-1}}$) is not maximal in $H_{v^{i-1}}$, find a maximal one, $M_{v^{i-1}}$, using the augmenting path algorithm, with the restriction that each augmented path always begins in a not-matched vertex $v^i \in H_{v^{i-1}} \cap V^i$. This last restriction will guarantee that if v^{i-1} was not matched in $H_{v^{i-1}}$, it remains unmatched in $H_{v^{i-1}}$.

3.2. Remove from M the pairs in $M|_{H_{v^{i-1}}}$ and add to M the pairs in $M_{v^{i-1}}$.

4. Update $\ell^i : V^i \to \{1, 2, \ldots, |K^i|\}$ satisfying P1, P2, P3, P4.

5. For each point $v^{i+1} \in V^{i+1}$, update weight $w^{i+1}(v^{i+1})$ if needed.

6. Construct $\ell^{i+1} : V^{i+1} \to \{1, 2, \ldots, |K^{i+1}|\}$ satisfying P1, P2, P4.

Observe that a labeling ℓ^i satisfying P1, P2, P3, P4 (Step 4 in the description of the algorithm above) can always be obtained. First, before updating, labeling ℓ^i satisfied P1, P2, P4 (Step 6). Second, to satisfy P3, we just interchange the labels between points in V^i with same weight, then the updated labeling ℓ^i will also P3 and also P1, P2 and P4.

Since the points $v^i \in V^i$ correspond to i-cells in K, $0 \leq i \leq n$, an order of all the points in the planar representation of the Hasse diagram of K provide a filter of K. Such an order O can be constructed as follows:

Algorithm 2. *Computing a filter F of K form a vertex-labeling and matching M of H obtained in Alg. 1.*

1. *Initially O is the ordered set of all the points in V^0 ordered by their labels.*
2. *For $i = 1$ to $i = n$ do:*
   (a) *For every point $v^i \in V^i$ matched with a point $v^{i-1} \in V^{i-1}$, insert v^i in O just after v^{i-1} (this way, the i-cell associated with the point v^i is added to the filter F just after its last face is added).*
   (b) *Add the rest of the points in V^i at the end of O ordered by their labels.*

Remark 1. The number of non-significant intervals in the filter F given by the order O coincides with the number of pairs in M.

Proposition 1. *Fixing a filter F of K up to dimension $i - 1$ (i.e., a filter of $\bigcup_{j=0}^{i-1} K^j$), Step 3 in the description of Alg. 1 produces a minimal i-barcode.*

Proof. Observe that fixing a labeling and matching up to level $i - 1$, Setp 3 of Alg. 1 produces a maximal matching between points at level $i - 1$, i and $i + 1$, with the condition that points in v^{i-1} already matched with points in V^{i-2} are not matched with any point in V^i. □

Implementation of the above algorithms is an ongoing work. Based on the previous proposition and some preliminary computations our conjecture is that the procedure explained above produces a filter with a minimal barcode.

5 Relations between Minimal Barcodes and Optimal Discrete Morse Function

Discrete Morse functions on cell complexes were defined by Forman in [5]. A function, $f : K \to \mathbb{R}$ is a discrete Morse function if for every $\sigma \in K$, f takes a value less than or equal to $f(\sigma)$ on at most one coface of σ and takes a value greater than or equal to $f(\sigma)$ on at most one face of σ. A cell σ is critical if all cofaces take strictly greater values and all faces are strictly lower. A discrete vector field V is a collection of pairs $(\sigma^i < \sigma^{i+1})$ of cells in K such that each cell is in at most one pair of V. A discrete Morse function defines a discrete vector field by pairing $\sigma^i < \sigma^{i+1}$ whenever $f(\sigma^i) \geq f(\sigma^{i+1})$. The critical cells are precisely those that do not appear in any pair. Discrete vector fields that arise from Morse functions are called gradient vector fields. A V-path is a sequence of cells: $\sigma_0^i, \sigma_0^{i+1}, \sigma_1^i, \sigma_1^{i+1}, \ldots \sigma_{r-1}^i, \sigma_{r-1}^{i+1}, \sigma_r^i$ where $(\sigma_t^i, \sigma_t^{i+1}) \in V$, $\sigma_t^{i+1} > \sigma_{t+1}^i$ and $\sigma_t^i \neq \sigma_{t+1}^i$. A V-path is a non-trivial closed V-path if $\sigma_r^i = \sigma_0^i$ for $r \geq 1$. Forman shows that a discrete vector field is the gradient vector field of a discrete Morse function if and only if there are no non-trivial closed V-paths.

There have been several works in the literature dealing with the problem of obtaining optimal discrete Morse functions (the function has the minimum possible number of critical cells in each dimension) and perfect Morse function (the number of critical i-cells coincides with the ith Betti number of the complex). In [13] it is shown that computing optimal Morse matchings in the setting of simplicial complexes is NP-hard. In [15], a linear algorithm to define optimal discrete Morse functions on discrete 2-manifolds is introduced. In [1] the authors establish conditions under which a 2-dim. simplicial complex admits a perfect discrete Morse function and conversely.

It is clear that our work presents similarities with the problem of the computation of optimal discrete Morse functions. But in our case, a fixed ordering on the 0-cells are given. Then, optimal discrete Morse matchings could not produce minimal barcodes (indeed, could not produce valid filters) and viceversa, the set of non-significant intervals in a minimal barcode could not be an optimal discrete Morse matching.

6   Conclusions and Future Work

In this paper, starting from a polyhedral complex K and an initial order of the set of vertices, we provide an algorithm for computing a filter on the Hasse diagram of the poset of faces of K such that the associated persistence barcode representation has a minimal number of significant intervals which correspond to homology classes with life-times longer than 1.

An idea to adapt the presented algorithm to compute minimal barcodes to the case in which significant intervals are intervals with length greater than k, for $k > 1$, could be: First, to compute a minimal barcode using the above algorithms. Observe that, in this case, the matched points are successively inserted in the filter whereas the non-matched points are successively added at the end. Second, modify the Hasse diagram pretending collapses of the pairs of cells associated with the non-significant intervals and apply the above algorithm again. Observe that in this step we only reorder the cells that have been added to the end. Third, repeat the process $k - 1$ times.

References

1. Ayala, R., Fernandez-Ternero, D., Vilches, J.A.: Perfect discrete Morse functions on 2-complexes. Pattern Recognition Letters 33, 1495–1500 (2012)
2. Carlsson, G., de Silva, V., Morozov, D.: Zigzag persistent homology and real-valued functions. In: Proc. 25th Annual Symposium on Computational Geometry (SoCG), pp. 247–256 (2009)
3. Conforti, M., Cornuejols, G., Zambelli, G.: Extended formulations in combinatorial optimization. 4OR: A Quarterly Journal of Operations Research 8(1), 1–48 (2010)
4. Edelsbrunner, H., Letscher, D., Zomorodian, A.: Topological persistence and simplification. In: Proc. 41st Annual Symposium on Foundations of Computer Science (FOCS 2000), pp. 454–463. IEEE Computer Society (2000)
5. Forman, R.: Morse theory for cell complexes. Advances in Mathematics 134, 90–145 (1998)
6. Gonzalez-Diaz, R., Real, P.: On the cohomology of 3D digital images. Discrete Applied Math. 147(2-3), 245–263 (2005)
7. Gonzalez-Diaz, R., Ion, A., Jimenez, M.J., Poyatos, R.: Incremental-Decremental Algorithm for Computing AT-Models and Persistent Homology. In: Real, P., Diaz-Pernil, D., Molina-Abril, H., Berciano, A., Kropatsch, W. (eds.) CAIP 2011, Part I. LNCS, vol. 6854, pp. 286–293. Springer, Heidelberg (2011)
8. Ghrist, R.: Barcodes: The persistent topology of data. Bulletin of the American Mathematical Society 45, 61–75 (2008)
9. Gutierrez, A., Monaghan, D., Jiménez, M.J., O'Connor, N.E.: Persistent Homology for 3D Reconstruction Evaluation. In: Ferri, M., Frosini, P., Landi, C., Cerri, A., Di Fabio, B. (eds.) CTIC 2012. LNCS, vol. 7309, pp. 139–147. Springer, Heidelberg (2012)
10. Kaibel, V.: Extended formulations in combinatorial optimization. Optima 85, 2–7 (2011)
11. Lamar-León, J., García-Reyes, E.B., Gonzalez-Diaz, R.: Human Gait Identification Using Persistent Homology. In: Alvarez, L., Mejail, M., Gomez, L., Jacobo, J. (eds.) CIARP 2012. LNCS, vol. 7441, pp. 244–251. Springer, Heidelberg (2012)
12. Hatcher, A.: Algebraic Topology. Cambridge University Press (2002)
13. Joswig, M., Pfetsch, M.E.: Computing Optimal Discrete Morse Funcions. Elec. Notes Disc. Math. 17, 191–195 (2004)
14. Kozlov, D.N.: Combinatorial Algebraic Topology. Springer (2008)
15. Lewiner, T., Lopes, H., Tavares, G.: Optimal Discrete Morse Functions for 2-manifolds. Comput. Geom. 26, 221–233 (2003)
16. Maver, J., Bajcsy, R.: Occlusions as a guide for planning the next view. IEEE Transactions on Pattern Analysis and Machine Intelligence 15(5), 417–433 (1993)
17. Munkres, J.: Elements of Algebraic Topology. Addison-Wesley Co. (1984)
18. Scott, W.R., Roth, G., Rivest, J.F.: View planning for automated three-dimensional object reconstruction and inspection. ACM Computing Surveys 35(1) (2003)
19. Tarabanis, K.A., Allen, P.K., Tsai, R.Y.: A survey of sensor planning in computer vision. IEEE Trans. on Robotics and Automation 11(1), 86–104 (1995)
20. Vanderbeck, F., Wolsey, L.A.: Reformulation and decomposition of integer programs. In: Junger, M., et al. (eds.) 50 Years of Integer Programming 1958-2008, pp. 431–502. Springer (2010)
21. Zomorodian, A., Carlsson, G.: Computing persistent homology. Discrete and Computational Geometry 33(2), 249–274 (2005)

A Fast Matching Algorithm
for Graph-Based Handwriting Recognition

Andreas Fischer[1], Ching Y. Suen[1],
Volkmar Frinken[2], Kaspar Riesen[3], and Horst Bunke[4]

[1] Centre for Pattern Recognition and Machine Intelligence, Concordia University
1455 de Maisonneuve Blvd West, Montreal, Quebec H3G 1M8, Canada
{an_fisch,suen}@encs.concordia.ca
[2] Computer Vision Center, Dept. of Computer Science,
Universitat Autònoma de Barcelona, Edifici O, 08193 Bellaterra, Spain
vfrinken@cvc.uab.cat
[3] Institute for Informations Systems, University of Applied Sciences and Arts
Northwestern Switzerland, Riggenbachstrasse 16, 4600 Olten, Switzerland
kaspar.riesen@fhnw.ch
[4] Institute of Computer Science and Applied Mathematics,
University of Bern, Neubrückstrasse 10, 3012 Bern, Switzerland
bunke@iam.unibe.ch

Abstract. The recognition of unconstrained handwriting images is usually based on vectorial representation and statistical classification. Despite their high representational power, graphs are rarely used in this
field due to a lack of efficient graph-based recognition methods. Recently,
graph similarity features have been proposed to bridge the gap between
structural representation and statistical classification by means of vector
space embedding. This approach has shown a high performance in terms
of accuracy but had shortcomings in terms of computational speed. The
time complexity of the Hungarian algorithm that is used to approximate
the edit distance between two handwriting graphs is demanding for a
real-world scenario. In this paper, we propose a faster graph matching
algorithm which is derived from the Hausdorff distance. On the historical
Parzival database it is demonstrated that the proposed method achieves
a speedup factor of 12.9 without significant loss in recognition accuracy.

1 Introduction

In many pattern recognition applications, graphs are the first choice to represent objects. Their ability to model different parts of an object as well as
their binary relations can be used to derive powerful representations of molecular compounds [1], computer networks [2], and symbols in digital images [3], to
name just a few. In the domain of handwriting recognition, graphs have found
widespread application for single character recognition, especially in the case of
Chinese characters that are composed of many complex strokes [4].

However, when it comes to the recognition of unconstrained handwriting images that contain complete sentences in natural language, graph-based representation is rarely used due to problems arising from the large variety of character

W.G. Kropatsch et al. (Eds.): GbRPR 2013, LNCS 7877, pp. 194–203, 2013.
© Springer-Verlag Berlin Heidelberg 2013

shapes, the large number of words in natural language, and the inability to segment connected handwriting into characters before recognition [5]. Available systems are usually based on vectorial pattern representation $\mathbf{x} \in \mathbb{R}^n$ and statistical classifiers, e.g., hidden Markov models [6] and recurrent neural networks [7]. They cannot be applied directly on graphs representing handwriting.

Recently, a general approach to bridging the gap between graph-based handwriting representation and statistical classification has been proposed in [8]. Based on dissimilarity space embedding [9,10], handwriting graphs are transformed into feature vectors by calculating their similarity to a set of character prototypes. Graph similarity is obtained by means of graph edit distance [11], an error-tolerant matching method that can be applied to any kind of graph. The similarities constitute the real-valued components of the feature vectors which can then be used in combination with any statistical classifier.

When compared with traditional statistical feature sets, the graph similarity features have shown a promising performance in terms of recognition accuracy [8]. However, although the Hungarian algorithm [12] was used to approximate the graph edit distance in cubic time with respect to graph size [13], the computational complexity remained high, resulting in slow execution speed.

In this paper, we derive a faster matching algorithm for graph-based handwriting recognition from the Hausdorff distance. The proposed method runs in quadratic time with respect to graph size and hence significantly reduces the complexity of the recognition process.

Experiments are conducted on the historical Parzival database [14] which includes images from a 13th century manuscript written in Old German. For a single word recognition task with hidden Markov models, it is demonstrated that the proposed matching algorithm achieves a speedup factor of 12.9 without significant loss in recognition accuracy.

The remainder of this paper is organized as follows. First, the graph similarity features are reviewed in Section 2. Next, the proposed matching algorithm is presented in Section 3. Then, experimental results are presented and discussed in Section 4. Finally, conclusions are drawn in Section 5.

2   Graph Similarity Features

In this section, we briefly review the graph similarity features that are described in detail in [8]. It is a general framework that allows the use of graph-based handwriting representation in combination with statistical classification by means of vector space embedding.

2.1   Handwriting Graphs

The handwriting graphs used in this paper are derived from handwriting skeleton images. An original image is shown in Figure 1a and a skeleton graph in Figure 1b. Image preprocessing includes binarization, correction of the baseline inclination, separation of the writing region into an upper, middle, and lower part, and thinning of the strokes to a width of one pixel.

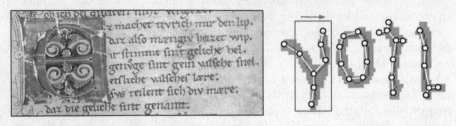

(a) original image   (b) skeleton graph, sliding window

Fig. 1. Handwriting graphs

Afterwards, a handwriting graph is constructed by adding endpoints, intersections, and the upper left pixel of circular structures to the set of nodes, labeled with their image position $(x, y) \in \mathbb{R}^2$. Then, further connection points at distance D along the skeleton are added as nodes. The connection point distance D is a parameter to be chosen by the user that determines the node density on the skeleton. Whenever two nodes are connected over the skeleton, they are linked with an undirected, unlabeled edge.

2.2   Vector Space Embedding and Recognition System

An intriguing challenge for connected handwriting recognition is the inability to segment the image into characters before recognition [5]. Instead, a common approach is to perform an oversegmentation with a sliding window moving from left to right over the image and extracting a sequence of feature vectors $x_1, \ldots, x_N$ with $x_i \in \mathbb{R}^n$. The segments are grouped into characters during recognition.

Handwriting graphs are transformed into feature vectors by means of dissimilarity space embedding [10]. First, prototype character graphs are selected, either manually or automatically [15]. Then, a sliding window is moved over the handwriting graph from left to right as illustrated in Figure 1b. At each position, the graph dissimilarity $d(g_1, g_2) \in \mathbb{R}$ between the subgraph g_1 in the window and the prototype graph g_2 is calculated for all prototypes. This results in a sequence of feature vectors $x_1, \ldots, x_N$ with $x_i = (d(g_{1,i}, g_{2,1}), \ldots, d(g_{1,i}, g_{2,n}))$ and a dimensionality $x_i \in \mathbb{R}^n$ equal to the number n of prototypes. The dissimilarity measure used is the graph edit distance which is discussed in detail in Section 3.1. An important property of the edit distance is that it can be applied to any kind of graph.

After embedding, the resulting feature vector sequence can be used for recognition with any statistical classifier. In this paper, we employ hidden Markov models (HMM) [6] for word recognition. For any further details on the recognition system as well as the graph similarity features in general, we refer to [8].

3   Fast Matching Algorithm

A shortcoming of the graph similarity features is their high computational time complexity for matching two handwriting graphs. For a median graph size of 30 nodes, the graph matching process takes about half a minute per word on a 2.66GHz personal computer (see Section 4). Considering a real-world scenario, for instance the daily processing of handwritten letters sent to a company or the processing of large collections of historical manuscripts for digital libraries, this computational speed is demanding in terms of hardware resources.

In this section, we derive a faster graph matching method from the Hausdorff distance. It preserves most properties of the formerly used approximate graph edit distance [13] which is based on a node assignment according to some edit cost. By allowing multiple node assignments for the proposed method, the time complexity is reduced from cubic to quadratic with respect to graph size.

In the following, the approximate graph edit distance is reviewed in Section 3.1, the Hausdorff distance is discussed in Section 3.2, and the proposed modified Hausdorff distance is introduced in Section 3.3.

3.1   Approximate Graph Edit Distance

To calculate the dissimilarity $d(g_1, g_2)$ between two graphs g_1 and g_2, representing the subgraph inside the sliding window and a character prototype (see Section 2.2, Figure 1b), the graph edit distance is used to derive graph similarity features [11]. This distance is given by the minimum cost of edit operations needed to transform g_1 into g_2. Possible edit operations include the substitution, deletion, and insertion of nodes and edges.

For the handwriting graphs under consideration (see Section 2.1), the Euclidean cost function is used with

- $c(n_1, n_2) = ||(x_1, y_1) - (x_2, y_2)||$ for node label substitution
- $c(n_1, \epsilon) = c(\epsilon, n_2) = C_n \geq 0$ for node deletion and insertion
- $c(e_1, \epsilon) = c(\epsilon, e_2) = C_e \geq 0$ for edge deletion and insertion

where (x_i, y_i) is the attribute vector associated with node n_i, representing the location of n_i in the two-dimensional plane. This definition ensures the edit distance to be metric [11]. The non-negative parameters C_n and C_e for deletion and insertion are optimized on a validation set to adapt the generic cost function to the graph data. As there are no edge labels, no edge label substitution cost need to be defined.

Usually, the edit distance is calculated with the A^* algorithm which performs a best-first tree search, possibly using a lower bound heuristic for the estimated future cost [11]. The A^* algorithm always finds the optimal solution but has an exponential time complexity with respect to the graph size. In order to match large handwriting graphs, an approximation is used to obtain a suboptimal edit distance in polynomial time [13]. The approximation reduces the edit distance to a node assignment problem which can then be solved in cubic time by Munkres'

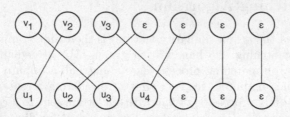

Fig. 2. Assignment problem

algorithm [12], also known as the Hungarian algorithm. Although the algorithm does not always find an optimal solution for the edit distance, it is reasonably accurate, especially for small distances among similar graphs which is important for the task of classification [13].

The edit distance can be formulated as a node assignment problem as illustrated in Figure 2. In this example, we consider a graph g_1 with three nodes v_1, v_2, v_3 (top row) that is matched with a graph g_2 having four nodes u_1, u_2, u_3, u_4 (bottom row). For each node in g_1, an ϵ-node is inserted in the bottom row, and for each node in g_2 an ϵ-node is inserted in the top row. Assignments between the top and the bottom row correspond with node edit operations, e.g., substitution (v_1, u_3), deletion (v_3, ϵ), and insertion (ϵ, u_2).

Finding a complete assignment with minimum cost[1] corresponds to finding an optimal solution for the edit distance if the edges of the graphs are ignored. By taking the implied cost of adjacent edge operations into account for each node assignment, the true edit distance can be approximated. The method is suboptimal because only local adjacent edge structures are matched instead of the global edge structure. Munkres' algorithm solves the assignment problem in $O(N^3)$ where N is the sum of the number of nodes in g_1 and g_2.

For graph similarity features, a normalization of the approximate graph edit distance with respect to its maximum value has proven beneficial. It is accomplished by $d_{max}(g_1, g_2) = N \cdot C_n + E \cdot C_e$ where N is the sum of the number of nodes in g_1 and g_2 and E is the sum of the number of edges. The maximum corresponds to the deletion of all nodes and edges in g_1 and the insertion of all nodes and edges in g_2. A similarity value is obtained by

$$\hat{s}(g_1, g_2) = 1 - \frac{d(g_1, g_2)}{d_{max}(g_1, g_2)} \tag{1}$$

Also, a normalization over all prototype characters $p \in P$ is performed at each sliding window position yielding the final graph similarity measure

$$s(g_1, g_2) = \frac{\hat{s}(g_1, g_2)^2}{\sum_P \hat{s}(g_1, p)} \tag{2}$$

---

[1] The assignment cost (ϵ, ϵ) is zero.

3.2 Hausdorff Distance

The Hausdorff distance is a distance measure between two subsets of a metric space. In case of finite subsets A and B the Hausdorff distance $H(A, B)$ is

$$H(A, B) = \max(\max_A \min_B d(a, b), \max_B \min_A d(a, b)) \qquad (3)$$

where $a \in A$, $b \in B$, and $d(a, b)$ is the underlying metric [16]. In Equation 3, $\min_B d(a, b)$ is the nearest neighbor distance of a in B, $\min_A d(a, b)$ is the nearest neighbor distance of b in A, and the Hausdorff distance corresponds to the maximum over all nearest neighbor distances.

The Hausdorff distance is widely used in the domain of image matching [16], for example to locate templates within target images. In its original definition, only the maximum over all nearest neighbor distances is taken into account. Hence, Hausdorff distance is sensitive to outliers in the data. A straight-forward modification that integrates all nearest neighbor distances can be achieved with

$$H'(A, B) = \sum_A \min_B d(a, b) + \sum_B \min_A d(a, b) \qquad (4)$$

Considering A as the nodes of graph g_1, B as the nodes of graph g_2, and $d(n_1, n_2) = c(n_1, n_2) = \|(x_1, y_1) - (x_2, y_2)\|$ as the underlying metric, i.e., the node substitution cost (see Section 3.1), the Hausdorff distance can be directly applied to the handwriting graphs. It ignores the edges of the graphs and can trivially be calculated in $O(NM)$ where N and M denote the number of nodes in g_1 and g_2, respectively.

Therefore the Hausdorff distance or its modification in Equation 4 can be used as a fast alternative for the approximate graph edit distance. In our experiments (see Section 4) a normalization with respect to all prototypes $p \in P$ has proven beneficial yielding the final graph dissimilarity measures

$$h(g_1, g_2) = \frac{H(g_1, g_2)}{\sum_P H(g_1, p)} \qquad (5)$$

$$h'(g_1, g_2) = \frac{H'(g_1, g_2)}{\sum_P H'(g_1, p)} \qquad (6)$$

3.3 Modified Hausdorff Distance

In this paper, we propose a novel modification of the Hausdorff distance that takes into regard not only substitution, but also deletion and insertion cost. It is defined as

$$H''(A, B) = \sum_A \min_B \bar{c}_1(a, b) + \sum_B \min_A \bar{c}_2(a, b) \qquad (7)$$

by replacing the metric d in Equation 4 with the cost functions $\bar{c}_1$ and $\bar{c}_2$. Again, A corresponds with the nodes of graph g_1 and B with the nodes of g_2. The cost functions $\bar{c}_1(n_1, n_2)$ and $\bar{c}_2(n_1, n_2)$ for matching node n_1 with node n_2 are

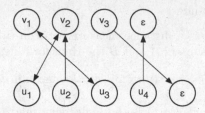

Fig. 3. Multiple assignments

$$\bar{c}_1(n_1, n_2) = \begin{cases} \frac{c(n_1, n_2)}{2}, & \text{if } c(n_1, n_2) < c(n_1, \epsilon) \\ c(n_1, \epsilon), & \text{otherwise} \end{cases} \tag{8}$$

$$\bar{c}_2(n_1, n_2) = \begin{cases} \frac{c(n_1, n_2)}{2}, & \text{if } c(n_1, n_2) < c(\epsilon, n_2) \\ c(\epsilon, n_2), & \text{otherwise} \end{cases} \tag{9}$$

with respect to the cost function c of the edit distance (see Section 3.1). That is, $\min_B \bar{c}_1(a, b)$ returns half of the node substitution cost $\frac{c(n_1, n_2)}{2}$ if the substitution (n_1, n_2) is preferred over deletion (n_1, ϵ) of n_1. Among all possible substitutions the one with the smallest cost is chosen. Otherwise, the deletion cost $c(n_1, \epsilon)$ is returned. Similarly, $\min_A \bar{c}_2(a, b)$ returns the best $\frac{c(n_1, n_2)}{2}$ if the substitution (n_1, n_2) is preferred over insertion (ϵ, n_2) of n_2.

The correspondence between the modified Hausdorff distance H'' and the approximate edit distance is illustrated in Figure 3 in analogy to Figure 2. Based on H'', each node of graph g_1 (top row) is either assigned to a node of graph g_2 (bottom row) if a substitution is preferred in $\bar{c}_1$ or to the ϵ node for deletion. Vice versa, each node of graph g_2 is either assigned to a node of graph g_1 if a substitution is preferred in $\bar{c}_2$ or to the ϵ node for insertion.

H'' equals the cost of all assignments. The cost of double assignments, e.g., (v_1, u_3) in Figure 3, is the full substitution cost $c(n_1, n_2) = \bar{c}_1(n_1, n_2) + \bar{c}_2(n_1, n_2)$. Deletions and insertions contribute their respective cost. The only difference to the assignment problem solved by the approximate edit distance is the possibility of multiple assignments, e.g., with v_2 in Figure 3. In such a case H'' is smaller than the approximate edit distance which is an upper bound of H''.

Finally, implied adjacent edge costs can be taken into account for the assignments in $\bar{c}_1$ and $\bar{c}_2$ in the same way as for the approximate edit distance. For small edit distances between two graphs, H'' is, indeed, expected to provide a good approximation of the edit distance in quadratic time.

For the graph similarity features, the usual normalization is applied (see Section 3.1). With $\hat{h}''(g_1, g_2) = 1 - \frac{H''(g_1, g_2)}{d_{max}(g_1, g_2)}$ the final graph similarity measure is obtained as

$$h''(g_1, g_2) = \frac{\hat{h}''(g_1, g_2)^2}{\sum_P \hat{h}''(g_1, p)} \tag{10}$$

4   Experimental Evaluation

Experiments are conducted on the historical Parzival database[2] [14] which includes images from a 13th century manuscript written in Old German. 11, 743
word images are considered that contain 3, 177 word classes and 87 characters.

For a single word recognition task with graph similarity features and HMM-
based recognition (see Section 2), the similarity function h'' obtained from the
modified Hausdorff distance is compared with the Hausdorff distance h, its
variant h', and the approximate edit distance s (see Section 3).

4.1   Setup

First, the word images are divided into three distinct sets for training, validation,
and testing. Half of the words are used for training and a quarter of the words
for validation and testing, respectively. For vector space embedding, 79 character
prototypes are used as in [8].

Parameters that are optimized with respect to the validation accuracy include
the connection point distance $D \in \{3, 5, 7, 9\}$ for graph-based representation, the
deletion and insertion cost $C_n, C_e \in \{0, 0.4D, 0.6D, \ldots, 1.4D\}$ of the graph edit
distance, and the number of Gaussian mixtures $G \in \{1, 2, \ldots, 30\}$ of the HMM.
Optimal parameter values are adopted from previous studies [8] conducted with
the approximate edit distance s. The same values are used for h, h', and h''.

Table 1. Word recognition accuracy on the test set in percentage

h	h'	h''	s
49.78	83.95	93.66	94.00

4.2   Results

The achieved word recognition accuracy on the test set is listed in Table 1. As
stated in [8], the best accuracy of 94.00% achieved with graph similarity features
and approximate edit distance s significantly outperforms traditional statistical feature sets which achieve a maximum accuracy of 90.49% [8]. This result
demonstrates the general effectiveness of the proposed graph-based approach to
handwriting recognition.

With an accuracy of 93.66% the proposed modified Hausdorff distance h''
achieves nearly the same performance as s. There is no statistically significant
difference between the results (t-test, $\alpha = 0.05$). That is, the improvement from
cubic to quadratic time complexity can be achieved without significant loss in
accuracy.

---

[2] http://www.iam.unibe.ch/fki/databases/iam-historical-document-database

Table 2. Runtime statistics. The median graph size in terms of number of nodes, the median number of graph matchings per word, the mean runtimes on a 2.66GHz processor for s and h'' per word in seconds, and the speedup factor.

Graph Size	Matchings	Runtime s	Runtime h''	Speedup
30	6162	33.24	2.57	12.9

When comparing the results of h and h', a very remarkable difference in performance is observed. This is possibly due to the fact that the Hausdorff variant h' is less sensitive to outliers. Still, both distance measures perform significantly worse than h'' and s.

The runtime statistics of h'' and s are listed in Table 2. Both methods are implemented in Java. For an optimal value $D_{opt} = 3$, the median number of graph nodes is 30 and the proposed algorithm achieves a speedup factor of 12.9.

An anomaly is observed for the cost function parameters $C_{n,opt} = 3$ and $C_{e,opt} = 0$. The edge cost parameter $C_{e,opt} = 0$ indicates that the use of edges in the chosen graph representation actually leads to worse results. We explain this anomaly by the fact that the handwriting images of historical manuscripts contain many broken characters due to binarization problems. Imposing an edge cost leads to stronger deviations from the clean prototype characters in this case.

5 Conclusions

Graph similarity features provide a general framework to combine graph-based representation and statistical classification for the recognition of handwritten text images. The framework proposed in this paper is based on vector space embedding of handwriting graphs with respect to a set of character prototypes. It showed a high recognition accuracy when compared with traditional statistical feature sets, but had shortcomings in computational speed when matching two handwriting graphs with an approximate graph edit distance.

In this paper, we propose a fast matching algorithm derived from the Hausdorff distance that reduces the complexity of the graph matching process from cubic to quadratic time with respect to graph size. The method retains most properties of the approximate edit distance but allows multiple node assignments. On the historical Parzival database it was demonstrated for an HMM-based word recognition task that a speedup factor of 12.9 could be achieved without significant loss in accuracy.

In the domain of handwriting recognition, future work includes the investigation of different graph-based representations of handwriting within the proposed framework. In the field of image matching, the proposed distance measure could be used as a variant of the Hausdorff distance in various applications, such as template location. Finally, for graph-based recognition in general, the algorithm offers a promising possibility to approximate the graph edit distance in quadratic time with respect to graph size. This issue needs to be verified on diverse graph data sets.

Acknowledgment. This work has been supported by the Swiss National Science Foundation fellowship project PBBEP2_141453, the Spanish project TIN2009-14633-C03-03, and the Spanish MICINN under the MIPRCV "Consolider Ingenio 2010" CSD2007-00018 project.

References

1. Mahé, P., Ueda, N., Akutsu, T., Perret, J., Vert, J.: Graph kernels for molecular structure-activity relationship analysis with support vector machines. Journal of Chemical Information and Modeling 45(4), 939–951 (2005)
2. Bunke, H., Dickinson, P.J., Kraetzl, M., Wallis, W.D.: A Graph-Theoretic Approach to Enterprise Network Dynamics. Progress in Computer Science and Applied Logic, vol. 24. Birkhäuser (2006)
3. Llados, J., Marti, E., Villanueva, J.: Symbol recognition by error-tolerant subgraph matching between region adjacency graphs. IEEE Trans. PAMI 23(10), 1137–1143 (2001)
4. Lu, S., Ren, Y., Suen, C.Y.: Hierarchical attributed graph representation and recognition of handwritten chinese characters. Pattern Recognition 24(7), 617–632 (1991)
5. Bunke, H., Varga, T.: Off-line Roman cursive handwriting recognition. In: Chaudhuri, B. (ed.) Digital Document Processing, pp. 165–173. Springer (2007)
6. Ploetz, T., Fink, G.A.: Markov models for offline handwriting recognition: A survey. Int. Journal on Document Analysis and Recognition 12(4), 269–298 (2009)
7. Graves, A., Liwicki, M., Fernandez, S., Bertolami, R., Bunke, H., Schmidhuber, J.: A novel connectionist system for improved unconstrained handwriting recognition. IEEE Trans. PAMI 31(5), 855–868 (2009)
8. Fischer, A., Riesen, K., Bunke, H.: Graph similarity features for HMM-based handwriting recognition in historical documents. In: Proc. 12th Int. Conf. on Frontiers in Handwriting Recognition, pp. 253–258 (2010)
9. Pekalska, E., Duin, R.: The Dissimilarity Representations for Pattern Recognition: Foundations and Applications. World Scientific (2005)
10. Riesen, K., Bunke, H.: Graph Classification and Clustering Based on Vector Space Embedding. World Scientific (2010)
11. Bunke, H., Allermann, G.: Inexact graph matching for structural pattern recognition. Pattern Recognition Letters 1(4), 245–253 (1983)
12. Munkres, J.: Algorithms for the assignment and transportation problems. Journal of the Society for Industrial and Applied Mathematics 5(1), 32–38 (1957)
13. Riesen, K., Bunke, H.: Approximate graph edit distance computation by means of bipartite graph matching. Image and Vision Computing 27(7), 950–959 (2009)
14. Fischer, A., Wüthrich, M., Liwicki, M., Frinken, V., Bunke, H., Viehhauser, G., Stolz, M.: Automatic transcription of handwritten medieval documents. In: Proc. 15th Int. Conf. on Virtual Systems and Multimedia, pp. 137–142 (2009)
15. Fischer, A., Bunke, H.: Character prototype selection for handwriting recognition in historical documents with graph similarity features. In: Proc. 19th European Signal Processing Conference, pp. 1435–1439 (2011)
16. Huttenlocher, D.P., Klanderman, G.A., Kl, G.A., Rucklidge, W.J.: Comparing images using the Hausdorff distance. IEEE Trans. PAMI 15, 850–863 (1993)

On the Evaluation of Graph Centrality for Shape Matching

Samuel de Sousa, Nicole M. Artner, and Walter G. Kropatsch

Vienna University of Technology
Pattern Recognition and Image Processing Group
Vienna, Austria
{sam,artner,krw}@prip.tuwien.ac.at

Abstract. Graph centrality has been extensively applied in Social Network Analysis to model the interaction of actors and the information flow inside a graph. In this paper, we investigate the usage of graph centralities in the Shape Matching task. We create a graph-based representation of a shape and describe this graph by using different centrality measures. We build a Naive Bayes classifier whose input feature vector consists of the measurements obtained by the centralities and evaluate the different performances for each centrality.

Keywords: centrality, shape matching, graph.

1 Introduction

Humans have the innate skill of recognizing objects by their appearance, shape, silhouettes, and contours. When this object recognition task is performed by machines, the shape representation is an important factor that needs to be taken into account, which might be considered as a key factor to obtain a good or bad recognition performance. We analyze the shape representation based on graph theory by abstracting pixels of an image as vertices and modeling their spatial relationship with edges.

Several implicit information can be extracted from graphs. For instance, centrality measurements of graphs or networks have been extensively explored in Social Network Analysis (SNA) [19] to understand the flow of information or to identify potential key actors inside the network. However, those measures could be also applied in a different context that may not be related to SNA and still be capable of achieving meaningful results. For instance, graph centralities could be used to represent a graph based on the distribution of centrality values over the vertices.

In this paper, we analyze the impact of using graph centralities in the modeling and description of shapes. To the best of our knowledge, there is no graph-based approach employing centrality measures for 2D shape matching. Hence, we represent a shape using a graph and calculate the following centrality measures: Degree, Betweenness, Closeness, PageRank, and Eigenvector. We divide a dataset of shapes into 8 classes and train classifiers to recognize those shapes

W.G. Kropatsch et al. (Eds.): GbRPR 2013, LNCS 7877, pp. 204–213, 2013.

by taking into consideration only the centralities of graphs. We add salt noise to the shapes in order to evaluate the impact that the absence of certain nodes and edge has in the recognition task. Our main contribution is a novel way of graph comparison and "matching" based on graph centralities, and the evaluation of their robustness under topological changes of the graph. Our results indicate that the closeness centrality is the most reliable centrality for the matching task under minor changes in the graph.

The remainder of this paper is organized as follows: Section 2 provides a literature review on graph centrality and graph-based shape representation. We introduce the centralities measurements and discuss our overall methodology in Section 3. Our experiments are explained in Section 4 narrowing our feature modeling and examining the results obtained by our classifiers. Finally, we present our conclusions and future work in Section 5.

2   Related Work

The concept of centrality has already been used by Bavelas in 1948 [1] to explain human behavior. It is frequently used in the analysis of different types of networks. Hence, many different measures of centrality have been proposed [4]. Borgatti et al. [4] try to give a graph-theoretic review of centrality measures, where measures are classified according to the features of their calculation. They focus on the three best-known measures of centrality: degree, closeness, and betweenness [9]. One of their findings is that there are four basic dimensions to distinguish between centrality measures: (i) the types of walks considered, (ii) the properties of walks measured, (iii) the type of nodal involvement, and (iv) the type of summarization.

Correa et al. [6] use derivatives of centrality in the visualization of social networks. The derivative of centrality informs how much a given node influences the importance of another node, even if they are not directly connected. They found out that derivatives of centrality are tool for analyzing social networks, they help to simplify the layout of complex networks and to visually measure the centralization degree of a network. Furthermore, they provide information to estimate other metrics like structural balance and uncertainty.

Mukherjee et al. [13] present an application of centrality in human action recognition. They employ centrality to create a compact codebook out of a large vocabulary of poses (bag-of-word approach). Cukierski et al. [7] use centrality to solve an open problem in the ISOMAP algorithm [12], which is a non-linear dimensionality reduction method. The ISOMAP algorithm computes geodesic distances between data points with the help of a neighborhood graph. Unfortunately, these graphs sometimes contain unwanted edges, which connect disparate regions of one or more manifolds and this leads to a distortion of the calculated geodesic distance matrix. This problem is where Cukierski et al. propose an edge-removal method based on graph betweenness centrality, which can robustly identify manifold-shorting edges.

In the literature, shock graphs are frequently employed in graph-based shape matching. Shock graphs are related to medial axis transform [3], but with a higher descriptive power. Each vertex in a shock graph is labeled by the shock type and the edges depend on the shock formation times. Siddiqi et al. [17] match 2D shapes based on directed, acyclic shock graphs. By employing a shock graph grammar the task of matching shock graphs can be reduced to matching shock trees, which can be done in polynomial time.

Sebastian et al. [15] match 2D shape outlines based on the edit distance between their corresponding shock graphs. They propose a novel framework, where they partition the shape space with the help of shock graph topology, discretize the space of deformations based of their shock graph transitions, and find the globally optimal sequence of transitions by employing a graph edit distance algorithm. Torsello et al. [18] present a geometric measure to determine the similarity between shapes calculated from the skeletons. This measure allows to distinguish between perceptually distinct shapes whose skeletons are ambiguous and to distinguish between the main skeletal structure and its ligatures.

Besides the works on shock graphs and skeletons, there are approaches which match shapes by their contour. Felzenszwalb and Schwartz [8] present a so-called shape-tree, which describes the boundary of shape at multiple levels of resolution. Their representation can be used to determine the similarity between two shapes and for matching a deformable shape model to a cluttered image. In [20], Zhu et al. propose a hierarchical deformable template, which describes an object by a hierarchical graph defined by parent-child relationships. In the top vertex the pose (position, orientation, and scale) of the center of the object is stored and in the child vertices the poses of points on the object boundary are described.

The review of the related work showed that, to the best of our knowledge, there is no graph-based approach employing centrality measures for 2D shape matching. Hence, this paper is the first attempt to use centrality in graph-based shape matching.

3   Graph-Based Shape Matching

Our methodology starts by building a graph-based representation of the shape and by calculating several centrality measures of this graph. The resulting centra-lity-based representation of the graph is further input into a classifier as the feature vector in order to distinguish different shapes by the centrality representation.

Given a graph $\mathcal{G} = (\mathcal{V}, \mathcal{E})$, a centrality can be interpreted as a function $f : \mathcal{V} \to \mathbb{R}$, which assigns a real value to each vertex $v \in \mathcal{V}$. In general, centrality measures the importance of a vertex within a graph or the importance of an actor in a social network [19]. We elucidate this concept using the graph displayed in Figure 1a. The measures of centrality computed for this undirected graph are listed in Figure 1b, where $b(v), c(v), d(v), e(v)$, and $r(v)$ stand for degree, closeness, between, eigenvector and rank of a vertex $v \in \mathcal{V}$, respectively. Those measures are explained in the following sections.

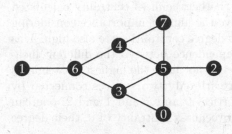

v	d	b	c	e	
0	2	0.00	0.53	0.58	0.09
1	1	0.00	0.38	0.18	0.06
2	1	0.00	0.43	0.33	0.05
3	3	4.00	0.63	0.72	0.14
4	3	4.00	0.63	0.72	0.14
5	5	9.50	0.70	1.00	0.23
6	3	6.50	0.58	0.55	0.15
7	2	0.00	0.53	0.58	0.09

(a) Sample graph $\mathcal{G} = (\mathcal{V}, \mathcal{E})$, $|\mathcal{V}| = 8$ and $|\mathcal{E}| = 10$.

(b) Centrality values obtained for the graph.

Fig. 1. For each vertex $v \in \mathcal{V}$, we compute the importance of the vertex according to a specific centrality

3.1 Degree Centrality

The degree of a vertex $d(v)$ is a measure that counts the number of edges incident to v [9]. By evaluating the degree centrality, one can compare the connectivity of vertices, but this measure does not tell how well-positioned a certain vertex is within the graph. In the graph of the example in Figure 1a, the highest degree is 5 and the lowest is 1.

The importance of a vertex with regard to degree centrality depends on the average degree in a graph. For instance, a degree 8 is considered high in a graph whose average degree is 2, but it is low in a graph whose average is 20. Hence, other centrality measures are capable of providing more detailed information about a vertex.

3.2 Betweenness Centrality

The communication between two non-adjacent vertices depends on the path between them. The main idea of betweenness centrality [19] is that vertices that lie on the geodesic path of many other vertices will possess great control over the information flow, due to the fact that they reside "between" others.

The betweenness centrality $b(v)$ of a vertex v is calculated as follows:

$$b(v) = \sum_{s \in V \setminus \{v\}} \sum_{t \in V \setminus \{s,v\}} \frac{\sigma_{st}(v)}{\sigma_{st}}, \qquad (1)$$

where σ_{st} is the number of geodesic paths between vertices s and t. The value $\sigma_{st}(v)$ stands for the number of geodesic paths between s and t via v. The computation of the betweenness centrality requires the determination of the geodesic paths by calculating the geodesic distances to all vertices in the graph. The lower complexity for this centrality is $O(V^2 \log V + VE)$ achieved by Johnson et al. [2].

The third column of Figure 1b shows the betweenness centrality calculated for the sample graph. Vertices 5 and 6 have the highest importance considering betweenness centrality. Furthermore, their degree centralities are also high. This shows that there is a certain degree of dependence between the different measures of centrality. A vertex with degree 1 will not have the highest betweenness centrality in a graph (except for a graph with only two vertices connected by an edge). Looking at the two pairs of vertices 0 and 7 and 1 and 2, one can see that even though all of them have betweenness centrality of 0, their degree centralities are different (1 and 2).

3.3 Closeness Centrality

Closeness centrality [9] is a measure that evaluates how close a certain vertex is to all other vertices in the graph. It computes the inverse of the shortest paths from a vertex to all other vertices. Let $d(s, v)$ be the shortest distance between vertices s and v, the closeness centrality $c(v)$ of a vertex v can be computed as follows:

$$c(v) = \sum_{s \in G \setminus \{v\}} \frac{1}{d(s, v)}. \tag{2}$$

In comparison to betweenness centrality, which can result in values equal to 0 (see vertices 0, 1, 2 and 7 in the example in Figure 1a), closeness centrality will only be 0 for vertices which are disconnected from the graph. In the sample graph in Figure 1a, vertex 5 obtained the highest closeness centrality $(c(v))$. However, the $c(v)$ of vertex 5 is not much higher than vertices 3 and 4 (in contrast to $b(v)$). In SNA it is known that if information is spread from the vertex with the highest $c(v)$, it will spread over the whole network in the shortest time possible [4].

3.4 Eigenvector Centrality

Given a graph $\mathcal{G} = (\mathcal{V}, \mathcal{E})$, the eigenvector centrality $e(v)$ corresponds to the eigenvector $X = (x_1, x_2, \ldots, x_{|\mathcal{V}|})^T$ associated with the highest eigenvalue λ of the graph adjacency matrix [14].

As proved by the Perron Frobenius theorem [10], a square matrix (such as the adjacency matrix of a graph) with positive values has a unique, largest eigenvalue with strictly, positive eigenvector entries.

3.5 PageRank Centrality

PageRank [5] is an algorithm for measuring the importance of a web page. According to [14], the page rank centrality $r(v)$ is a variant of the eigenvector centrality and it can be determined by the following equation:

$$\mathbf{R} = \frac{1 - d}{n} . \mathbf{1} + d\mathbf{L}\mathbf{R}, \tag{3}$$

where $\mathbf{R} = (r_1, r_2, \cdots, r_n)^T$ is the page rank vector, where r_i is the page rank of vertex (webpage) i and n is the total number of vertices (webpages), d is a damping factor with d = 0.85, $\mathbf{1}$ is a column vector, and $\mathbf{L}$ is a modified adjacency matrix (for details on the computation see [14]).

3.6   Centrality Shape Descriptor

We propose to describe 2D shapes by histograms of their centrality measures, which are calculated on a 8-neighborhood graph. Each centrality measure describes different properties of a shape. Figure 2 displays the same graph with different representations of centralities.

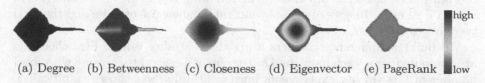

(a) Degree   (b) Betweenness   (c) Closeness   (d) Eigenvector   (e) PageRank

Fig. 2. Visualization of the centrality values of a shape. The centrality values are color coded from low to high values.

The degree centrality (Fig. 2a) shows high values in the center of the shape and low values in the boundaries. Hence, the degree histogram will simply count how many vertices exist of the two groups: (i) vertices inside the shape with 8-connected neighboring vertices, and (ii) vertices on the boundary. The betweenness centrality (Fig. 2b) shows high values for vertices that were frequently traversed by geodesic paths of the graph. Closeness representation (Fig. 2c) contains higher values for vertices that are "closer" to all other vertices and low values for "distant" ones. The eigenvector centrality (Fig. 2d) shows high values in the "center" of the shape and those values decrease towards the boundaries. Finally, PageRank (Fig. 2e) results show high values in the center of the shape and small values on the boundary. However, those values on the boundary are slightly different from the degree results, they are not as uniform as the degree centrality.

4   Experiments

The histograms of the graph centralities are used to create feature vectors for a Naive Bayes classifier. We have used the Kimia's Shape 99 database [16] which contains 11 images of 9 classes. Figure 3 displays one image of each class. Considering that once the graph is built, we do not use any image information such as boundaries, curvature. One way to evaluate the robustness of each centrality when the topology of graph changes, i.e. when the number of vertices and edges of the graph change. To achieve that we add different percentages of random

| plane | animal | dude | tool | shark | scribble | rabbit | stingray | hand |

Fig. 3. Sample images of classes used for training

noise to the images of the database. As we are working with binary images and the foreground (shape) is black, we add random "salt" noise (white pixels) which will cause missing nodes when the graph is constructed. The first three images of Figure 3 do not show any noise. The following two figures (tool and shark) show 1% of noise. Images of scribble and rabbit show 5% of noise and the final two images show 10% of noise.

We build six different classifiers using the centrality values. Five classifiers are computed using each individual histogram of a centrality measure with fixed number of bins (in our experiments bins = 40, value was chosen empirically). One extra classifier (referred from now on as "all") combines all five histograms as a feature vector of the training data.

Our code was developed in python and the centralities were calculated using the igraph library[1]. We train our six classifiers using the original dataset without any noise. We classify the original training set and corrupted versions of it with 1%, 5%, and 10% noise as previously described. The classification of the training data aims to evaluate the correctness of the classifier, It demonstrates, for instance, how well the histograms of each centrality differ from each class and thus allow correct classification of new data. Figure 4 shows the impact of 10% of noise in computation of the centrality values. Closeness centrality (Fig. 4c) shows robust results against noise whereas other centralities suffer a great impact from additive noise.

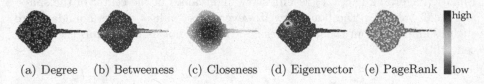

(a) Degree   (b) Betweeness   (c) Closeness   (d) Eigenvector   (e) PageRank

Fig. 4. Visualization of the centrality values of a shape under 10% of noise. The centrality values are color coded from low to high values.

The results of classification are displayed in Table 1. First column displays the classifier evaluated. Second column (no noise) displays the result of the classification on the training data, which shows that the classifier using all centralities

---

[1] http://igraph.sourceforge.net

Table 1. Results of classification using different centrality measures. The closeness centrality achieved the best results under random noise.

classifier	No noise	1% noise	5% noise	10% noise
All	**0.98**	0.85	0.65	0.46
Betweenness	0.82	0.67	0.51	0.29
Closeness	0.94	**0.93**	**0.92**	**0.83**
PageRank	0.81	0.32	0.11	0.11
Eigenvector	0.89	0.69	0.27	0.26
Degree	0.81	0.37	0.17	0.14

obtained the best results (98% of correctness). Most of the individual centralities were able to perform well (with at least 81% of correctness). However, when 1% of noise is added, the closeness centrality starts to obtain the best performance. The other centrality measures are more sensitive to noise than closeness and their results drop remarkably: the addition of random noise had the highest impact on the results of PageRank, Degree, and Eigenvector. We observe that closeness

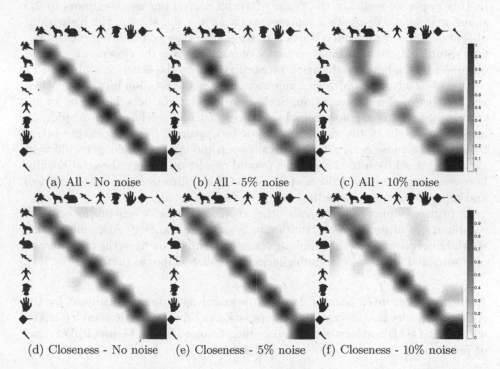

(a) All - No noise   (b) All - 5% noise   (c) All - 10% noise

(d) Closeness - No noise   (e) Closeness - 5% noise   (f) Closeness - 10% noise

Fig. 5. Confusion Matrices of classification. First row shows the combination of all histograms. Second row displays the closeness centrality. Each matrix column shows the real class and the row shows the predicted class.

centrality remains stable in the under the missing nodes in the graph. Even when 10% of noise is added, it still obtains better results than other centralities on the training data.

It can be concluded that closeness centrality is the most robust centrality measure for our shape matching and that the simple combination of all centralities did not obtain the best result but it was still considerably better than the other isolated centralities. Furthermore, the performance of the degree centrality changes severely as under "salt" noise due to the fact that the vertices inside the shape have degrees smaller than eight and therefore the whole histogram distribution of degree is significantly different from the training data.

Figure 5 shows the confusion matrices of the classification results for classifiers *all* and *closeness*. For instance, Figure 5d shows that the classifier of closeness centrality misclassified *airplanes*, *dogs*, *sharks*, and *dudes* in the training data. This might have occurred due to the sharp edges of the pictures leading to similar histograms. However, the per-class classification of closeness is still very high after 10% of noise.

5   Conclusion

In this paper we evaluate the usage of graph centralities as descriptors in 2D shape matching. We create a representation for the shape using the histograms of centralities and we train a Naive Bayes classifier based on that representation. Our results indicate that in the presence of random noise, the closeness centrality obtained the highest classification rates compared to the other centralities.

One clear advantage of such a representation is the rotation invariance achieved by the graph representation. Scale invariance might be achieved if, for instance, histograms are normalized. However, if based solely on the histogram, important details of the shape might not be captured, *i.e.* some shapes might result in the same centrality histogram even though they belong to different classes. One solution to this problem could consist on adding the relationships between mountains and valleys of the shape in one direction (such as clockwise) and integrating this relationship in the feature modeling.

In future, we plan to investigate other representations for centralities or other combinations of multiple centralities such as using boosting. Also, integrating spatial information into the graph representation might increase the performance of results and help in the disambiguation of similar shapes as performed by [11].

Acknowledgments. Samuel de Sousa acknowledges financial support by the Austrian Agency for International Cooperation in Education & Research (OeAD) within the OeAD Sonderstipendien program, financed by the Vienna PhD School of Informatics.

References

1. Bavelas, A.: A mathematical model for group structure. Human Organization 7(3), 16–30 (1948)
2. Black, P.E.: Johnson's algorithm. Dictionary of Algorithms and Data Structures (2004)
3. Blum, H.: Biological shape and visual science (part I). Journal of Theoretical Biology 38, 205–287 (1973)
4. Borgatti, S.P., Everett, M.G.: A graph-theoretic perspective on centrality. Social Networks 28(4), 466–484 (2006)
5. Brin, S., Page, L.: The anatomy of a large-scale hypertextual web search engine. Computer Networks and ISDN Systems 30(1-7), 107–117 (1998)
6. Correa, C., Ma, K.-L.: Visualizing social networks. In: Aggarwal, C.C. (ed.) Social Network Data Analytics, pp. 307–326. Springer, US (2011)
7. Cukierski, W., Foran, D.: Using betweenness centrality to identify manifold short-cuts. In: IEEE International Conference on Data Mining Workshops, ICDMW 2008, pp. 949–958 (December 2008)
8. Felzenszwalb, P., Schwartz, J.: Hierarchical matching of deformable shapes. In: IEEE Conference on Computer Vision and Pattern Recognition, CVPR 2007, pp. 1 –8 (June 2007)
9. Freeman, L.: Centrality in social networks: Conceptual clarification. Social Networks 1(3), 215–239 (1979)
10. Frobenius, G.: Über Matrizen aus nicht negativen Elementen. In: Sitzungsberichte Königlich Preussichen Akademie der Wissenschaft, pp. 456–477 (1912)
11. Iglesias-Ham, M., García-Reyes, E., Kropatsch, W., Artner, N.: Convex deficiencies for human action recognition. Journal of Intelligent & Robotic Systems 64, 353–364 (2011)
12. de Silva, V., Tenenbaum, J.B., Langford, J.C.: A global geometric framework for nonlinear dimensionality reduction. Science 290, 2319–2323 (2000)
13. Mukherjee, S., Biswas, S., Mukherjee, D.: Recognizing human action at a distance in video by key poses. IEEE Transactions on Circuits and Systems for Video Technology 21(9), 1228–1241 (2011)
14. Okamoto, K., Chen, W., Li, X.-Y.: Ranking of closeness centrality for large-scale social networks. In: Preparata, F.P., Wu, X., Yin, J. (eds.) FAW 2008. LNCS, vol. 5059, pp. 186–195. Springer, Heidelberg (2008)
15. Sebastian, T., Klein, P., Kimia, B.: Recognition of shapes by editing their shock graphs. IEEE Transactions on Pattern Analysis and Machine Intelligence 26(5), 550–571 (2004)
16. Sharvit, D., Chan, J., Tek, H., Kimia, B.B.: Symmetry-based indexing of image databases. Journal of Visual Communication and Image Representation 9(4), 366–380 (1998)
17. Siddiqi, K., Shokoufandeh, A., Dickinson, S., Zucker, S.: Shock graphs and shape matching. International Journal of Computer Vision 35, 13–32 (1999)
18. Torsello, A., Hancock, E.R.: A skeletal measure of 2d shape similarity. Computer Vision and Image Understanding 95(1), 1–29 (2004)
19. Wasserman, S., Faust, K.: Social Network Analysis. Methods and Applications. Cambridge University Press, New York (1994)
20. Zhu, L., Chen, Y., Yuille, A.: Learning a hierarchical deformable template for rapid deformable object parsing. IEEE Transactions on Pattern Analysis and Machine Intelligence 32(6), 1029–1043 (2010)

Shape Recognition as a Constraint Satisfaction Problem

Aline Deruyver[1] and Yann Hodé[2]

[1] ICube Laboratory, BFO Team, UMR 7005, 67 412 ILLKIRCH, France
`aline.deruyver@unistra.fr`
[2] Centre Hospitalier, G08, 68 250 Rouffach, France

Abstract. This article proposes a new way to modelize spatial constraints in order to recognize shapes in the context of constraints satisfaction problems (CSP). The proposed spatial constraints take into account not only distances and orientations but also equational properties of the border lines of segmented regions (some characteristic points are defined). This approach is used to build graphs of constraints and the arc-consistency of these graphs is checked by using the AC_{BC} algorithm. The experimentations show the efficiency of this approach to recognize geometrical shapes such as circle and ring in over-segmented images.

1   Introduction

Recognizing objects with a specific shape is a current challenge of image interpretation. Among the most popular segmentation methods, many of them are composed of the two following steps: The first one consists in splitting the image in small homogeneous regions where pixels are so similar that they are supposed to belong to the same object. The second one consists in merging these regions to retrieve the object seen in the image.

We focus our work on this two steps approach. Although the assumption underlying the first step is not always valid (sometimes the boundaries between an object and its background or between two objects are not materialized by a local change of pixel values and are build by the brain [1]), it holds in many cases. A very popular algorithm to obtain an initial set of small homogeneous regions from a grey scale image is the watershed algorithm [2]. It has the advantage of not necessary requiring interactive information as seeds put in targeted regions and it may be used without notifying any threshold value which is very interesting for building an automatic segmentation method.

The merging process of the second step follows rules related with local or regional assumptions about what makes two connected regions similar or not. The most obvious criterion is the grey level value. It may be assumed that two connected regions belonging to the same object, have more similar grey levels values, than two connected regions belonging to two different objects [3], [4]. However this simple idea is far from being always true. Moreover, this approach assumes implicitly a bottom-up process. It means that information is processed

W.G. Kropatsch et al. (Eds.): GbRPR 2013, LNCS 7877, pp. 214–223, 2013.

sequentially with increasing complexities and that the observed shapes are the result of this data driven process [1]. In this framework, one looks for the laws that determine how a combination of information at a given level produces new information at a higher level. But are we sure that these laws that would drive the signal transformation and interpretation are the key of pattern recognition? The possibility of a mixed bottom-up and top down strategy has been hypothesized in human, according to perceptual experiments on normal subjects and on subjects with a lesioned brain ([5],[6]). Then, it is possible for example that the brain has a library of basic shapes and that he tries to match the image content with this library. If we accept this hypothesis, some top-down process may occur. The problem is what formalism may be used to match high level representation (knowledge) and low level representation (set of pixels) and how to match them. We have proposed [7] a way to manage this two levels with the graph formalism: Region Adjacency Graph (RAG) for the low level and semantic graph for the high level. The two structures are matched thanks to arc-consistency checking algorithms. The aim is

- either to control the merging process of the nodes of the RAG as far as the merged nodes may be matched with the nodes of the semantic graph such that it stays arc-consistant [7],
- either to univocally match the nodes of the RAG with a node of the semantic graph thanks to its constraints. This way of doing is relevant only when univocal matching is possible for a majority of the segmented regions. Faithfully, this property can be observed in many cases [8].

In both cases, the more the semantic graph imposed constraints, the more the solution will be univocal and relevant. Semantic graphs are very convenient to represent spatial relationships between different parts of an object. To increase the level of constraints, it could be useful to describe as precisely as possible the shape of each part of an object. A shape may be completely and very precisely described with a set of equations, each equation describing a part of the shape between two of its characteristic points. We propose in this paper, a way to integrate these equational constraints into the semantic graph.

We first define which parameters have to be used to describe morphologic properties along the border lines of studied regions of pixels. Next, we show how the matching between the low level data and the high level model of representation can be done. The definition of a finite domain constraint satisfaction problem with bilevel constraints are recalled and the definition of hyper-Arc consistency problems with bilevel constraints is given. In section 4, we show how to build constraints able to express some properties of border lines of the shape with respect to equations describing the edges. We show, in section 5, how this approach can be used to recognise simple shapes such as circles. Finally some experiments are presented showing the relative noise robustness of our algorithm. In these experiments we try to retrieve the ring and the center of a water meter.

2   Characteristic Points of a Segmented Region

To know if a segmented region is compatible with equational constraints, it is necessary to check if the edge pixels of this region are compatible with the equational constraints. Storing the coordinates of all the edge pixels of a segmented region is not very convenient and some characteristic pixels of the segmented region may summarize the whole contour of this region.

In order to define points of interest describing the relations between two regions in a given direction, we define the interface curve of a region A connected to a region B as the set of pixels of the region A that are connected to pixels of region B. Among the different characteristic points of this curve, we choose the following points:

- The convex extremity of the interface curve (red circles in Figure 1.a). In fact there is three values to store corresponding to the coordinates of the extremum segment. Indeed sometimes we can have several extremum points having the same abcisse coordinate as we can see in Figure 1.b.
- The extremum points: Points corresponding to extremums of the interface curve (yellow circles in Figure 1.a) in terms of coordinates.

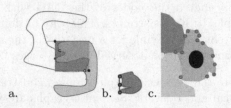

Fig. 1. Characteristic points of a shape

3   Constraint Satisfaction Problems with Bilevel Constraints: Arc-Consistency and Hyper-Arc Consistency

In a previous work, the Connectivity-Direction-Metric Formalism was defined and was used to recognize objects (cerebral anatomy, faces, flowers, ...) made up of several compounds, spatially organized in the context of a constraint satisfaction problem with bilevel constraints. [8,9]. For each type of object, a graph of constraints was built to describe the spatial constraints between each compound of the object to recognize. The $AC4_{BC}$ Algorithm was applied to check if the spatial organization of the different segmented parts of the object satisfy the set of constraints imposed by the graph. In this context, the aim was to find a given spatial organization between the different compounds of the object and not to retrieve an object with a specific shape.

New constraints introducing the characteristic points defined in the previous section will be used in the same context. However, to check if a shape respects morphological properties described by mathematical equations, it is sometimes necessary to impose constraints between more than two compounds. For example, to check if three extremum points are on a given arc of circle, we need to build a constraint between these three points. In this case it will be necessary to build an hyper-arc. Moreover, it is also necessary to check if the sequence of regions making up the edge of an object are continuously linked. The notion of intra-node constraint will be used to perform this checking. Then for a better understanding of the paper, it is necessary to recall the definitions of a constraint satisfaction problem with bilevel constraints and of a arc-consistency problem with bilevel constraints. The definition of hyper-arc consistency problems with bilevel constraints is given as well.

3.1 Finite-Domain Constraint Satisfaction Problem with Bilevel Constraints ($FDCSP_{BC}$)

In a $FDCSP_{BC}$, we define two kinds of constraint: the binary inter-node constraints C_{ij} between two nodes and the binary intra-node constraints $Cmpi$ between two values that could be associated with the node i (i.e. compatible). In the framework of image analysis, these constraints are mainly spatial constraints between regions. Let be D_i the domain associated to a node i.

Definition 1. *Let $Cmpi$ be a compatibility relation, such that $(a, b) \in Cmpi$ iff a and b are compatible. Let C_{ij} be a constraint between i and j. Let us consider a pair S_i, S_j such that $S_i \subset D_i$ and $S_j \subset D_j$, $S_i, S_j \models C_{ij}$ means that (S_i, S_j) satisfies the oriented constraint C_{ij}. Within the image analysis framework, the sets S_i and S_j contain sets of segmented regions.*

$$S_i, S_j \models C_{ij} \Leftrightarrow \begin{cases} \forall a_i \in S_i, \exists (a'_i, a_j) \in S_i \times S_j, \\ such\ that\ (a_i, a'_i) \in Cmpi \\ and\ (a'_i, a_j) \in C_{ij} \end{cases}$$

The sets $\{S_1...S_n\}$ satisfy $FDCSP_{BC}$ iff $\forall C_{ij}\ S_i, S_j \models C_{ij}$.

We associate a graph G to a constraint satisfaction problem in the following way: **(1)** G has a node i for each variable i. **(2)** A directed arc (i, j) is associated with each constraint C_{ij}. **(3)** Arc(G) is the set of arcs of G and e is the number of arcs in G. **(4)** Node(G) is the set of nodes of G and n is the number of nodes in G.

3.2 Arc-Consistency Problem with Bilevel Constraints (AC_{BC}) associated with the $FDCSP_{BC}$

Definition 2. *Let $(i, j) \in arc(G)$. Let $\mathcal{P}(D_i)$ be the set of sub parts of the domain D_i. Arc (i,j) is arc consistent with respect to $\mathcal{P}(D_i)$ and $\mathcal{P}(D_j)$ iff $\forall S_i \in \mathcal{P}(D_i)\ \exists S_j \in \mathcal{P}(D_j)$ such that $\forall v \in S_i\ \exists t \in S_i,\ \exists w \in S_j\ C_{mpi}(v, t)$ and $C_{ij}(t, w)$. (v and t could be identical)*

The definition of an arc consistent graph becomes:

Definition 3. *Let $\mathcal{P}(D_i)$ be the set of sub parts of the domain D_i. Let $P=\mathcal{P}(D_1)$ $\times$ .... $\times \mathcal{P}(D_n)$. A graph G is arc-consistent with respect to P iff $\forall(i,j) \in arc(G)$: (i,j) is arc-consistent with respect to $\mathcal{P}(D_i)$ and $\mathcal{P}(D_j)$.*

The purpose of an arc-consistency algorithm with bilevel constraints is, given a graph G and a set P, to compute P', the largest arc-consistent domain with bilevel constraints for G in P. The $AC4_{BC}$ derived from the AC_4 algorithm proposed by Mohr and Henderson in 1986 [10,11] solve the AC_{BC} problem (See [8] and [9] for the details of the algorithm).

3.3   Hyper-Arc Consistency Problem with Bilevel Constraints (HAC_{BC}) associated with the $FDCSP_{BC}$

In order to build more detailed constraints, it is sometimes necessary to manage n-ary constraints. It is quite easy to generalize the notion of bilevel constraints to n-ary constraints. In this case we have an hyper-graph. The definition of the hyper-arc consistency with bilevel constraints is:

Definition 4. *Let be a CSP (X,D,C) with X a set of nodes, D a set of domains and C a set of constraints, and a constraint $c \in C$ applied on the node x1, x2,...,xn with their domains $D_{x1}, D_{x2}, ..., D_{xn}$ respectively such that $c \subseteq D_{x1} \times D_{x2} \times ... \times D_{xn}$. Let Cmp_{xi} be a compatibility relation (intra-node constraint) associated with the node xi, $i \in [1..n]$ such that $(a,b) \in Cmp_{xi}$ if and only if a and b are compatible. Then c is hyper-arc consistent with two levels of constraints if for each $i \in [1..n]$ and $a \in D_{xi}, \exists a' \in D_{xi}$ and $\exists$ an n-uplet $d \in c$ with $a \in d$ and $(a,a') \in Cmp_{xi}$ and d satisfies the constraint c.*

A CSP is hyper-arc consistent if all its constraints are hyper-arc consistent.

4   Rules to Retrieve a Curve with Local Constraints

Let C be a continuous part of a curve defined by a function $y = f(x)$ between two values x_{min} and x_{max}. As we work on an over-segmented image, we can consider that the derivative of the small part C of the curve is monotonic. Let two regions named head and tail be such that $(x_{min}, f(x_{min}))$ is a characteristic point of the head region, and $(x_{max}, (f(x_{max}))$ is a characteristic point of the tail region. Let P be a set of connected regions forming a chain between the head region and the tail region. Let Sh_{in} the inner part of a shape and Sh_{out} the outer part of the shape. A region $R \in P$ must have the 4 following properties:

1. Rule 1: R has to be "internal edge compatible" with the shape: at least a characteristic point should be compatible with the equational constraint of the edge of the shape and all the other characteristic points should be compatible with the equational constraint defining the edge or the inner part of the shape. Moreover, $\forall R' \in P$, $R' \in D_i$ where D_i is the Domain of the node i corresponding to the edge of the shape.

2. Rule 2: All the regions connected to R have to be compatible with the shape and have to belong to the node of the shape or to the node of regions which are adjacent to the shape. It means that, if p_c and $p\prime_c$ are characteristic points of regions, $\forall p_c \in R$, if $p_c \in Sh_{in}$, $p\prime_c \in Sh_{out} \Rightarrow p\prime_c \notin R$ and $\forall p_c \in R$, if $p_c \in Sh_{out}$, $p\prime_c \in Sh_{in} \Rightarrow p\prime_c \notin R$ (See Figure 2a).

3. Rule 3: $\forall R \in P$, it exists an "edge continuity" between R and another connected regions $R' \in P$. It means that the two regions have to be linked at the level of the pixels belonging to the edge (See Figure 2b). This constraint depends on the mathematical equation discribing the edge that we look for. This rule can be easily implemented thanks to the intra-node constraints ($\mathcal{C}mpi$) defined in section 2.4. See [7] for the implementation of $\mathcal{C}mpi$.

4. Rule 4: If we consider that the derivative of C is monotonic, C must be concave or convex with respect to a segment whose the extremities are two characteristic points belonging to a same border interface between R and a connected region.

a.   b.

Fig. 2. a. The external edge of the circle is made up of regions 1, 2, 3 and 4 (their characteristic points are outside of the diameter of the circle defined by south and north extremum regions (black points). However, the region 1 does not belong to the external edge because it has a characteristic point inside the circle and not only on the border. b. In the left drawing, the connected regions A and B satisfy the edge continuity, in the right drawing A and B do not satisfy the edge continuity as the location of the connection between the two regions is not on the theoretical edge defined by the shape equation.

Thanks to these four properties we can state the following theorem:

Theorem 1. *The satisfaction of the four rules defined previously is enough to guarantee that $\forall p \in C, \exists R \in P$ such that p belongs to the edge pixels of R.*

This theorem assures that working with characteristic points of regions is enough to check that all the edge pixels of the regions of P follow C.

Proof. Given the four previous rules, we want to show that all the pixels of the curve C oriented in a given direction (for example the North direction) belong to the edges of segmented regions of the image.

Initial Constraint (1): We can state from the rules 1, 2, 3 and 4 that all the characteristic points of all the regions are compatible with the curve and that no region has characteristic points both inside and outside the curve.

Let suppose that a pixel p of the curve (yellow point in Figure 3b) is internal to a region R which is a part of the curve thanks to its characteristic points. Then the region R breaks the continuity of the curve. It exists a north projection (P_N) and a south projection (P_S) of this pixel on the north and the south edges of this region (red points in Figure 3c). By definition, for this region R, it exists a north extremum $(extr_N)$ and a south extremum $(extr_S)$ among its characteristic points and these extremums are such that, according to a north/south axis, $extr_N < P_N < p < P_S < extr_S$. As we have assumed that the pixel p is on the curve C and that the derivative of C is monotonic, and thanks to the rule 4, it is possible to show that the north extremum is necessarily external to the shape defined by the curve C and the south extremum is necessarily internal to the shape defined by the curve C. Then the region R has characteristic points internal and external to the shape, which is impossible with respect to the initial constraint (1). Then, it does not exist any pixel of the curve that can be internal to the region R which none characteristic pixels are on the curve. Then all the pixels of the curve C are on edge pixels of the regions. In conclusion the curve C is entirely described by a sub-part of the set of edge pixels of a set of regions.

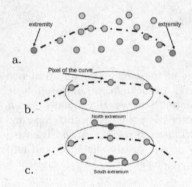

Fig. 3. a. Curve defined by its equation and its characteristic points (in particular, its extremities belonging respectively to head region and tail region of P). Pink points: characteristic points external to the curve belonging to regions of $P : y \leq f(x)$. Green points: characteristic points internal to the curve belonging to regions of $P : y \geq f(x)$ b. We suppose that the yellow pixel is internal to a region R. c. The only possible kind of configuration, thanks to the rule 4. In red, the north projection and the south projection of the yellow pixel on the north and the south edges of R.

5   Application to the Retrieval of a Circle

We suppose that we want to retrieve a circle with a given radius $r \pm \varepsilon$ with the equation $(x - a)^2 + (y - b)^2 = r^2$. The coordinates of the center (a, b) are deduced during the reasoning. The compatibility of the points with the equation is checked thanks to the property defined by the previous theorem. The design of a graph of constraints describing a circle has to be made in nine steps:

- Creating the nodes of the graph:
  - The nodes corresponding to the convex salient points of the circle in the vertical and horizontal directions with respect to an orthonormal coordinate system.
  - The nodes corresponding to the edge of the object (circle) and to the edge of the background connected to the object.
  - The nodes corresponding to the non edge part of the object and to the non edge part of the background.
  - The node of the whole object (union of the edge and non edge nodes of the circle and union of the edge and non edge nodes of the background)
- Determining the:
  - binary constraints between each couple of salient points nodes to check if regions of pixels are compatible with equational contraints.
  - n-ary (hyperarcs) constraints linking more than two salient point nodes to check if regions are compatible with equational constraints (the n-ary constraints are checked after binary constraints to save time-computing).
  - n-ary constraints linking the salient point nodes with the edge node of the object and the edge node of the background according to the equational constraints.
  - n-ary constraints linking the salient point nodes with the edge node of the object which ensures the four properties defined in section 5. The path between the salient point nodes and a region of the edge is only made up of region belonging to the edge. The notion of intra-node constraint defined in section 2.4 ($\mathcal{C}mpi$) is used to define these contraints.
  - union constraints between edge and non edge nodes and their union node.

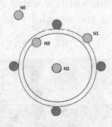

Fig. 4. Nodes to describe a circle. In red, extremum points (N4, N5, N6, N7).

In the case of a circle the graph contains 10 nodes (see Figure 4) with N0: non edge part of the background, N1: background edge, N2: non edge part of the circle, N3: circle edge, N4: Est extremum, N5: west extremum, N6: south extermum, N7: north extremum N8: whole circle (union of N2 and N3) N9: whole background (union of N0 and N1). See the constraints in the Figure 5.

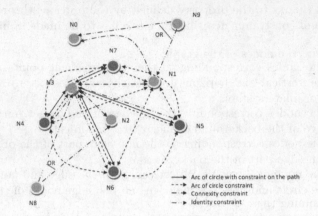

Fig. 5. The graph describing the spatial constraints between each node. N4, N7 and N6 are linked with an hyper-arc constraint checking if it satisfies the property of an arc of circle. The same hyper arc is imposed on N5, N7 and N6. The identity constraints imposed on N3 are linked with a logical OR operator (by using the notion of quasi-arc consistency [12]). The same operator OR is applied on the identity constraints imposed on N8 and N9. The other arcs are linked with a logical AND operator.

6   Experiments

In this application the aim is to localize the water meter in the image in order to detect if it is not broken, to recognize the type of water meter (analogical or numerical) and to read the numerical value displayed on it if there is one. These images are very noisy and the grey level values are not a relevant information to recognize the frame and the center of the water meter. The only way to make a correct interpretation is to use the spatial relations and the morphological

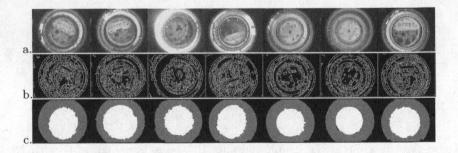

Fig. 6. Interpretation of water meter images. a: original images, b: segmented images with a watershed algorithm c: detection of the frame and the center of the water meter. (The original images are supplied by the company "Véolia").

characteristics of the object subparts. Our approach has been applied to a set of 14 images : 7 images containing true water meters and 7 containing broken water meters. The aim was to localize the frame and the center of the water meter. The images not containing a water meter are detected. In this case the graph is not consistent. Figure 6 presents the 7 labeled images.

7   Conclusion and Discussion

Thanks to the new constraints, the detection of the center of the water meters which is a circle is quite good and precise. The results are much better than those obtained in the previous works which does not include these constraints. More experiments have to be made on other kind of shapes. However the framework proposed in this paper is very generic and can be adapted in a quite natural way to other shapes that can be described with mathematical equations. The Hough transform is another approach using mathematical equations as well, but one interest of our approach is the checking of the constraint of continuity on the curve which is not guaranteed with the Hough transform. Moreover it gives the possibility to look for more complex shapes like diamond or any kind of polygon.

References

1. Marr, D.: Vision. Freeman and Company, New York (1982)
2. Meyer, F.: Un algorithme optimal de partage des eaux. In: Proceeding 8th Congress AFCET, Lyon-Villeurbanne, vol. 2, pp. 847–859 (1992)
3. Lallich, S., Muhlenbachand, F., ·Jolion, J.: A test to control a region growing process within a hierarchical graph. Pattern Recognition Letter 36(10), 2201–2211 (2003)
4. Jolion, J.: Stochastic pyramid revisited. Pattern Recognition Letter 24, 1035–1042 (2003)
5. Beck, D.M., Kastner, S.: Top-down and bottom-up mechanisms in biasing competition in the human brain. Vision Res. 49(10), 1154–1165 (2009)
6. Kveraga, K., Ghuman, A.S., Bar, M.: Top-down predictions in the cognitive brain. Brain Cogn. 65(2), 145–168 (2007)
7. Deruyver, A., Hodé, Y., Jolion, J.-M.: Graph consistency checking: A tool to check the semantic consistency of a segmentation. International Journal of Semantic Computing 5, 179–210 (2011)
8. Deruyver, A., Hodé, Y.: Constraint satisfaction problem with bilevel constraint: application to interpretation of over segmented images. Artificial Intelligence 93, 321–335 (1997)
9. Deruyver, A., Hodé, Y.: Qualitative spatial relationships for image interpretation by using a conceptual graph. Image and Vision Computing 27, 876–886 (2009)
10. Mohr, R., Henderson, T.: Arc and path consistency revisited. Artificial Intelligence 28, 225–233 (1986)
11. Bessière, C.: Arc-consistency and arc-consistency again. Artificial Intelligence 65, 179–190 (1994)
12. Deruyver, A., Hodé, Y., Brun, L.: Image interpretation with a conceptual graph: labeling over-segmented images and detection of unexpected objects. Artificial Intelligence 173, 1245–1265 (2009)

Gaussian Wave Packet on a Graph

Furqan Aziz, Richard C. Wilson, and Edwin R. Hancock*

Department of Computer Science, University of York, YO10 5GH, UK
{furqan,wilson,erh}@cs.york.ac.uk

Abstract. The wave kernel provides a richer and potentially more expressive means of characterising graphs than the more widely studied wave equation. Unfortunately the wave equation whose solution gives the kernel is less easily solved than the corresponding heat equation. There are two reasons for this. First, the wave equation can not be expressed in terms of the familiar node-based Laplacian, and must instead be expressed in terms of the *edge-based Laplacian*. Second, the eigenfunctions of the edge-based Laplacian are more complex than those of the node-based Laplacian. This paper presents the solution of a wave equation on a graph. Wave equation provides an interesting alternative to the heat equation defined using the Edge-based Laplacian. This provides the prerequisites for deeper analysis of graphs and their characterisation. For instance it potentially allows the study of non-dispersive solutions or solitons. In this paper we give a complete solution of the wave equation for a Gaussian wave packet. To simulate the equation on a graph, we assume the initial distribution be a Gaussian wave packet on a single edge of the graph. We show the evolution of this Gaussian wave packet with time on some synthetic graphs.

Keywords: Edge-based Laplacian, Wave Equation, Gaussian wave packet.

1 Introduction

Traditional graph theory defines a discrete Laplacian, Δ, as an operator which is defined only on the vertices of a graph. This Laplacian has found application in many areas like computer vision, machine learning and pattern recognition. For example Fiedler [1] has used the eigenvector corresponding to second smallest eigenvalue of the Laplacian for graph partitioning. Xiao et al [2] have used heat kernel, which is derived from graph Laplacian, to embed the nodes of a graph in Euclidean space. Zhang et al[3] have used the heat kernel for anisotropic image smoothing. The graph Laplacian was used by Coifman and Lafon[4] for dimensionality reduction of data.

The discrete Laplacian defined over the vertices of a graph, however, cannot link most results in analysis to a graph theoretic analogue. For example the wave equation $u_{tt} = \Delta u$, defined with discrete Laplacian, does not have finite speed of propagation. In [5,6], Friedman and Tillich develop a calculus on graph which

---

* Edwin Hancock was supported by a Royal Society Wolfson Research Merit Award.

W.G. Kropatsch et al. (Eds.): GbRPR 2013, LNCS 7877, pp. 224–233, 2013.

provides strong connection between graph theory and analysis. Their work is based on the fact that graph theory involves two different volume measures. i.e., a "vertex-based" measure and an "edge-based" measure. This approach has many advantages. It allows the application of many results from analysis directly to the graph domain.

While the method of Friedman and Tillich leads to the definition of both a divergence operator and a Laplacian (through the definition of both vertex and edge Laplacian), it is not exhaustive in the sense that the edge-based eigenfunctions are not fully specified. In a recent study we have fully explored the eigenfunctions of the edge-based Laplacian and developed a method for explicitly calculating the edge-interior eigenfunctions of the edge-based Laplacian [7]. This reveals a connection between the eigenfunctions of the edge-based Laplacian and both the classical random walk and the backtrackless random walk on a graph. The eigensystem of the edge-based Laplacian contains eigenfunctions which are related to both the adjacency matrix of the line graph and the adjacency matrix of the oriented line graph.

As an application of the edge-based Laplacian, we have recently presented a new approach to characterizing points on a non-rigid three-dimensional shape[8]. This is based on the eigenvalues and eigenfunctions of the edge-based Laplacian, constructed over a mesh that approximates the shape. This leads to a new shape descriptor signature, called the Edge-based Heat Kernel Signature (EHKS). The EHKS was defined using the heat equation, which is based on the edge-based Laplacian. This has applications in shape segmentation, correspondence matching and shape classification.

Wave equation provides potentially richer characterisation of graphs than heat equation. Initial work by ElGhawalby and Hancock [9] has revealed some if its potential uses. They have proposed a new approach for embedding graphs on pseudo-Riemannian manifolds based on the wave kernel. However, there are two problems with the rigourous solution of the wave equation; a) we need to compute the edge-based Laplacian, and b) the solution is more complex than the heat equation.

In this paper we present a solution of the edge-based wave equation on a graph. We assume a Gaussian wave packet on one of the edge of the graph, and see its evolution over time. The remainder of this paper is organized as follows. We commence by introducing graphs and some definitions. In section 3, we introduce the eigensystem of the edge-based Laplacian. In section 4, we give a general solution of the wave equation, and in section 5 we give the solution for the Gaussian wave packet as initial condition. Finally we show simulation of our work on some synthetic graphs.

2   Graphs

A *graph* $G = (V, E)$ consists of a finite nonempty set V of *vertices* and a finite set E of unordered pairs of vertices, called *edges*. A *directed graph* or *digraph* $D = (V_D, E_D)$ consists of a finite nonempty set V_D of vertices and a finite set E_D of ordered pairs of vertices, called *arcs*. So a digraph is a graph with an orientation on each edge. A digraph D is called *symmetric* if whenever (u, v) is an arc of D, (v, u) is also an arc of D. There is a one-to-one correspondence between the set of symmetric digraphs and the set of graphs, given by identifying an edge of the graph with an arc and its inverse arc on the digraph on the same vertices. We denote by $D(G)$ the symmetric digraph associated with the graph G.

The *line graph* $L(G) = (V_L, E_L)$ is constructed by replacing each arc of $D(G)$ by a vertex. These vertices are connected if the head of one arc meets the tail of another. Therefore

$$V_L = \{(u, v) \in D(G)\}$$

$$E_L = \{((u, v), (v, w)) : (u, v) \in D(G), (v, w) \in D(G)\}$$

The *oriented line graph* $OL(G) = (V_O; E_O)$ is constructed in the same way as the $L(G)$ except that reverse pairs of arcs are not connected, i.e. $((u, v), (v, u))$ is not an edge. The vertex and edge sets of $OL(G)$ are therefore

$$V_L = \{(u, v) \in D(G)\}$$

$$E_L = \{((u, v), (v, w)) : (v, w)), (u, v) \in D(G), (v, w) \in D(G), u \neq w\}$$

Figure 1(a) shows a simple graph, 1(b) its digraph, and 1(c) the corresponding oriented line graph.

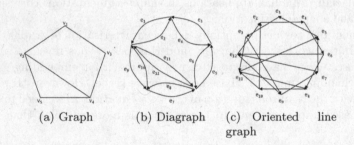

(a) Graph   (b) Diagraph   (c) Oriented   line
                                            graph

Fig. 1. Graph, its digraph, and its oriented line graph

3   Edge-Based Eigensystem

In this section we review the eigenvalues and eigenfunctions of the edge-based Laplacian[5][7]. Let $G = (V, E)$ be a graph with a boundary ∂G. Let $\mathcal{G}$ be the geometric realization of G. The geometric realization is the metric space consisting of vertices V with a closed interval of length l_e associated with each edge $e \in \mathcal{E}$. We associate an edge variable x_e with each edge that represents the standard coordinate on the edge with $x_e(u) = 0$ and $x_e(v) = 1$. For our work, it will suffice to assume that the graph is finite with empty boundary (i.e., $\partial G = 0$) and $l_e = 1$.

3.1   Vertex Supported Edge-Based Eigenfunctions

The vertex-supported eigenpairs of the edge-based Laplacian can be expressed in terms of the eigenpairs of the normalized adjacency matrix of the graph. Let A be the adjacency matrix of the graph G, and $\tilde{A}$ be the row normalized adjacency matrix. i.e., the $(i, j)th$ entry of $\tilde{A}$ is given as $\tilde{A}(i, j) = A(i, j)/\sum_{(k,j)\in E} A(k, j)$. Let $(\phi(v), \lambda)$ be an eigenvector-eigenvalue pair for this matrix. Note $\phi(.)$ is defined on vertices and may be extended along each edge to an edge-based eigenfunction. Let ω^2 and $\phi(e, x_e)$ denote the edge-based eigenvalue and eigenfunction. Here $e = (u, v)$ represents an edge and x_e is the standard coordinate on the edge (i.e., $x_e = 0$ at v and $x_e = 1$ at u). Then the vertex-supported eigenpairs of the edge-based Laplacian are given as follows:

1. For each $(\phi(v), \lambda)$ with $\lambda \neq \pm 1$, we have a pair of eigenvalues ω^2 with $\omega = \cos^{-1} \lambda$ and $\omega = 2\pi - \cos^{-1} \lambda$. Since there are multiple solutions to $\omega = \cos^{-1} \lambda$, we obtain an infinite sequence of eigenfunctions; if $\omega_0 \in [0, \pi]$ is the principal solution, the eigenvalues are $\omega = \omega_0 + 2\pi n$ and $\omega = 2\pi - \omega_0 + 2\pi n, n \geq 0$. The eigenfunctions are $\phi(e, x_e) = C(e) \cos(B(e) + \omega x_e)$ where

$$C(e)^2 = \frac{\phi(v)^2 + \phi(u)^2 - 2\phi(v)\phi(u)\cos(\omega)}{\sin^2(\omega)}$$

$$\tan(B(e)) = \frac{\phi(v)\cos(\omega) - \phi(u)}{\phi(v)\sin(\omega)}$$

   There are two solutions here, $\{C, B_0\}$ or $\{-C, B_0 + \pi\}$ but both give the same eigenfunction. The sign of $C(e)$ must be chosen correctly to match the phase.

2. $\lambda = 1$ is always an eigenvalue of $\tilde{A}$. We obtain a principle frequency $\omega = 0$, and therefore since $\phi(e, x_e) = C \cos(B)$ and so $\phi(v) = \phi(u) = C \cos(B)$, which is constant on the vertices.

3.2   Edge-Interior Eigenfunctions

The edge-interior eigenfunctions are those eigenfunctions which are zero on vertices and therefore must have a principle frequency of $\omega \in \{\pi, 2\pi\}$. Recently we have shown that these eigenfunctions can be determined from the eigenvectors of the adjacency matrix of the oriented line graph[7]. We have shown that the eigenvector corresponding to eigenvalue $\lambda = 1$ of the oriented line graph provides a solution in the case $\omega = 2\pi$. In this case we obtain $|E| - |V| + 1$ linearly independent solutions. Similary the eigenvector corresponding to eigenvalue $\lambda = -1$ of the oriented line graph provides a solution in the case $\omega = \pi$. In this case we obtain $|E| - |V|$ linearly independent solutions. This comprises all the principal eigenpairs which are supported on the vertices.

3.3   Normalization of Eigenfunctions

Note that although these eigenfunctions are orthogonal, they are not normalized. To normalize these eigenfunctions we need to find the normalization factor corresponding to each eigenvalue. Let $\rho(\omega)$ denotes the normalization factor corresponding to eigenvalue ω. Then

$$\rho^2(\omega) = \sum_{e \in \mathcal{E}} \int_0^1 \phi^2\left(e, x_e\right) dx_e$$

Evaluating the integral, we get

$$\rho(\omega) = \sqrt{\sum_{e \in \mathcal{E}} C(e)^2 \left[\frac{1}{2} + \frac{\sin\left(2\omega + 2B(e)\right)}{4\omega} - \frac{sin(2B(e))}{4\omega}\right]}$$

Once we have the normalization factor to hand, we can compute a complete set of orthonormal bases by dividing each eigenfunction with the corresponding normalization factor. Once normalized, these eigenfunctions form a complete set of orthonormal bases for $L^2(\mathcal{G}, \mathcal{E})$.

4   General Solution of the Wave Equation

Let a graph coordinate $\mathcal{X}$ defines an edge e and a value of the standard coordinate on that edge x. The eigenfunctions of the edge-based Laplacian are

$$\phi_{\omega,n}(\mathcal{X}) = C(e, \omega) \cos\left(B(e, \omega) + \omega x + 2\pi n x\right)$$

The edge-based wave equation is

$$\frac{\partial^2 u}{\partial t^2}(\mathcal{X}, t) = \Delta_E u(\mathcal{X}, t)$$

We look for separable solutions of the form $u(\mathcal{X}, t) = \phi_{\omega,n}(\mathrm{X})g(t)$. This gives

$$\phi_{\omega,n}(\mathcal{X})g''(t) = g(t)\left(\omega + 2\pi n\right)^2 \phi(\omega, n)$$

which gives a solution for the time-based part as

$$g(t) = \alpha_{\omega,n} \cos\left[(\omega + 2\pi n)t\right] + \beta_{\omega,n} sin\left[(\omega + 2\pi n)t\right]$$

By superposition, we obtain the general solution

$$u(\mathcal{X}, t) = \sum_{\omega} \sum_{n} C(e, \omega) \cos\left[B(e, \omega) + \omega x + 2\pi n x\right]$$
$$\left\{\alpha_{\omega,n} \cos\left[(\omega + 2\pi n)t\right] + \beta_{\omega,n} sin\left[(\omega + 2\pi n)t\right]\right\}$$

4.1   Initial Conditions

Since the wave equation is second order partial differential equation, we can impose initial conditions on both position and speed

$$u(\mathcal{X}, 0) = p(\mathcal{X})$$

$$\frac{\partial u}{\partial t}(\mathcal{X}, 0) = q(\mathcal{X})$$

and we obtain

$$p(\mathcal{X}) = \sum_\omega \sum_n \alpha_{\omega,n} C(e, \omega) \cos\left[B(e, \omega) + \omega x + 2\pi n x\right]$$

$$q(\mathcal{X}) = \sum_\omega \sum_n \beta_{\omega,n}(\omega + 2\pi n) C(e, \omega) \cos\left[B(e, \omega) + \omega x + 2\pi n x\right]$$

We can obtain these coefficients using the orthogonality of the eigenfunctions. So we get

$$\alpha_{\omega,n} = \sum_e C(e, \omega) \frac{1}{2}\left[F_{\omega,n} + F_{\omega,n}^*\right]$$

where

$$F_{\omega,n} = e^{iB} \int_0^1 dx\, p(e, x) e^{i\omega x} e^{i2\pi n}$$

similarly

$$\beta_{\omega,n}(\omega + 2\pi n) = \sum_e C(e, \omega) \frac{1}{2}\left[G_{\omega,n} + G_{\omega,n}^*\right]$$

where

$$G_{\omega,n} = e^{iB} \int_0^1 dx\, q(x, e) e^{i(\omega + 2\pi n)x} = e^{iB} \int_0^1 dx\, p'(x, e) e^{i(\omega + 2\pi n)x}$$

5   Gaussian Wave Packet

Let the initial position be a Gaussian wave packet $p(e, x) = e^{-a(x-\mu)^2}$ on one particular edge and zero everywhere else. Then we have

$$F_{\omega,n} = e^{iB} \int_0^1 dx\, e^{-a(x-\mu)^2} e^{i\omega x} e^{i2\pi n x}$$

$$= e^{iB} e^{i\mu\omega} e^{-\frac{\omega^2}{4a}} \int_0^1 dx\, e^{-a\left(x - \mu - \frac{i\omega}{2a}\right)^2} e^{i2\pi n x}$$

Let the Gaussian is fully contained on one edge. i.e., $p(x, e)$ is only supported on this edge, then

$$F_{\omega,n} = e^{iB} e^{i\mu\omega} e^{-\frac{\omega^2}{4a}} \int_{-\infty}^{\infty} dx\, e^{-a\left(x - \mu - \frac{i\omega}{2a}\right)^2} e^{i2\pi n x}$$

Solving, we get

$$F_{\omega,n} = \sqrt{\frac{\pi}{a}} e^{i[B+\mu(\omega+2\pi n)]} e^{-\frac{1}{4a}(\omega+2n\pi)^2}$$

Similarly we obtain

$$F_{\omega,n}^* = \sqrt{\frac{\pi}{a}} e^{-i[B+\mu(\omega+2\pi n)]} e^{-\frac{1}{4a}(\omega+2n\pi)^2}$$

and so

$$\alpha_{\omega,n} = \sqrt{\frac{\pi}{a}} e^{-\frac{1}{4a}(\omega+2n\pi)^2} C(e,\omega) \cos[B+\mu(\omega+2\pi n)]$$

Since $p(x,e)$ is zero at both ends the coefficients β can be found straightforwardly.

$$\beta_{\omega,n} = \sqrt{\frac{\pi}{a}} e^{-\frac{1}{4a}(\omega+2n\pi)^2} C(e,\omega) \sin[B+\mu(\omega+2\pi n)]$$

5.1   Complete Reconstruction

Let f be the edge on which the initial function is non-zero. Let the Gaussian is fully contained on one edge. Then

$$u(\mathcal{X},t) = \sum_{\omega} \sqrt{\frac{\pi}{a}} C(\omega,e) C(\omega,f) \sum_{n} e^{-\frac{1}{4a}(\omega+2\pi n)^2}$$
$$\cos\left[B(\omega,e) + \omega x + 2\pi n x\right] \cos\left[B(\omega,f) + (\omega+2\pi n)(t+\mu)\right]$$

For a particular sequence with principal eigenvalue ω, we need to calculate

$$u_\omega = \sum_{n} \sqrt{\frac{\pi}{a}} e^{-\frac{1}{4a}(\omega+2\pi n)^2} \cos\left[B(\omega,e) + \omega x + 2\pi n x\right] \cos\left[B(\omega,f) + (\omega+2\pi n)(t+\mu)\right]$$

Writing the cosine in exponential form, we obtain

$$u_w = \sum_{n} \sqrt{\frac{\pi}{a}} e^{-\frac{1}{4a}(\omega+2\pi n)^2}$$
$$\times \frac{1}{4} \left[e^{i[B(e,\omega)+B(f,\omega)]} e^{i(\omega+2\pi n)(x+t+\mu)} + e^{-i[B(e,\omega)+B(e,\omega)]} e^{-i(\omega+2\pi n)(x+t+\mu)} \right.$$
$$\left. + e^{i[B(e,\omega)-B(f,\omega)]} e^{i(\omega+2\pi n)(x-t-\mu)} + e^{-i[B(e,\omega)-B(e,\omega)]} e^{-i(\omega+2\pi n)(x-t-\mu)} \right]$$

We need to evaluate terms like terms like $\sum_n \frac{\pi}{a} e^{-\frac{1}{4a}} e^{i[B(e,\omega)+B(f,\omega)]} e^{i(\omega+2\pi n)(x+t+\mu)}$, where the values of ω and n depend on the particular eigenfunction sequence under evaluation.

Let $\mathcal{W}(z)$ be z wrapped to the range $[-\frac{1}{2}, \frac{1}{2})$, i.e.,

$$\mathcal{W}(z) = z - \left\lfloor z + \frac{1}{2} \right\rfloor$$

Solving for all cases, the complete solution becomes

$$u(\mathcal{X}, t) = \sum_{\omega \in \Omega_a} \frac{C(\omega, e)C(\omega, f)}{2} \left(e^{-a\mathcal{W}(x+t+\mu)^2} \cos\left[B(e, \omega) + B(f, \omega) + \omega \left\lfloor x + t + \mu + \frac{1}{2} \right\rfloor \right] \right.$$

$$\left. + e^{-a\mathcal{W}(x-t-\mu)^2} \cos\left[B(e, \omega) - B(f, \omega) + \omega \left\lfloor x - t - \mu + \frac{1}{2} \right\rfloor \right] \right)$$

$$+ \frac{1}{2|E|} \left(\frac{1}{4} e^{-a\mathcal{W}(x+t+\mu)^2} + \frac{1}{4} e^{-a\mathcal{W}(x-t-\mu)^2} \right)$$

$$+ \sum_{\omega \in \Omega_c} \frac{C(\omega, e)C(\omega, f)}{4} \left(e^{-a\mathcal{W}(x-t-\mu)^2} - e^{-a\mathcal{W}(x+t+\mu)^2} \right)$$

$$+ \sum_{\omega \in \Omega_c} \frac{C(\omega, e)C(\omega, f)}{4} \left((-1)^{\lfloor x-t-\mu+\frac{1}{2} \rfloor} e^{-a\mathcal{W}(x-t-\mu)^2} \right.$$

$$\left. - (-1)^{\lfloor x+t+\mu+\frac{1}{2} \rfloor} e^{-a\mathcal{W}(x+t+\mu)^2} \right)$$

where Ω_a represents the set of vertex-supported eigenvalues and Ω_b and Ω_c represent the set of edge-interior eigenvalues respectively. i.e., π and 2π.

6   Experiments

In this section, we show the evolution of Gaussian wave packet on some simple graphs. Figure 2 shows the result for a graph with five nodes and seven edges

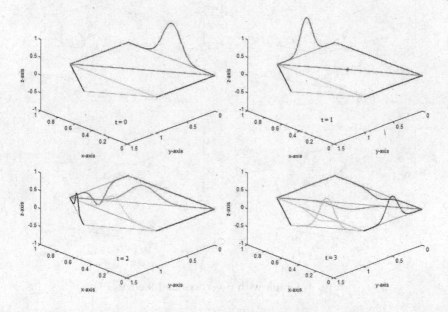

Fig. 2. Graph with 5 vertices and 7 edges

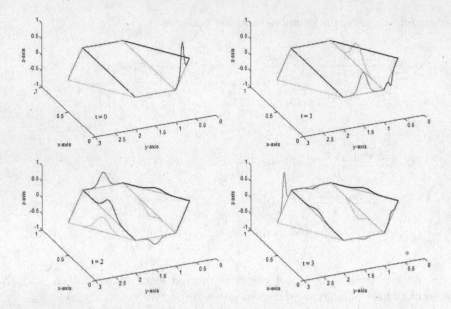

Fig. 3. Graph with 6 vertices and 8 edges

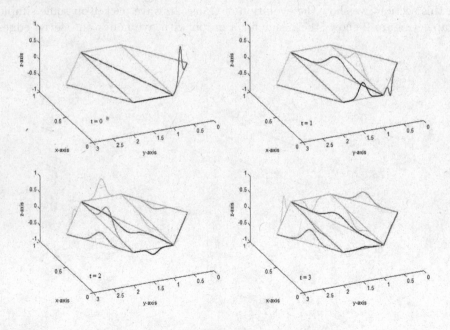

Fig. 4. Graph with 6 vertices and 9 edges

for time $t = 0$, $t = 1$, $t = 2$ and $t = 3$. Note that when the wave packet hits a node with degree greater than 2, some part of the packet is reflected back while the other part is equally distributed to the connecting edges. A similar result is shown for graph with six nodes and eight edges in Figure 3, and for a graph with six nodes and nine edges in Figure 4.

7 Conclusion and Future Work

In this paper we have developed a complete solution of the wave equation on a graph which is based on the edge-based Laplacian of a graph. We assume the initial distribution be a Gaussian wave packet and shown its evolution with time on different graphs. The advantage of using the edge-based Laplacian over vertex-based Laplacian is that it allows the direct application of many results from analysis to graph theoretic domain. For example it allows the study of non-dispersive solutions or solitons. In future our goal is to use the solution of the wave equation and other equations defined using the edge-based Laplacian for characterizing graphs with higher accuracy.

References

1. Fiedler, M.: A property of eigenvectors of non-negative symmetric matrices and its application to graph theory. Czechoslovak Mathematics Journal (1975)
2. Xiao, B., Yu, H., Hancock, E.R.: Graph matching using manifold embedding. In: Campilho, A.C., Kamel, M.S. (eds.) ICIAR 2004. LNCS, vol. 3211, pp. 352–359. Springer, Heidelberg (2004)
3. Zhang, F., Hancock, E.R.: Graph spectral image smoothing using the heat kernel. Pattern Recognition, 3328–3342 (2008)
4. Coifman, R.R., Lafon, S.: Diffusion maps. Applied and Computational Harmonic Analysis, 5–30 (2006)
5. Friedman, J., Tillich, J.P.: Wave equations for graphs and the edge based laplacian. Pacific Journal of Mathematics, 229–266 (2004)
6. Friedman, J., Tillich, J.P.: Calculus on graphs. CoRR (2004)
7. Wilson, R.C., Aziz, F., Hancock, E.R.: Eigenfunctions of the edge-based laplacian on a graph. Linear Algebra and its Applications (2013)
8. Aziz, F., Wilson, R.C., Hancock, E.R.: Shape signature using the edge-based laplacian. In: International Conference on Pattern Recognition (2012)
9. ElGhawalby, H., Hancock, E.R.: Graph embedding using an edge-based wave kernel. In: Hancock, E.R., Wilson, R.C., Windeatt, T., Ulusoy, I., Escolano, F. (eds.) SSPR & SPR 2010. LNCS, vol. 6218, pp. 60–69. Springer, Heidelberg (2010)

Exact Computation of Median Surfaces
Using Optimal 3D Graph Search[*]

Zhengwang Wu[1], Xiaoyi Jiang[2], Nanning Zheng[1],
Yuehu Liu[1], and Dachuan Cheng[2]

[1] Institute of Artificial Intelligence and Robotics, Xi'an Jiaotong University, China
[2] Department of Mathematics and Computer Science,
University of Münster, Germany
[3] Department of Biomedical Imaging, China Medical University, Taiwan

Abstract. In this paper we formulate the generalized median surface
problem and present its exact solution by means of an optimal 3D graph
search algorithm. In addition to the general interest in median surface
computation our work is also motivated by the task of parameter space
exploration without ground truth, which is an effective means of dealing
with the difficult parameter problem. A concrete application in this context will be demonstrated on artery boundary detection in ultrasound
data. It will be shown that the median computation can not only avoid
the parameter training, but also potentially achieve even better results
than with trained parameters. Particularly in situations with no available
ground truth, the median-based approach can thus be a good alternate.

1 Introduction

Median computation has turned out to be a useful concept in pattern recognition [1]. Given an object set S in space U, the generalized median is defined
by $x \in U$ which minimizes the sum of distances to all objects in S and can
be considered as a good representative of the given set. Another motivation of
median computation is to eliminate some erroneous objects by averaging over
all objects. Generally, the median concept is motivated by well established results from supervised classifier combination: By averaging the results of several
classifiers a more reliable classification can be achieved [2].

The median concept has been concretized to a lot of domains including vectors
[3], strings [4], graphs [5], clusterings [6], and segmentations [7]. In [8] the 2D
median contour problem is investigated. In this work we consider the related 3D
median surface problem.

There exist only very few general frameworks for median computation. One
such framework described in [5] is based on an embedding into the vector
space. The median vector is computed by means of the Weiszfeld algorithm
[3] and inversely transformed to the original space. Another general framework
[9] computes the weighted mean of a pair of objects in an evolutionary scheme.

---

[*] This work is supported by NSFC, Grant No. 90920008.

W.G. Kropatsch et al. (Eds.): GbRPR 2013, LNCS 7877, pp. 234–243, 2013.

Both frameworks are approximative only and therefore suitable for those median problems with inherently high computational complexity. Indeed, they have been applied to computing generalized median of strings [4,9] and graphs [5], both being of $\mathcal{NP}$-hard problems.

Many median computation algorithms have been developed for specific domains to integrate as much as possible domain-specific knowledge in order to obtain possibly exact solutions in an efficient way. For instance, the generalized median string problem is $\mathcal{NP}$-hard for the edit distance, but simplified histogram-based distances enable low-order polynomial time [10]. For 2D contours dynamic programming can be used to determine the optimal median contour in a time linear to the image size [8]. In this work we will show that for the class of so-called *terrain-like surfaces* (to be formally defined later) considered here, we can apply an optimal 3D graph search algorithm to exactly and efficiently solve median surface problem.

In addition to the general interest in median surface computation this work is also motivated by our recent work on exploring the parameter space of segmentation algorithms without ground truth. Ensemble techniques similar to multiple classifier systems should be developed to achieve the best possible segmentation result on a per-image basis.

The outline of this paper is as follows. In Section 2 we define the median surface problem under consideration, which is further motivated by segmentation parameter exploration in Section 3. The median surface problem will be exactly solved by applying an optimal 3D graph search algorithm (Section 4). We report experimental results to demonstrate the usability of median surface computation in the context of segmentation parameter exploration in Section 5. Finally, some discussions in Section 6 conclude the paper.

2   Median Surface Problem

The surfaces of concern in this paper are terrain-like (height-field) as specified in Definition 1.

Definition 1. *A terrain-like surface is a function:* $f : X \times Y \to Z$ *with* $X = \{1, 2, \ldots, M\}$, $Y = \{1, 2, \ldots, N\}$, *and* $Z = \{1, 2, \ldots, L\}$. *In order to guarantee surface connectivity in 3D, an additional smoothness constraint requires* $|f(x + 1, y) - f(x, y)| \leq \Delta_x$ *and* $|f(x, y + 1) - f(x, y)| \leq \Delta_y$ *for small positive constants* Δ_x *and* Δ_y.

In the following we will use the term "surface" only for the sake of simplicity. This class of surfaces are very common in image analysis. In 3D biomedical volume datasets an important task is to detect such terrain-like surfaces, possibly in an optimal manner. Stacking 2D images along the time axis also results in 3D volume datasets and related terrain-like surface detection tasks.

Given a set S of K surfaces $\{S_1, S_2, \ldots, S_K\}$ and a distance function $d()$ which measures the dissimilarity of two surfaces, the general median surface is defined by:

$$\overline{S} \;=\; \arg\min_{s\in U_S} \sum_{i=1}^{K} d(s, S_i) \tag{1}$$

where U_S represents the space (universe) of all potential solutions, i.e. surfaces within the volume $X \times Y \times Z$.

The distance function is defined by:

$$d(s, S_i) \;=\; \sum_{x=1}^{M} \sum_{y=1}^{N} w_{xy} \cdot \rho(s(x,y), S_i(x,y)) \tag{2}$$

where ρ is a dissimilarity function for scalar values. Any function suitable for a certain application, e.g. L_p, can be used for this purpose. In particular, those from robust statistics [11] may help to achieve improved performance against outliers in the input surface data. In the simplest case the weight w_{xy} can be set to be constant for all (x, y) positions. But in general, a location-sensitive weight gives us more flexibility to incorporate application-specific knowledge to a largest extent. For our segmentation parameter exploration we will fully utilize this flexibility.

3 Motivation

One motivation of median surface computation is exploring segmentation parameter space without ground truth. Segmentation algorithms mostly have some parameters and their optimal setting is not a trivial task. In automatic parameter training a training image set with (manual) ground truth segmentation is assumed to be available. Then, a subspace of the parameter space is explored to find out the best parameter setting. For each parameter setting candidate a performance measure is computed in the following way:

- Segment each image of the training set based on the parameter setting;
- Compute a performance measure by comparing the segmentation result and the corresponding ground truth;
- Compute the average performance measure over all images of the training set.

The optimal parameter setting is given by the one with the largest average performance measure. Since fully exploring the subspace can be very costly, space subsampling [12] and genetic search [13] have been proposed. While this approach is reasonable and has been successfully practiced in several applications, its fundamental disadvantage is the need of ground truth segmentation. The manual generation of ground truth is always painful and thus a main barrier of wide use in many situations.

Recently, it is proposed to apply the concept of generalized median for implicitly exploring the parameter space without the need of ground truth segmentation. Assuming a reasonable subspace of the parameter space (i.e. a lower and upper bound for each parameter), it is sampled into a finite number $\mathcal{M}$ of parameter settings. Then, the segmentation procedure is run for all the $\mathcal{M}$ parameter settings and the generalized median of the $\mathcal{M}$ segmentation results is computed. The rationale here is that the median segmentation tends to be a good one within the explored parameter subspace, as successfully demonstrated for 2D contour detection [8] and region segmentation [7]. Segmentation of terrain-like surfaces is one of the most important problems in (biomedical) image analysis. Thus, median surface computation can help to alleviate the segmentation parameter problem in 3D surface segmentation as well.

Another situation is segmentation of 2D images along the time axis. Many algorithms from the literature, e.g. [14], perform the segmentation independently on all images and thus cannot guarantee a continuous segmentation over time, which is highly desired when working with high-speed imaging devices. If the parameter space exploration technique described above is applied in this case to the 3D volumes by stacking all image-wise segmentations along the time axis, we obtain a continuous temporal segmentation without any extra effort as a nice spinoff of handling the parameter problem. For doing this, we certainly need to relax the input for the median surface computation to potentially include discontinuous surfaces. But this is not a problem at all.

In summary median surface computation is not only an interesting topic in its own right but also of substantial practical value. This motivates us to find an efficient way for exact median surface computation.

4   Exact Computation by Optimal 3D Graph Search

In this section we show that the median surface problem defined in Eq. (1) can be transformed into an optimal 3D surface detection problem, which is solvable by an optimal 3D graph search algorithm in low-order polynomial time.

First, we reformulate Eq. (1) as follows.

$$
\begin{aligned}
\overline{S} &= \arg\min_{s \in U_S} \sum_{i=1}^{K} d(s, S_i) \\
&= \arg\min_{s \in U_S} \sum_{i=1}^{K} \sum_{x=1}^{M} \sum_{y=1}^{N} w_{xy} \cdot \rho(s(x,y), S_i(x,y)) \\
&= \arg\min_{s \in U_S} \sum_{x=1}^{M} \sum_{y=1}^{N} w_{xy} \cdot \underbrace{\sum_{i=1}^{K} \rho(z = s(x,y), S_i(x,y))}_{c_{xyz}} \\
&= \arg\min_{s \in U_S} \underbrace{\sum_{x=1}^{M} \sum_{y=1}^{N} c_{xyz}}_{C(s)} \\
&= \arg\min_{s \in U_S} C(s)
\end{aligned}
\tag{3}
$$

A candidate solution surface $s \in U_S$ is characterized by the z-value $s(x, y)$ for each position (x, y) on the grid $X \times Y$. We assign each point (x, y, z) in the volume $X \times Y \times Z$ a cost c_{xyz}, which is determined by its deviations (in z-direction) from the K input surfaces $S_i(x, y)$ and the position-specific weight w_{xy}. Then, the goodness of a candidate solution surface s can be measured by $C(s)$, i.e. summing up the costs of all positions. Therefore, the median surface is simply the optimal surface with minimal cost from the solution space U_S (consisting of all terrain-like surfaces within the volume $X \times Y \times Z$).

The discussion above leads to the following new optimization problem. We first compute a cost c_{xyz} for each point (x, y, z) in the volume $X \times Y \times Z$. Then, the median surface is determined by finding the terrain-like surface within the volume with the minimal sum of costs.

It is important to notice that we cannot solve this optimization problem by computing the optimal z-value for each of the $M \times N$ positions (x, y) *independently*, which could be done, for instance, by enumerating all z-values out of Z and minimizing c_{xyz}. Doing it this way, we may encounter the trouble of generating a discontinuous resultant surface. Only for simple cases (e.g. constant weight w_{xy} and $\rho = L_2$) the simple position-wise optimization is guaranteed to deliver an optimal continuous resultant surface. But in general, a global optimization approach is needed.

For the special case $N = 1$ (i.e. the y-axis vanishes), the 3D optimal surface segmentation is reduced to a 2D optimal contour detection problem. This simplified problem was solved in [8] by a highly efficient dynamic programming algorithm. Unfortunately, there is no direct way of extending the dynamic programming solution to the 3D problem at hand.

Fortunately, the optimal 3D graph search algorithm described in [15] solves exactly our 3D optimal surface detection problem. In the following we briefly present the most important steps of this algorithm and the readers are referred to [15] for more details. A node-weighted directed graph $G = (V, E, W)$ is constructed as follows. For each point (x, y, z) in the volume $X \times Y \times Z$ a corresponding node $V(x, y, z)$ is defined in G, whose weight $W(x, y, z)$ is assigned according to:

$$W(x, y, z) = \begin{cases} c_{xyz} - c_{xy,z-1} & z > 1 \\ c_{xyz} & z = 1 \end{cases} \tag{4}$$

where c_{xyz} is the cost defined in Eq. (3). G contains two types of arcs: $E = E^a \cup E^r$. The set E^a of intraposition arcs rules the connections within the same position (x, y). Each node $V(x, y, z)$ ($z > 1$) has a direct arc to the node $V(x, y, z - 1)$ below it, i.e.,

$$E^a = \{< V(x, y, z), V(x, y, z - 1) > \mid z > 1\}$$

The set E^r of interposition arcs rules the connections of adjacent positions and is defined by:

$$E^r = \begin{cases} \{< V(x,y,z), \ V(x+1,y,\max(0,z-\Delta_x)) > \\ \quad \mid x \in \{1,\dots,M-1\}, \ z \in Z\} \ \cup \\ \{< V(x,y,z), \ V(x-1,y,\max(0,z-\Delta_x)) > \\ \quad \mid x \in \{2,\dots,M\}, \ z \in Z\} \ \cup \\ \{< V(x,y,z), \ V(x,y+1,\max(0,z-\Delta_y)) > \\ \quad \mid y \in \{1,\dots,N-1\}, \ z \in Z\} \ \cup \\ \{< V(x,y,z), \ V(x,y-1,\max(0,z-\Delta_y)) > \\ \quad \mid y \in \{2,\dots,N\}, \ z \in Z\} \end{cases}$$

Given the constructed digraph G, a closed set C is a subset of nodes such that all successors of any nodes in C are also contained in C. The cost of a closed set is the total cost of all its nodes. In [15] it was shown that the original optimal surface detection problem is equivalent to finding a minimum nonempty closed set in G. This is a well studied problem in graph theory and can be solved by computing a minimum $s-t$ cut in a related graph G_{st} (see [15] for the details of constructing G_{st} from G). In our implementation the Boykov-Kolmogorov algorithm [16] is applied to compute the minimum $s-t$ cut. For a graph with n nodes and m arcs, the theoretical worst-case time complexity for this algorithm is $\mathcal{O}(n^2mc)$, where c is the cost of the minimum cut.

5   Experimental Results

In this section we demonstrate a practical use of median surface computation by applying the algorithm described above to segmentation parameter exploration on ultrasound data.

5.1   Ultrasound Image Data and Experimental Settings

The task considered here is the extraction of artery boundaries from ultrasound videos. An artery has a near wall and a far wall, as illustrated in Fig. 1(b). Along with the time axis the 2D images can be regarded a 3D volume, see Fig. 1(a). Three ultrasound videos from patients were used in our experiments. For these videos, a ground truth of the arterial walls (golden standard) was labeled manually.

The algorithm from [14] was applied to detect the two contours in each 2D image. These contours are stacked to build 3D, possibly discontinuous, surfaces. For each video, 100 different parameter settings were used to generate 100 near wall surfaces and 100 far wall surfaces. Then, a median surface was computed from each of the two surface ensembles. For this computation a position-wise weight w_{xy}, see Eq. (2), is needed to measure the dissimilarity between the z-values of a candidate median surface and an input surface at (x,y). In our current implementation this was done in the following way. A normal distribution is estimated using all z-values of the 100 input surfaces at (x,y). Then, w_{xy} is chosen to be [1.0 - density at $z = s(x,y)$].

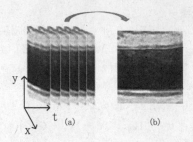

Fig. 1. Ultrasound video. (a) Along with the time axis the 2D images build a 3D volume. (b) The near (blue) and far (red) wall of the artery in a single image.

For comparison purpose the best-performing one among the 100 parameter settings was determined for each video by comparing with the ground truth. In all our tests the comparison between two surfaces, e.g. a segmented surface and a ground truth, was done by computing the average L_1 deviation in z per (x, y) position.

5.2   Comparison with the Best Parameter Setting

Totally we have 6 test cases (near and far wall, 3 videos). For each of the test cases the median surface was computed from the 100 input surfaces and compared with the surface from the best-performing parameter setting, which was determined by a per-video basis. The results are shown in Table 1, which are further detailed in Figures 2-4 with the average deviation per image and the distribution of deviations.

These results indicate that basically no real performance differences exist between the best-performing parameter setting and our approach of parameter space exploration by means of median surface computation. This fact is particularly remarkable since the parameter optimization was done on a per-video basis in contrast to the popular practice of using training images. In the latter case the trained best parameters can be expected to achieve good results on additional test images, but in general not the best result per image. Overall, it can be concluded that without using any ground truth information, the generalized median technique is able to produce segmentations of identical quality as the training approach.

5.3   Comparison with the Ground Truth

Since all the real data include manually labeled ground truth (GT), we compared our median result with the ground truth. The average L_1 deviation in z is shown in Table 2. In addition, the results from the best-performing parameter setting (BP) were also compared with GT. Some results are given in Figure 5 for illustration purpose. As can be seen in Table 2, these results turn out to be inferior to our median segmentation results. Using our median surface algorithm thus can not only avoid the parameter training (which is only possible with existing ground truth), but also potentially achieve even better segmentation results than with

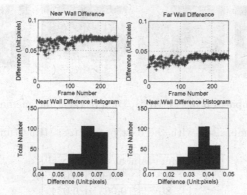

Fig. 2. Comparison with the best parameter setting: video 1

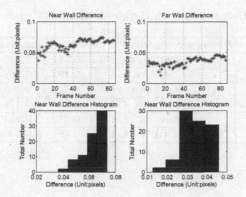

Fig. 3. Comparison with the best parameter setting: video 2

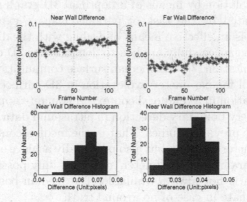

Fig. 4. Comparison with the best parameter setting: video 3

the best parameters. This fact is clearly due to the ensemble nature of the median surface computation.

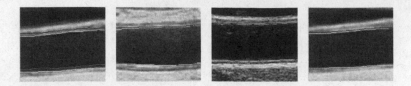

Fig. 5. Comparison of GT (red), BP (yellow), and median (blue)

Table 1. Comparison with the best parameter setting (unit: pixels)

video	#images	near wall	far wall
1	251	0.066	0.036
2	86	0.063	0.034
3	111	0.065	0.036

Table 2. Comparison with the ground truth GT (unit: pixels)

video	#images	comparison type	near wall	far wall
1	251	median vs. GT	0.341	0.336
		BP vs. GT	0.492	0.388
2	86	median vs. GT	0.321	0.365
		BP vs. GT	0.504	0.478
3	111	median vs. GT	0.312	0.334
		BP vs. GT	0.481	0.436

6   Conclusion

In this paper we have formulated the generalized median surface problem and presented its exact solution by means of an optimal 3D graph search algorithm. This work is motivated by the task of parameter space exploration without ground truth, which is an effective means of dealing with the difficult parameter problem and has been successfully applied to domains like 2D contour detection [8] and region segmentation [7]. Our median surface computation algorithm thus provides a useful tool for parameter exploration in 3D surface segmentation or 2D contour segmentation in a temporal context. A concrete application has been demonstrated on artery boundary detection in ultrasound data, which confirmed the findings from the previous studies. That is, the median computation can not only avoid the parameter training, but also potentially achieve even better results than with trained parameters. Parameter training is only possible with existing ground truth, which is not always available. The median-based approach can thus be a good alternate in case of no ground truth.

The optimal 3D graph search algorithm in [15] designed for terrain-like surface detection has several interesting extensions. One extension is for simultaneously detecting multiple surfaces subject to certain spatial constraints. In addition, the algorithm can be applied to tube-like, or more generally star-shaped, surface segmentation based on transforming the initial image data to another space. These extensions allow us to study the median surface problem for a broader range of surface classes and will be investigated in future.

References

1. Jiang, X., Bunke, H.: Learning by generalized median concept. In: Pattern Recognition and Machine Vision, pp. 1–16. River Publishers (2010)
2. Rokach, L.: Pattern Classification Using Ensemble Methods. World Scientific (2010)
3. Weiszfeld, E., Plastria, F.: On the point for which the sum of the distances to n given points is minimum. Annals of Operations Research 167, 7–41 (2009)
4. Jiang, X., Wentker, J., Ferrer, M.: Generalized median string computation by means of string embedding in vector spaces. Pattern Recognition Letters 33, 842–852 (2012)
5. Ferrer, M., Karatzas, D., Valveny, E., Bardají, I., Bunke, H.: A generic framework for median graph computation based on a recursive embedding approach. Computer Vision and Image Understanding 115, 919–928 (2011)
6. Vega-Pons, S., Ruiz-Shulcloper, J.: A survey of clustering ensemble algorithms. Int. Journal of Pattern Recognition and Artificial Intelligence 25, 337–372 (2011)
7. Franek, L., Abdala, D.D., Vega-Pons, S., Jiang, X.: Image segmentation fusion using general ensemble clustering methods. In: Kimmel, R., Klette, R., Sugimoto, A. (eds.) ACCV 2010, Part IV. LNCS, vol. 6495, pp. 373–384. Springer, Heidelberg (2011)
8. Wattuya, P., Jiang, X.: A class of generalized median contour problem with exact solution. In: Yeung, D.-Y., Kwok, J.T., Fred, A., Roli, F., de Ridder, D. (eds.) SSPR & SPR 2006. LNCS, vol. 4109, pp. 109–117. Springer, Heidelberg (2006)
9. Franek, L., Jiang, X.: Evolutionary weighted mean based framework for generalized median computation with application to strings. In: Gimel'farb, G., Hancock, E., Imiya, A., Kuijper, A., Kudo, M., Omachi, S., Windeatt, T., Yamada, K. (eds.) SSPR & SPR 2012. LNCS, vol. 7626, pp. 70–78. Springer, Heidelberg (2012)
10. Solnon, C., Jolion, J.-M.: Generalized vs set median strings for histogram-based distances: Algorithms and classification results in the image domain. In: Escolano, F., Vento, M. (eds.) GbRPR 2007. LNCS, vol. 4538, pp. 404–414. Springer, Heidelberg (2007)
11. Stewart, C.: Robust parameter estimation in computer vision. SIAM Reviews 41, 513–537 (1999)
12. Min, J., Powell, M.W., Bowyer, K.W.: Automated performance evaluation of range image segmentation algorithms. IEEE Transactions on Systems, Man, and Cybernetics, Part B 34, 263–271 (2004)
13. Pignalberi, G., Cucchiara, R., Cinque, L., Levialdi, S.: Tuning range image segmentation by genetic algorithm. EURASIP J. Adv. Sig. Proc. 2003, 780–790 (2003)
14. Cheng, D., Jiang, X.: Detections of arterial wall in sonographic artery images using dual dynamic programming. IEEE Trans. Information Technology in Biomedicine 12, 792–799 (2008)
15. Li, K., Wu, X., Chen, D., Sonka, M.: Optimal surface segmentation in volumetric images-a graph-theoretic approach. IEEE Trans. on Pattern Analysis and Machine Intelligence (PAMI) 28, 119–134 (2006)
16. Boykov, Y., Kolmogorov, V.: An experimental comparison of min-cut/max-flow algorithms for energy minimization in vision. IEEE Trans. on Pattern Analysis and Machine Intelligence (PAMI) 26, 1124–1137 (2004)

Estimation of Distribution Algorithm for the Max-Cut Problem

Samuel de Sousa, Yll Haxhimusa, and Walter G. Kropatsch

Vienna University of Technology
Pattern Recognition and Image Processing Group
Vienna, Austria
{sam,yll,krw}@prip.tuwien.ac.at

Abstract. In this paper, we investigate the MAX-CUT problem and propose a probabilistic heuristic to address its classic and weighted version. Our approach is based on the Estimation of Distribution Algorithm (EDA) that creates a population of individuals capable of evolving at each generation towards the global solution. We have applied the MAX-CUT problem for image segmentation and defined the edges' weights as a modified function of the L2 norm between the RGB values of nodes. The main goal of this paper is to introduce a heuristic for MAX-CUT and additionally to investigate how it can be applied in the segmentation context.

Keywords: max-cut, graph cut, eda, segmentation.

1 Introduction

Many problems in computer vision end up by assigning a certain label (corresponding to a class) to a pixel or a region in the image. Therefore, it is required to choose a proper representation in order to assign such label. Many algorithms that are suitable for graph theoretical problems can also be applied in the computer vision domain if the problem is modeled using the graph formulation. Thus, the choice of representing images as graphs has several advantages over other approaches.

A graph theoretical clustering algorithm consists of searching for a certain combinatorial structure in the edge weighted graph, such as the minimum spanning tree [9,16] or normalized cut [26,28]. Among those methods, the complete linkage clustering algorithm [20] reduces the search to the problem of finding a complete subgraph (i.e. the maximal clique [24]) in the image. Also, graph-based spectral methods have been successfully used for clustering [21] as well.

Given a graph $\mathcal{G} = (\mathcal{V}, \mathcal{E})$, MAX-CUT is the problem of finding a partition $(T, \bar{T})$ of the nodes V that maximizes the number of edges between T and its complement set $\bar{T}$. This problem belongs to the class $\mathcal{NP}$-Hard [11], therefore, no polynomial time algorithm is able to solve MAX-CUT for any arbitrary class of graphs, although several approximations have been proposed. In fact, for planar graphs, it is possible to compute the maximum cut in polynomial time [14].

W.G. Kropatsch et al. (Eds.): GbRPR 2013, LNCS 7877, pp. 244–253, 2013.

In this paper, we propose a heuristic for the MAX-CUT problem for any arbitrary class of graphs and we model the weighted MAX-CUT as the problem of maximizing the sum of weighted edges between two sets of nodes. Nodes of the same set connected by an edge should be merged into one single cluster and nodes of different sets connected by a bichromatic edge should remain separated. Our probabilistic heuristic is based on the Estimation of Distribution Algorithm (EDA) [1]. Moreover, we have used the segmentation task to show the applicability of the problem in the pattern recognition domain. We address the weighted version of MAX-CUT whose weights belong to $\mathbb{R}$.

The remainder of this paper is organized as follows: Section 2 provides a literature review on graph-based segmentation. The MAX-CUT problem is introduced and explained in Section 3. Our heuristic is disclosed in Section 4. We map the theoretical graph problem into image segmentation in Section 5. Our experiments are described in Section 6. Finally, we present our conclusions and future directions in Section 7.

2   Related Work

Early graph-based clustering methods [29] use fixed thresholds and local measures in computing a cluster, i.e. the minimum spanning tree (MST) is computed. The clustering criterion is to break the MST edges with the largest weight. The work of Urquhart [27] attempts to overcome the problem of fixed threshold by normalizing the weight of an edge using the smallest weight incident on the vertices touching that edge. The methods in [9,16] use an adaptive criterion that depend on local properties rather than global ones and have the minimum spanning tree as the base algorithm. It is shown in [7] that minimum spanning tree clustering technique, although unsupervised one, approaches the performance of 'Bayes classifier', as the number of sample points from each class increases.

The methods based on minimum cuts [4,6] in graph are designed to minimize the similarity between pixels that are being split [28,26]. Authors in [28] define a cut criterion, but it was biased toward finding small components. Shi and Malik [26] developed the normalized cut criterion to address this bias, which takes into consideration self-similarity of regions. These cut-criterion methods capture the non-local properties of the image, in contrast with the simple graph-based methods such as breaking edges in the MST. However they provide only a characterization of such cut rather than of final segmentation as it is provided by Felzenszwalb [9]. Shi and Malik [26] developed an approximation method for computing the minimum normalized cut, closely related to spectral graph methods, e.g [10].

The minimal spanning tree and the minimum cut are explicitly defined on weighted edge graph, whereas the concept of a maximal clique is defined on unweighted edge graphs. As a consequence, maximal clique based clustering algorithms work on unweighted graphs derived from the edge weighted graphs by means of thresholding [17]. Pavan and Pelillo [24] generalized the concept of maximal clique to weighted graphs.

Markov Random Field (MRF) has been used for clustering [12]. However the use of MRF for image clustering usually leads to $\mathcal{NP}$-Hard problems. The graph-based approximation method for MRF problems [5] yields practical solution, if the number of labels for the pixel is small, which limits these methods for use in segmentation and clustering.

A disadvantage of graph theoretical approaches for image segmentation, i.e. clustering, is that these algorithms in some real-time applications are very time consuming.

3   Max-Cut

Given an undirected graph $\mathcal{G} = (\mathcal{V}, \mathcal{E})$, a cut in the graph is a partition of the vertices $\mathcal{V}$ into T and $\bar{T}$. Let $\bar{T} = \mathcal{V} \setminus T$ be the complement set of T and $E(T, \bar{T})$ be the set of edges connecting a vertex in T with another in $\bar{T}$. The MAX-CUT problem consists of finding the cut that maximizes $|E(T, \bar{T})|$. It is one of the problems of Karp [19] and it belongs to the $\mathcal{NP}$-Hard class. An example of a maximum cut is shown in Figure 1a.

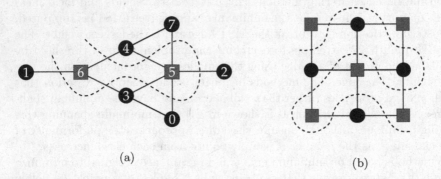

(a)   (b)

Fig. 1. (a) The cut that maximizes the number of bichromatic edges between the T (red squared nodes) and $\bar{T}$ (black circular nodes). (b) The Maximum Cut of a 4-connected representation.

Considering the complexity of $\mathcal{NP}$-Hard problems, common approaches consist of creating ρ-approximated algorithms, i.e. polynomial algorithms whose solution is ρ times the optimal solution [13]. There is a vast and growing amount of algorithms to deal with $\mathcal{NP}$-Hard problems. Researchers seek to find approaches that are capable of achieving better approximation rates as well as they attempt to demonstrate that there are no better approximations above a certain threshold. For instance, Goemans-Williason [13] proposed an approximated algorithm to the MAX-CUT problem whose rate is close to:

$$\alpha = min_{0 \leq \theta \leq \pi} \frac{2}{\pi} \frac{\theta}{1 - \cos\theta} > 0.87856. \tag{1}$$

According to the authors, the approach was a substantial improvement in nearly twenty years. Subsequently, Håstad [15] sets a barrier that unless $\mathcal{P} = \mathcal{NP}$, MAX-CUT can not be approximated by a deterministic algorithm that adheres to a rate strictly exceeding 16/17 [15,18].

Finally, Kaporis et al. [18] proposed a deterministic algorithm in polynomial time that approximates almost all instances of MAX-CUT with a rate above the Håstad threshold. Their solution became the first improvement of MAX-CUT after a decade [18]. Thus, seeking to break the barrier imposed by Håstad, Kaporis et al. use two strategies: Assuming that the maximum cut is not known, it becomes necessary to (i) find an upper bound for the MAX-CUT as well as (ii) to improve significantly the known lower bounds.

4   Estimation of Distribution Algorithm

Evolutionary Algorithms (EA) and more specifically Estimation of Distribution Algorithms (EDA) consist of an ensemble of individuals (agents) sampling the search space for potential solutions of a given problem. Those individuals have a knowledge about the laws of the environment and a quality measurement that represents how able those individuals are to solve the problem [2]. Those solutions are created based on chromosomes and each chromosome has a probability p of being chosen. Evolutionary Algorithms have been applied to the Max-Cut problem before such as in [8], where the authors propose a hybrid evolutionary algorithm using Variable Neighborhood Search and Memetic Algorithm.

In this paper, chromosomes are represented by the nodes of the graph and the nodes should follow a probability distribution, such as the uniform distribution in which all nodes are likely to be chosen with the same probability.

We apply the Population Based Incremented Learning (PBIL) algorithm [3] which takes a vector of probabilities $P = \{p(v_1), p(v_2), \ldots, p(v_n)\}$ associated with how capable a chromosome (v_i) is to provide a solution for the problem. We create a population $S = \{s_1, s_2, \ldots, s_n\}$ of individuals that will choose a subset of nodes to compose the cut T. The best individual (s_{best}^g) of generation g is selected to survive and it is added into generation $g+1$. Hence, we guarantee that $s_{best}^{g+1} \geq s_{best}^g$. The probability of $p^g(v_i)$ is updated as follows:

$$p^{g+1}(v_i) = (1 - \alpha) \times p^g(v_i) + \alpha \times \frac{\sum_{k=1}^{|S|}[v_i \in s_k]}{|S|}, \tag{2}$$

where $\alpha \in [0, 1]$ is a learning rate parameter that weights the impact of both terms of the formula. By using a high α, we decrease the impact of the probability in the previous generation $((1 - \alpha) \times p^g(v_i))$ and we increase the impact of having this node chosen by many individuals. In our experiments, $\alpha = 0.5$, which means we balance equally the importance of both terms. In this algorithm, each individuals choose one or more nodes to belong to the solution, the number of nodes chosen by each individual is computed randomly in such a way to follow the probability distribution of the nodes.

Data: $\mathcal{G}(\mathcal{V}, \mathcal{E}), P, G$
Result: S_{best}

```
 1  begin
 2  │   S_best = ∅
 3  │   for  i = 1 to G do
 4  │   │   Sⁱ ← population(G(V, E), P);
 5  │   │   Sⁱ ← Sⁱ ∪ {S_best};
 6  │   │   Sⁱ_best ← evaluate(Sⁱ);
 7  │   │   if Sⁱ_best > S_best then
 8  │   │   │   S_best ← Sⁱ_best;
 9  │   │   end
10  │   end
11  end
```

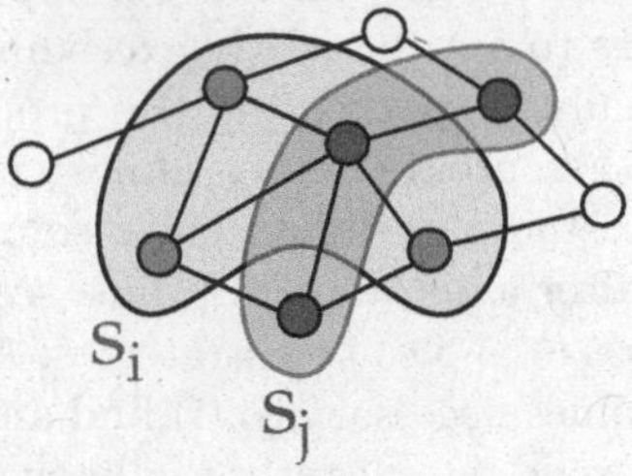

S_i

S_j

(a) Estimation of Distribution Algorithm: Individuals (S) evolve through generations G until the best individual (s_{best}) is found.

(b) Each individual (e.g. s_i, s_j) selects a cut T of the graph. Probability of nodes in generation $g+1$ is updated using s_{best}^g.

Fig. 2. Estimation of Distribution Algorithm

Figure 2a shows the Estimation of Distribution Algorithm. Line 4 creates a population based on the current graph and on the nodes' probabilities P. We add our best individual S_{best} into the current population which ensures that the next generation will produce as good results as the current one. Figure 2b shows an example of two individuals selecting a subset of nodes of the graph as their candidate solution for the cut. The best individual survives the current generation and evolves.

5   Max-Cut-Based Image Segmentation

In order to segment the image, we first create a graph representation. One approach could consider on assigning each pixel of the image as a node and using a 4-connected neighborhood to create the edges. However, considering that Max-Cut tries to maximize the bichromatic edges, by using this representation, we might end up with a cut such that all edges are bichromatic as displayed in Figure 1b.

Our graph representation is built as follows: Given a non visited region in the image, we add a seed to that region and we grow this seed by adding pixels whose absolute difference to the seed does not exceed a certain threshold t (in our experiments $t = 40$). For each region we average the intensity components of the RGB of all pixels belonging to that region and we define the weight of an edge (e_w) between two nodes (v_i, v_j) as a modified function of the $L2$ norm between the two regions:

$$e_w(v_i, v_j) = (2 \times [||rgb(v_i) - rgb(v_j)||_2 > t] - 1) \times ||rgb(v_i) - rgb(v_j)||_2. \quad (3)$$

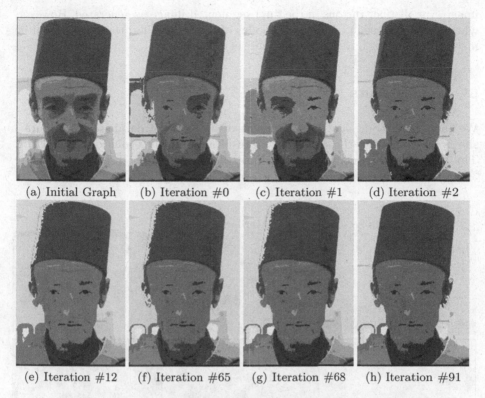

(a) Initial Graph   (b) Iteration #0   (c) Iteration #1   (d) Iteration #2

(e) Iteration #12   (f) Iteration #65   (g) Iteration #68   (h) Iteration #91

Fig. 3. Evolution of the segmentation through the EDA generations

This equation states that based on a threshold t, an edge between two nodes might be either positive or negative. This negative weight assumption prevents the EDA of choosing certain edges in the segmentation process. In the classic MAX-CUT problem, an edge has the weight of 1. But in the weighted version we maximize $\sum e_w(v_i, v_j); v_i \in T, v_j \in \bar{T}; e_w(v_i, v_j) \in \mathbb{R}$ which is the sum of the bichromatic edges' weights.

During the construction of the graph, each pixel in the boundary of two regions will produce an edge, which means that nodes modeling bigger regions will have more edges. In this way, our implementation of graph allows multiple edges between two nodes in order to give preference to bigger regions in the segmentation process. The addition of negative edges work as a mechanism to prevent Max-Cut of choosing those edges in the search for the global maxima. The negative edges penalize the cost function in such a way that if the EDA chooses one such edge, the cost will be smaller than not choosing that edge. Hence, we still try to maximize the sum of $e_w(v_i, v_j)$, however, some edges will not be added into the final cut.

We map the nodes' cut into the region segmentation as follows: whenever there is a bichromatic edge in the graph, there will be isolated regions in the image, *i.e.* distinct regions in the final segmented image as generated. However, when there is an edge between two nodes of the same color, those nodes will be merged into a single region in the final segmentation.

6   Experiments

We have applied our algorithm on images of the Berkeley database [22]. Figure 3 shows an example of the segmentation results during the evolution of the EDA. The initial graph generated by the watershed technique is displayed in Figure 3a and final result of segmentation is available in Figure 3h.

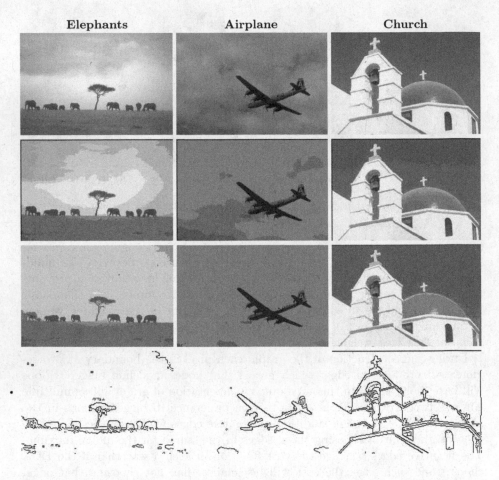

Fig. 4. Segmentation of images from the Berkeley database [22] using the EDA. First row displays the input image. The second row displays the initial graph representation. Our segmentation obtained by Max-Cut is displayed in the third row. We show the edges between regions in the fourth row.

As aforementioned, this paper attempts to apply the MAX-CUT into the segmentation problem. We have computed the results for other images (Elephants, Airplane, Church) in Figure 4. For all images, nodes that were merged into a single node succeeded to do so, due to the fact that the distance between regions in the RGB space are relatively small. Hence, those edges were modeled as negative edges and were not selected by the EDA to compose the final result because the addition of negative edges would penalize the cost.

For instance, the segmentation result obtained in the Elephants' picture did not merge the blue sky in the right upper corner considering that the distance between those two regions in RGB was higher than the threshold used during the optimization. On the other hand, the left-most elephant was merged with a piece of sky, which clearly does not produce a correct result. However, many segmentation algorithms, including many graph cuts use interaction with users by adding manual scribbles or regions containing the object of interest to help the segmentation procedure [23,25]. The negative edges' assumption is an attempt to improve the segmentation results by computing a similarity measure between regions. However, other mechanisms such as brushing or scribbling could be used to map MAX-CUT in the segmentation. One could add some knowledge about the spatial location of the regions to the cost function as an attempt to bring color information and spatial configuration together to improve the results.

7   Conclusions

Graph-based representation of an image has many advantages over other representations due to the fact that many problems can be posed in graph theoretical manner. In this paper we investigate the MAX-CUT problem which belongs to the class of problems called $\mathcal{NP}$-Hard and use the Algorithm of Estimation of Distribution to compute a solution for it.

The focus of this paper is to show a heuristic for the MAX-CUT problem and to show that it can be applied in the segmentation task by assuming, for instance, the negative edges' concept. We are continuously investigating how this problem could be better explored for segmentation.

Acknowledgments. Samuel de Sousa acknowledges financial support by the Austrian Agency for International Cooperation in Education & Research (OeAD) within the OeAD Sonderstipendien program, financed by the Vienna PhD School of Informatics.

References

1. Estimation of Distribution Algorithms: A New Tool for Evolutionary Computation (Genetic Algorithms and Evolutionary Computation). Springer (October 2001)
2. Bäck, T.: Evolutionary algorithms in theory and practice: evolution strategies, evolutionary programming, genetic algorithms. Oxford University Press, Oxford (1996)

3. Baluja, S.: Population-based incremental learning: A method for integrating genetic search based function optimization and competitive learning (1994)
4. Boykov, Y., Lea, G.F.: Graph Cuts and Efficient N-D Image Segmentation. Int. J. Comput. Vision 70(2), 109–131 (2006)
5. Boykov, Y., Veksler, O., Zabih, R.: Markov random fields with efficient approximations. In: Proceedings of IEEE Conference on Computer Vision and Pattern Recognition, USA, pp. 648–655. IEEE Computer Society (1998); Also as Cornell CS technical report TR97-1658, December 3 (1997)
6. Boykov, Y.Y., Jolly, M.P.: Interactive graph cuts for optimal boundary & region segmentation of objects in N-D images, vol. 1, pp. 105–112 (2001)
7. Chowdhury, N., Murhty, C.: Minimal spanning tree based clustering technique: Relationship whith bayes classifier. Pattern Recognition 30(11), 1919–1929 (1997)
8. Duarte, A., Sánchez, A., Fernández, F., Cabido, R.: A low-level hybridization between memetic algorithm and vns for the max-cut problem. In: Proceedings of the 2005 Conference on Genetic and Evolutionary Computation, GECCO 2005, pp. 999–1006. ACM, New York (2005)
9. Felzenszwalb, P.F., Huttenlocher, D.P.: Efficient graph-based image segmentation. International Journal of Computer Vision 59(2), 167–181 (2004)
10. Fiedler, M.: A property of eigenvectors of nonnegative symmetric matrices and its application to graph theory. Checz Mathematical Journal 25(100), 619–633 (1975)
11. Garey, M.R., Johnson, D.S.: Computers and Intractability: A Guide to the Theory of NP-Completeness. W. H. Freeman & Co., New York (1979)
12. Geman, S., Geman, D.: Stochastic relaxation, gibbs distribution, and the bayesian restoration of images. IEEE Transactions on Pattern Analysis and Machine Intelligence 6, 721–741 (1984)
13. Goemans, M.X., Williamson, D.P.: Improved approximation algorithms for maximum cut and satisfiability problems using semidefinite programming. J. ACM 42(6), 1115–1145 (1995)
14. Hadlock, F.: Finding a Maximum Cut of a Planar Graph in Polynomial Time. SIAM Journal on Computing 4(3), 221–225 (1975)
15. Håstad, J.: Some optimal inapproximability results. J. ACM 48(4), 798–859 (2001)
16. Haxhimusa, Y., Kropatsch, W.: Segmentation Graph Hierarchies. In: Fred, A., Caelli, T.M., Duin, R.P.W., Campilho, A.C., de Ridder, D. (eds.) SSPR&SPR 2004. LNCS, vol. 3138, pp. 343–351. Springer, Heidelberg (2004)
17. Jain, A.K., Dubes, R.: Algorithms for Clustering Data. Prentice Hall, Berlin (1988)
18. Kaporis, A.C., Kirousis, L.M., Stavropoulos, E.C.: Approximating almost all instances of MAX-CUT within a ratio above the håstad threshold. In: Azar, Y., Erlebach, T. (eds.) ESA 2006. LNCS, vol. 4168, pp. 432–443. Springer, Heidelberg (2006)
19. Karp, R.M.: Reducibility among combinatorial problems. In: Miller, R.E., Thatcher, J.W. (eds.) Complexity of Computer Computations, pp. 85–103. Plenum Press (1972)
20. Lance, J., Williams, W.: A general theory of classificatory sorting strategies: I hierarchical systems. Journal on Computing 9, 373–380 (1967)
21. Luo, B., Wilson, R.C., Hancock, E.R.: Spectral feature vectors for graph clustering. In: Caelli, T.M., Amin, A., Duin, R.P.W., Kamel, M.S., de Ridder, D. (eds.) SSPR & SPR 2002. LNCS, vol. 2396, pp. 83–93. Springer, Heidelberg (2002)
22. Martin, D., Fowlkes, C., Tal, D., Malik, J.: A database of human segmented natural images and its application to evaluating segmentation algorithms and measuring ecological statistics. In: Proc. 8th Int'l Conf. Computer Vision, vol. 2, pp. 416–423 (July 2001)

23. Noma, A., Graciano, A.B., Cesar Jr., R.M., Consularo, L.A., Bloch, I.: Interactive image segmentation by matching attributed relational graphs. Pattern Recognition 45(3), 1159–1179 (2012)
24. Pavan, M., Pelillo, M.: Graph-theoretic approach to clustring and segmentation. In: Proceedings of IEEE Conference on Computer Vision and Pattern Recognition, vol. 1, pp. 145–152. IEEE Computer Society (2003)
25. Rother, C., Kolmogorov, V., Blake, A.: "grabcut": interactive foreground extraction using iterated graph cuts. ACM Trans. Graph. 23(3), 309–314 (2004)
26. Shi, J., Malik, J.: Normalized Cuts and Image Segmentation. IEEE Transactions on Pattern Analysis and Machine Intelligence 22(8), 888–905 (2000)
27. Urquhart, R.: Graph theoretical clustering based on limited neighborhood sets. Pattern Recognition 15(3), 173–187 (1982)
28. Wu, Z., Leahy, R.M.: An optimal graph theoretic approach to data clustering: Theory and its application to image segmentation. IEEE Transactions on Pattern Analysis and Machine Intelligence 15(11), 1101–1113 (1993)
29. Zahn, C.: Graph-theoretical methods for detecting and describing gestal clusters. IEEE Transaction on Computing 20, 68–86 (1971)

Author Index